BIOSYNTHESIS

Molecular and Cell
BIOCHEMISTRY

BIOSYNTHESIS

SMITH AND WOOD

CHAPMAN & HALL
University and Professional Division
London · New York · Tokyo · Melbourne · Madras

UK	Chapman & Hall, 2–6 Boundary Row, London SE1 8HN
USA	Chapman & Hall, 29 West 35th Street, New York NY10001
JAPAN	Chapman & Hall Japan, Thomson Publishing Japan, Hirakawacho Nemoto Building, 7F, 1-7-11 Hirakawa-cho, Chiyoda-ku, Tokyo 102
AUSTRALIA	Chapman & Hall Australia, Thomas Nelson Australia, 102 Dodds Street, South Melbourne, Victoria 3205
INDIA	Chapman & Hall India, R. Seshadri, 32 Second Main Road, CIT East, Madras 600 035

First edition 1992

© 1992 Chapman & Hall

Typeset in 10/11½pt Palatino by EJS Chemical Composition,
Midsomer Norton, Bath, Avon
Printed in Hong Kong

ISBN 0 412 40760 4

British Library Cataloguing in Publication Data

Biosynthesis. – (Molecular and cell biochemistry)
I. Smith, C.A.   II. Wood, E.J.   III. Series
574.19

ISBN 0–412–40760–4

Library of Congress Cataloging-in-Publication Data

Biosynthesis/[edited by] C.A. Smith and E.J. Wood
      p.   cm.—(Molecular and cell biochemistry)
      ISBN 0–412–40760–4
      1. Biosynthesis.   I. Smith, C.A. (Chris A.)   II. Wood, E.J.
(Edward J.), 1941–   .   III. Series.
QP517.B57B56   1991
574.19'29—dc20   91-9916
                                                                                  CIP

Copy Editors: Sara Firman and Judith Ockenden
Sub-editor: Simon Armstrong
Production Controller: Marian Saville
Layout Designer: Geoffrey Wadsley (after an original design by Julia Denny)
Illustrators: Ian Foulis and Associates
Cover design: Amanda Barragry

Contents

Editors'
foreword

This book is one of a series of brief fundamental texts for junior under-graduates and diploma students in the biological sciences. The series, Molecular and Cell Biochemistry, covers the whole of modern biochemistry, integrating animal, plant and microbial topics. The intention is to give the series special appeal to the many students who read biochemistry for only part of their course and who are looking for an all-encompassing and stimulating approach. Although all books in the series bear a distinct family likeness, each stands on its own as an independent text.

Many students, particularly those with less numerate backgrounds, find elements of their biochemistry courses daunting, and one of our principal concerns is to offer books which present the facts in a palatable style. Each chapter is prefaced by a list of learning objectives, with short summaries and revision aids at the ends of chapters. The text itself is informal, and the incorporation of marginal notes and information boxes to accompany the main text give a tutorial flavour, complementing and supporting the main narrative. The marginal notes and boxes relate facts in the text to applicable examples in everyday life, in industry, in other life sciences and in medicine, and provide a variety of other educational devices to assist, support, and reinforce learning. References are annotated to guide students towards effective and relevant additional reading.

Although students must start by learning the basic vocabulary of a subject, it is more important subsequently to promote understanding and the ability to solve problems than to present the facts alone. The provision of imaginative problems, examples, short answer questions and other exercises are designed to encourage such a problem-solving attitude.

A major challenge to both teacher and student is the pace at which biochemistry and molecular biology are advancing at the present time. For the teacher and textbook writer the challenge is to select, distill, highlight and exemplify, tasks which require a broad base of knowledge and indefatigable reading of the literature. For the student the challenge is not to be overwhelmed, to understand and ultimately to pass the examination! It is hoped that the present series will help by offering major aspects of biochemistry in digestible portions.

This vast corpus of accumulated knowledge is essentially valueless unless it can be used. Thus these texts carry frequent, simple exercises and problems. It is expected that students will be able to test their acquisition of knowledge but also be able to use this knowledge to solve problems. We believe that only in this way can students become familiar and comfortable with their knowledge. The fact that it is useful to them will mean that it is retained, beyond the last examination, into their future careers.

This series was written by lecturers in universities and polytechnics who have many years of experience in teaching, and who are also familiar with current developments through their research interests. They are, in addition, familiar with the difficulties and pressures faced by present-day students in the biological sciences area. The editors are grateful for the co-operation of all their authors in undergoing criticism and in meeting requests to re-write (and sometimes re-write again), shorten or extend what they originally wrote. They are also happy to record their grateful thanks to those many individuals who very willingly supplied illustrative material promptly and generously. These include many colleagues as well as total strangers whose response was positive and unstinting. Special thanks must go to the assessors who very carefully read the chapters and made valuable suggestions which gave rise to a more readable text. Grateful thanks are also due to the team at Chapman & Hall who saw the project through with good grace in spite, sometimes, of everything. These include Dominic Recaldin, Commissioning Editor, Jacqueline Curthoys, formerly Development Editor, Simon Armstrong, Sub-editor, and Marian Saville, Production Controller.

Finally, though, it is the editors themselves who must take the responsibility for errors and omissions, and for areas where the text is still not as clear as students deserve.

Contributors

DR B. CATLEY   *Department of Biochemistry, Heriot-Watt University, Edinburgh, UK.   Chapter 4.*

DR J.J. GAFFNEY   *Department of Biological Sciences, Manchester Polytechnic, Manchester, UK.   Chapter 8.*

DR G. HARTMANN   *Department of Biochemistry, University of Surrey, Guildford, UK.   Chapter 7.*

DR P.J. LARGE   *Department of Applied Biology, University of Hull, Hull, UK. Chapter 6.*

DR J.D. McGIVAN   *Department of Biochemistry, University of Bristol, Bristol, UK.   Chapter 3.*

DR P.A. MILLNER   *Department of Biochemistry, University of Leeds, Leeds, UK.   Chapter 1.*

DR N.M. PACKTER   *Department of Biochemistry, University of Leeds, Leeds, UK.   Chapter 9.*

DR K. SNELL   *Department of Biochemistry, University of Surrey, Guildford, UK.   Chapter 7.*

DR C.A. SMITH   *Department of Biological Sciences, Manchester Polytechnic, Manchester, UK.   Chapters 1 and 8.*

DR L. STEVENS   *Department of Biological and Molecular Sciences, University of Stirling, Stirling, UK.   Chapter 5.*

DR E.J. WOOD   *Department of Biochemistry, University of Leeds, Leeds, UK. Chapter 1.*

Preface

Not much more than a hundred years ago it was believed that organic compounds could be synthesized only by organisms. Although this was shown subsequently not to be the case by the explosive development of organic chemistry as a discipline and, indeed, as an industry, the ability of most organisms to synthesize all the chemical components of which they are composed remains truly astonishing.

As a result of many studies carried out largely between the 1920s and the 1960s, we now have a very good map of the pathways of metabolism and the enzymic mechanisms of all biochemical cellular transformations. With the discovery of the true nature of polymeric molecules of life, the proteins, polysaccharides and nucleic acids, a second tier of understanding was added: not only do the monomer building blocks need to be biosynthesized, they also have to be put together in a particular order specified by a genetic blueprint. These processes, too, we now understand very well, although lacunae still remain in our understanding.

This volume gives an account to some of the key processes in the field of biosynthesis. The opening chapter seeks to tease out the common features of biosynthetic mechanisms and the strategies employed to build macromolecules from simple units. The ensuing chapters consider the making of monomer units such as carbohydrates and nucleotides; the joining together of monomers to make polysaccharides; and the means by which useable energy is obtained and deployed in synthetic reactions.

Chapter 2 looks at life's fundamental energy source, the process of photosynthesis. Pioneering research in the 1950s led to spectacular advances in our understanding of the so called 'dark reactions' of photosynthesis yet, despite unremitting studies since then, we still have only sketchy ideas about the precise mechanisms of light trapping and energy conversion. The following two chapters look at the processes by which carbohydrates and polysaccharides are synthesized from the simpler sugars manufactured by photosynthesis, and of how glucose, life's major fuel, is recycled in vertebrate systems.

All life depends on the remarkable ability of a few groups of microorganisms to capture nitrogen from the air and incorporate it into organic compounds. That they can do this in aqueous solution at atmospheric pressure and a few degrees above freezing is more remarkable still. Nitrogen fixation, the subject of Chapter 5, is exciting great interest among molecular geneticists who hope to locate the genes for nitrogen fixation proteins and transfer them into plants of potential economic importance.

In Chapter 6 we examine the manufacture of amino acids, the monomer units of the proteins. The details of the pathways here are well established and, although they are complex, the text seeks to stress the common means of achieving biosynthetic ends. The details of how nucleic acids and proteins are built up from nucleotides and amino acids, respectively, have been described in the first five chapters of *Molecular Biology and Biotechnology*.

Although the story of these syntheses form important elements in the theme of biosynthesis, it was thought to be more appropriate to deal with them in the context of molecular biology. Chapter 7 describes in outline how the purine and pyrimidine bases of nucleic acids are formed.

This book sets out to identify the broad general principles by which organisms make more of themselves. Despite the apparent bewildering array of biochemicals and the reaction pathways needed to make them, the primary purpose is to simplify the story by continually stressing common principles. If this book succeeds in helping students to distinguish the themes and variations behind the rich orchestration, it will have served its purpose.

Abbreviations

A	adenine (alanine)
ACP	acyl carrier protein
ACTH	adrenal corticotrophic hormone
ADP	adenosine diphosphate
Ala, A	alanine
AMP	adenosine monophosphate
cAMP	adenosine 3′,5′-cyclic monophosphate
Arg, R	arginine
Asn, N	asparagine
Asp, D	aspartic acid
ATP	adenosine triphosphate
ATPase	adenosine triphosphatase
C	cytosine (cysteine)
CDP	cytidine diphosphate
CMP	cytidine monophosphate
CTP	cytidine triphosphate
CoA, CoASH	coenzyme A
CoQ, Q	coenzyme Q, ubiquinone
Cys, C	cysteine
d-	2-deoxy-
D	aspartic acid
d-Rib	2-deoxyribose
DNA	deoxyribonucleic acid
cDNA	complementary DNA
e-	electron
E	glutamic acid
E	oxidation–reduction potential
F	phenylalanine
F	the Faraday (9.648×10^4 coulomb mol^{-1})
FAD	flavin adenine dinucleotide
Fd	ferredoxin
fMet	N-formyl methionine
FMN	flavin mononucleotide
Fru	fructose
g	gram
g	acceleration due to gravity
G	guanine (glycine)
G	free energy
Gal	galactose
Glc	glucose

Gln, Q	glutamine
Glu, E	glutamic acid
Gly, G	glycine
GDP	guanosine diphosphate
GMP	guanosine monophosphate
GTP	guanosine triphosphate
H	histidine
H	enthalpy
Hb	haemoglobin
His, H	histidine
Hyp	hydroxyproline (HOPro)
I	isoleucine
Ig G	immunoglobulin G
Ig M	immunoglobulin M
Ile, I	isoleucine
ITP	inosine triphosphate
J	Joule
K	degrees absolute (Kelvin)
K	lysine
L	leucine
Leu, L	leucine
ln x	natural logarithm of x $= 2.303 \log_{10} x$
Lys, K	lysine
M	methionine
M_r	relative molecular mass, molecular weight
Man	mannose
Mb	myoglobin
Met, M	methionine
N	asparagine
N	Avogadro's number (6.022×10^{23})
N	any nucleotide base (e.g. in NTP for nucleotide triphosphate)
NAD$^+$	nicotinamide adenine dinucleotide
NADP$^+$	nicotinamide adenine dinucleotide phosphate
P	proline
P_i	inorganic phosphate
PP_i	inorganic pyrophosphate
Phe, F	phenylalanine
Pro, P	proline
Q	coenzyme Q, ubiquinone
Q	glutamine
R	arginine
R	the gas constant ($8.314 \, J \, K^{-1} \, mol^{-1}$)
Rib	ribose
RNA	ribonucleic acid

mRNA	messenger RNA
rRNA	ribosomal RNA
tRNA	transfer RNA
s	second
s	sedimentation coefficient
S	svedberg unit (10^{-13} seconds)
S	serine
SDS	sodium dodecylsulphate
Ser, S	serine
T	thymine
T	threonine
Thr, T	threonine
TPP	thiamine pyrophosphate
Trp, W	tryptophan
TTP	thymidine triphosphate (dTTP)
Tyr, Y	tyrosine
U	uracil
UDP	uridine diphosphate
UDP-Glc	uridine diphosphoglucose
UMP	uridine monophosphate
UTP	uridine triphosphate
V	valine
V	volt
Val, V	valine
W	tryptophan
Y	tyrosine

Greek alphabet

A	α	alpha		N	ν	nu
B	β	beta		Ξ	ξ	xi
Γ	γ	gamma		O	o	omicron
Δ	δ	delta		Π	π	pi
E	ε	epsilon		P	ϱ	rho
Z	ζ	zeta		Σ	σ	sigma
H	η	eta		T	τ	tau
Θ	θ	theta		Y	υ	upsilon
I	ι	iota		Φ	ϕ	phi
K	κ	kappa		X	χ	chi
Λ	λ	lambda		Ψ	w	psi
M	μ	mu		Ω	ω	omega

Basic principles of biosynthesis

Objectives

After reading this chapter, you should be able to:

☐ outline the overall strategies of biosyntheses used by cells to make more of their own substance;

☐ describe the general ways in which small organic molecules are made from inorganic precursors (CO_2, H_2O, NH_3);

☐ explain how monomer units are linked together by condensation reactions to form macromolecules;

☐ appreciate the importance of biological membranes in many aspects of biosynthesis.

1.1 Introduction

As a result of a great deal of biochemical research over the last 60 to 70 years, it is now understood, at least in principle, how a cell produces or obtains the specific molecules it needs to maintain its structure and to reproduce itself. The term **biosynthesis** is synonymous with **anabolism**; a building up from relatively simple molecules to more complex ones, which will form the structure and carry out the functions of the cell.

Some organisms carry out the whole process. Starting with CO_2 and H_2O and an inorganic source of nitrogen they build all the cellular materials they need using the energy of sunlight. Other organisms obtain many of the basic building blocks they need, organic compounds such as sugars and amino acids and fats for example, and use these, oxidizing some of them to obtain energy and using others to make their cellular macromolecules such as polysaccharides, proteins and nucleic acids.

See *Energy in Biological Systems*

The life processes of cells and organisms depend crucially on the properties of proteins, nucleic acids and polysaccharides, which are large and complicated molecules. However, for all their complexity, they are manufactured by joining together a variety of simpler units (called **monomers**) in various permutations. The process does not stop at **macromolecule** formation for these molecules may be assembled in further levels of hierarchy to make cellular units such as membranes, organelles and chromosomes, and these ultimately form whole cells (Fig. 1.1). The majority of these structures are self-assembling. This assembly process may be looked on as an extension of biosynthesis, although for the most part no additional energy has to be supplied to make it happen. The macromolecules are so exquisitely designed that they tend to self-associate in the appropriate way spontaneously under physiological conditions.

Macromolecule: *means large molecule, but what does that mean? The division is hard to define. Most people would call insulin, a very small protein of* M_r *about 6000, a macromolecule. Anything smaller they would not.*

Reference Yudkin, M. and Offord, R. (1973) *Comprehensible Biochemistry*, p. 45ff, Longman, London. A rather old text now, but one that lives up to its title. Good for reading about the reasons why things are as they are.

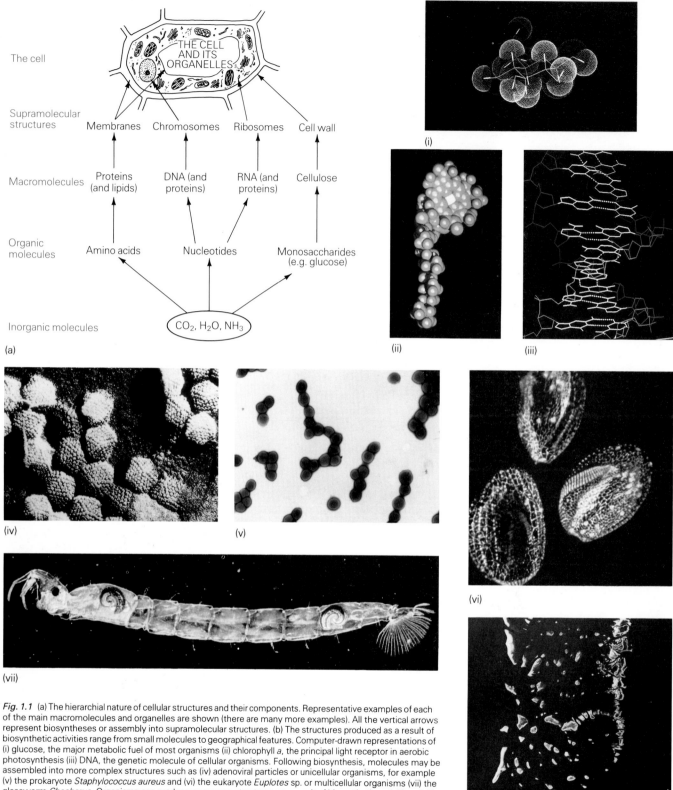

Fig. 1.1 (a) The hierarchial nature of cellular structures and their components. Representative examples of each of the main macromolecules and organelles are shown (there are many more examples). All the vertical arrows represent biosyntheses or assembly into supramolecular structures. (b) The structures produced as a result of biosynthetic activities range from small molecules to geographical features. Computer-drawn representations of (i) glucose, the major metabolic fuel of most organisms (ii) chlorophyll *a*, the principal light receptor in aerobic photosynthesis (iii) DNA, the genetic molecule of cellular organisms. Following biosynthesis, molecules may be assembled into more complex structures such as (iv) adenoviral particles or unicellular organisms, for example (v) the prokaryote *Staphylococcus aureus* and (vi) the eukaryote *Euplotes* sp. or multicellular organisms (vii) the glassworm *Chaobarus*. Organisms can produce enormous structures as a result of biosynthetic activities; (viii) shows a 200-mile length of the Great Barrier Reef photographed from a satellite. The whole reef is 1260 miles long and is the product of coral polyps. Photographs courtesy (i)–(iii) Dr C. Freeman, Polygen, University of York, UK, (iv) Dr M.V. Nermut, National Institute for Medical Research, London, UK, (v) Dr A. Curry, Public Health Laboratory, Withington Hospital, Manchester, UK. (vi) Tracor Northern, Milton Keynes, UK, (vii) M.J. Hoult, Department of Biological Sciences, Manchester Polytechnic, Manchester, UK, (viii) NASA, USA.

Stanozolol

Testosterone

Steroids are derived from naturally occurring male sex-hormones. The male sex-hormone testosterone is responsible for initiating the development of male secondary sexual characteristics (its **androgenic** action). This hormone also promotes, in males and females, protein synthesis (its **anabolic** action). A variety of steroid drugs have been developed which, compared with testosterone, have an increased anabolic action but reduced androgenic stimulation, hence the term **anabolic steroids**. These drugs have some clinical uses and are widely used in veterinary practice in countries outside the EEC to promote rapid increase in bulk in livestock. It is such veterinary preparations that are available to athletes, although the extent to which they are abused in athletics is the subject of controversy.

Drug-taking in organized sport came to prominent public attention during the 1950s. More recently, the stripping of the 1988 Olympic 100 m Gold Medal from Ben Johnson, for taking the banned anabolic steroid stanazolol, drew immense media attention to the uses of anabolic steroids by athletes.

The use of anabolic steroids probably offers some advantages for weight-lifting, and 'explosive' sprinting events. They increase muscle development by stimulating protein synthesis. Anabolic steroids also promote aggressive behaviour and allow athletes to train harder and recover from training and competitive events more quickly. It appears, however, that increases in strength are only certain to occur where the use of anabolic steroids is associated with regular training exercises.

A number of side-effects are associated with the use of anabolic steroids. For example, they cause masculinization in women, the symptoms of which include growth of facial hair and deepening of the voice. Other problems include acne, menstrual irregularities, and clitoral hypertrophy. In men, fertility can be affected, and cardiovascular disease and liver damage, including liver cancer, are also known side-effects. It is likely that the full range of pathologies will not be recognized until the 1990s when athletes who took large doses of anabolic steroids in the 1960–1970s, before the introduction of controlling legislation, reach late middle-age.

This chapter gives an outline of biosynthesis, concentrating especially on the common themes that are found in all life forms. Thus, the ways in which a nucleic acid or a protein molecule are synthesized in a bacterium or in a human cell are identical in all major aspects. Furthermore, the processes of biosynthesis, both of organic compounds and of macromolecules, are catalysed by enzymes which are similar in their range of activities and specificities in all life forms.

The key processes in building new cellular material are obtaining energy, enzyme action, the production of small organic molecules from simple (inorganic) precursors, the production of macromolecules, the control of these processes (and especially the *informational* input) and, finally, self-assembly to organelles and then cells. These processes will be examined in turn.

☐ The chemical composition of all cells is remarkably similar. Cells contain about 70% water, and all contain proteins, nucleic acids, carbohydrates and lipids in roughly the same proportions. The different types of these classes of compounds give cells their individuality, but the processes by which the molecules are made are invariably very similar if not identical.

1.2 Energy

The maintenance of cellular structures, which are far more organized than those of the environment, and the production of new cellular material, require energy. Green plants and photosynthetic bacteria obtain this energy from the radiant energy of the sun. Animals, and other organisms such as the bacteria of decay, obtain their energy by degrading organic molecules that have been formed by other organisms. Practically all organisms are, therefore, ultimately dependent upon the sun's energy.

☐ Carbohydrate is sometimes stated to be the product of photosynthesis and it is true that it is usually the primary product. However, *all* the compounds in a plant are ultimately formed from CO_2, H_2O and NH_3. These, apart from minerals, are the only materials green plants require.

See Chapter 2 and *Energy in Biological Systems*, Chapters 1–3

Exercise 1

How do plants survive during the hours of darkness?

In animals, plants and microorganisms, energy is used for the synthesis or interconversions of the constituents of macromolecules (amino acids, nucleotides, sugars, etc.), and of lipids, and also for the synthesis of macromolecules from precursor monomer units.

With an enormous number of ways of obtaining energy and an enormous number of ways of using energy in any given cell, it is expedient to have a common **energy currency**, and this intermediary takes the form of a nucleotide triphosphate, usually adenosine triphosphate (ATP), although other nucleotide triphosphates are frequently used.

ATP

Processes that supply energy produce ATP and processes that require energy utilize ATP. The synthesis of ATP requires energy and its hydrolysis releases energy. Thus, there is a **coupling** between energy-producing and energy-requiring reactions, *via* the intermediary ATP (Fig. 1.2). This coupling may be understood in terms of thermodynamics: only reactions that release energy occur spontaneously. However, a reaction that will not proceed because its energy change is unfavourable under a given set of conditions may be made by proceed by coupling it with a reaction for which the energy change is favourable, such that the *overall* reaction has a favourable change.

The following is an example of coupling exergonic and endergonic reactions such that the overall reaction proceeds. (Note that the free energy changes given are *standard* free energy changes. Energetically favourable reactions have *negative* free energy changes.)

Formation of glucose 6-phosphate, a sugar phosphate ester, by condensation (the hydrolysis of this compound is favoured):

$$\text{Glucose} + \text{phosphate} \rightleftharpoons$$
$$\text{Glucose 6-phosphate} + H_2O \quad \Delta G^{0'} = +13.8 \text{ kJ mol}^{-1}$$
$$\text{ATP} + H_2O \rightleftharpoons \text{ADP} + \text{phosphate} \quad \Delta G^{0'} = -30.7 \text{ kJ mol}^{-1}$$

$$\overset{\text{hexokinase}}{\text{ATP} + \text{Glucose} \rightleftharpoons}$$
$$\text{ADP} + \text{Glucose 6-phosphate} \quad \Delta G^{0'} = -16.8 \text{ kJ mol}^{-1}$$

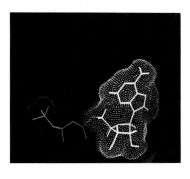

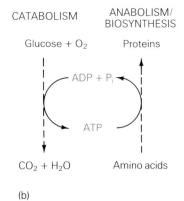

CATABOLISM   ANABOLISM/ BIOSYNTHESIS

Glucose + O_2 Proteins

ADP + P_i

ATP

$CO_2 + H_2O$ Amino acids

(a)   (b)

Fig. 1.2 (a) Molecular model of ATP. Courtesy Dr C. Freeman, Polygen, University of York, UK. (b) The ADP–ATP cycle couples energy-yielding processes to energy-requiring ones. Note that the catabolism of glucose, like most catabolic pathways, is oxidative and initially yields reducing equivalents whose energy is used to produce ATP in the election transport chain. In contrast, the assembly of strings of amino acids to make proteins is a series of condensation reactions (see Fig. 1.3) not involving a change in oxidation state.

Exercise 2

What do you understand by the term 'currency'?

Reference McGilvery, R.W. (1975) *Biochemical Concepts*, W.B. Saunders, Philadelphia. Another rather old text, but it gives different view on energy relationships.

Reference Hinkle, P.C. and McCarty, R.C. (1978) How cells make ATP, *Scientific American*, **238(3)**, 104–23. A splendid overview with beautiful diagrams which aims to unify ideas about ATP biosynthesis in different organisms.

Oxidation and reduction versus condensation and hydrolysis

The breakdown of carbohydrates to CO_2 and H_2O is an oxidative process (Fig. 1.2). In contrast, the biosynthesis of proteins from amino acids does not involve a change in oxidation state but is a **condensation**. Condensation is the reverse of a hydrolysis: the equivalent of a molecule of water is removed as two units are joined (Fig. 1.3). Many of the reactions that produce macromolecules are condensations. Proteins are one example: another is the formation of oligo- and polysaccharides from monosaccharides. The reaction requires the removal of the elements of a water molecule and the hydrolysis of ATP to ADP and inorganic phosphate (P_i). The reaction is thermodynamically favoured.

Cells have evolved ways of producing ATP as a result of oxidative reactions. The breakdown of carbohydrates, and indeed of most food materials, are oxidative processes. Reducing equivalents are produced which are subsequently used to drive the biosynthesis of ATP from ADP and P_i in

☐ ATP may be looked upon as a 'dehydrating reagent' to promote condensation. Its hydrolysis, $ATP + H_2O \rightarrow ADP + P_i$, is energetically favourable and consumes a molecule of water. It is important not to carry this analogy too far because the condensations proceed in solution, with a high concentration of water present throughout.

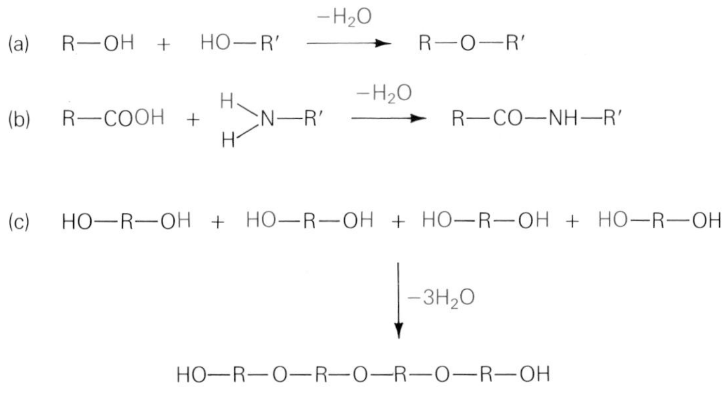

(a) $\quad$ R—OH $\;+\;$ HO—R' $\;\xrightarrow{-H_2O}\;$ R—O—R'

(b) $\quad$ R—COOH $\;+\;$ $\overset{H}{\underset{H}{}}$N—R' $\;\xrightarrow{-H_2O}\;$ R—CO—NH—R'

(c) $\quad$ HO—R—OH $\;+\;$ HO—R—OH $\;+\;$ HO—R—OH $\;+\;$ HO—R—OH

$\qquad\qquad\qquad\qquad \downarrow -3H_2O$

$\qquad\qquad$ HO—R—O—R—O—R—O—R—OH

Fig. 1.3 In condensation reactions, the elements of water are removed when two organic compounds combine: the reverse is a hydrolysis. Where molecules have more than one function groups, (c), chains may be formed.

Exercise 3

Can you think of other dehydrating reagents used in chemistry?

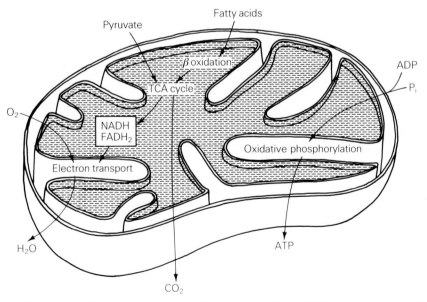

Fig. 1.4 The mitochondria have a central role in aerobic respiratory metabolism. Catabolism of pyruvate resulting from the glycolysis of glucose or of fatty acids generates reducing equivalents (NADH, $FADH_2$). Subsequent re-oxidation of these using molecular oxygen generates ATP for driving energy-requiring processes such as biosynthesis (see Fig. 1.2).

Table 1.1 *Life-styles – the metabolic classification of organisms*

Organism	Carbon source	Energy source	Electron donors	Examples
Photolithotroph	CO_2	Sunlight	H_2O, H_2S, S	Green plants, photosynthetic bacteria, cyanobacteria
Photo-organotroph	Organic compounds	Sunlight	Organic compounds	Non-sulphur purple bacteria
Chemolithotroph	CO_2	Oxidation–reduction reactions	H_2S, H_2S, Fe (II), NH_3	Hydrogen, sulphur, iron and denitrifying bacteria
Chemo-organotroph	Organic compounds	Oxidation–reduction reactions	Organic compounds e.g. glucose	Higher animals, most micro-organisms, non-photosynthetic plant cells

membrane-bound systems such as mitochondria (Fig. 1.4). Therefore, ATP forms a vital link between oxidative processes and condensation reactions.

Trapping light energy

See *Energy in Biological Systems*, Chapter 3

The fundamental energy-trapping activity that forms the basis of all life is that of capturing light from the sun. This process, the so-called *light reaction of photosynthesis*, has been studied extensively although it is still incompletely understood. The details appear elsewhere: the key point here is that a significant amount of the trapped energy ultimately appears at ATP. So, here again, this key intermediary is produced and is then used to drive energy-requiring biosynthetic reactions in plants and photosynthetic bacteria.

Some of the light energy trapped appears not as ATP but as 'reducing power' in the form of NADPH. The conversion of CO_2 to carbohydrate, (CH_2O), requires the **reduction** of CO_2, hence the need for reducing equivalents.

—— *Exercise 4* ——

Explain why photosynthesis needs 'reducing power' or 'reducing equivalents'.

□ In green plants H_2O is used as the 'reducing agent', but in photosynthetic bacteria other compounds are used, such as H_2S or a wide variety of organic compounds ('H_2A') (see Chapter 2).

Life-styles

An understanding of the need for ATP and its various production pathways offer insights into the 'life-styles' of different types of organism (Table 1.1). Green plants and photosynthetic bacteria use the energy from light to build up organic molecules from CO_2, H_2O and nitrogenous compounds: ATP drives the process. All other organisms take in preformed organic molecules. They break down some of these, mostly oxidatively, to release energy, which is trapped as ATP. Some of the organic molecules are used as such. Others are modified and ultimately incorporated into macromolecules and cell structures. ATP or other nucleoside triphosphates supply the energy to drive these processes.

1.3 *The importance of enzymes*

Life is possible only because the reactions in cells are catalysed and controlled by biological catalysts called **enzymes**. The process of biosynthesis must clearly involve formation of carbon–carbon and other bonds, and these reactions are catalysed by enzymes. All enzymes are macromolecules and the majority of them are proteins.

Enzymes can catalyse only simple transformations

Enzymes catalyse reactions with legendary efficiency. However, enzymes are classified into only six categories according to the type of reaction catalysed (Table 1.2). The initial reaction of someone looking at a metabolic pathways chart for the first time is usually one of horror at the complexity (Fig. 1.5), and

See *Biological Molecules*, Chapters 4 and 5

The light reaction of photosynthesis: that part of the photosynthetic process whereby light is trapped. It certainly consists of more than a single reaction. The light reaction may be distinguished experimentally from all the subsequent reactions, called the dark reactions.

Reference Metzler, D. (1977) *Biochemistry: The Chemical Reactions of Living Cells*, pp. 631–58, A useful chapter on the chemistry of many biosynthetic reactions and pathways.

Table 1.2 *The six classes of enzymes*

Class	Name	Catalytic activity
1	Oxidoreductases	Change of oxidation state, e.g. —CHOH to —C=O
2	Transferases	Transfer of a group, e.g. transfer of C_1 units such as methyl
3	Hydrolases	Hydrolysis, e.g. of esters or of peptides
4	Lyases	Bond-breaking reactions, e.g. carbon–nitrogen lyase
5	Isomerases	Isomerizations, e.g. *cis–trans* epimerizations
6	Ligases	Bond-forming reactions, e.g. carbon–carbon bonds

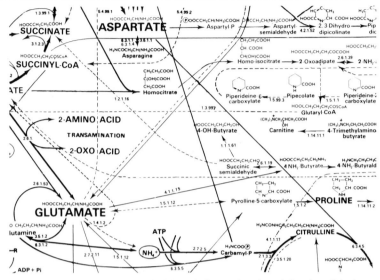

Fig. 1.5 Part of a metabolic pathways chart. Such 'metabolic maps' are intended to chart all the intermediates and reactions going on in a cell. Because many thousands of reactions are proceeding all at once the charts are inevitably complex. Such charts do not easily show which routes involve major throughput of materials and which are quantitatively minor, nor do they show how pathways are controlled. In contrast with a real map, such as that for the London Underground System, it is the compounds that move between enzymes, although the compounds appear in the positions of 'stations' linked by arrows! Reproduced by permission of Dr D.E. Nicholson.

yet individual steps in metabolic pathways, whether biosynthetic or degradative, are all relatively simple transformations. As an illustration, to split a glucose molecule (C_6) into two pyruvate molecules (C_3) in the course of **glycolysis** takes a dozen steps.

Box 1.2
Industrial diamonds

Industrial diamonds. Courtesy of de Beers, UK.

Diamonds are the hardest substance known. They have a variety of industrial uses in drilling, sawing, and as abrasives. Something like 40% of industrial diamonds are produced by heating cheap graphite to 3000 K at a pressure of 12 625 kPa (125 atm) in the presence of transition metal catalysts. The extreme conditions induce rearrangements of carbon–carbon bonds in the graphite, resulting in the formation of diamond.

In contrast, organisms can synthesize and rearrange carbon–carbon bonds at 310 K (37°C) at a pressure of 101 kPa (1 atm) using enzymes at catalysts! However, this is perhaps an unfair comparison, since organisms break and rearrange these bonds by concentrating on a restricted number of bonds.

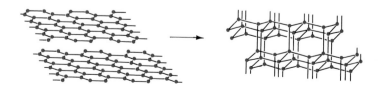

Glycolysis: *the process by which sugars, usually glucose, are catabolized to provide energy. The process may take place either aerobically or anaerobically.*

Reference A number of companies supply 'metabolic maps' or metabolic pathways charts. The one by D E Nicholson (currently in its 16th edition) is available from the Sigma Chemical Company. Since the first edition over 28 years ago, more than 800 000 have been printed. Dr Nicholson wrote an article entitled 'Some Reflections on Metabolic Cartography' published in *Biochemical Education* in 1972 (Vol. 1, p. 6).

Yields need to be very high in biochemical processes

In metabolic reactions, there must be as near to a 100% yield as possible. Organic chemists accept that 100% yields are impossible, and in an organic preparation the chemist will expect to have to purify compounds *en route*. Cells have no way of carrying out purifications and cannot easily discard unwanted by-products. The enzymes catalysing metabolic transformations have evolved to a peak of perfection. Reaction pathways are controlled to such an exquisite degree that yields are extremely high.

OPTICAL PURITY. This high degree of perfection applies nowhere more significantly than in the optical purity of naturally produced materials. Many organic molecules with asymmetric carbon atoms can exist as optical isomers. The synthesis of just one of two possible optical isomers by synthetic means is notoriously difficult. Usually, a mixture of isomers needs to be 'resolved' at the end of a synthesis in a separate step. In contrast, optical purity is the rule in organisms. This is because enzymes, and all the other macromolecules responsible for biochemical transformations, operate by a process of three-dimensional recognition. Complementary shapes must match, as they do in the combination of an enzyme's active site with its substrate. Only the 'correct' isomer is recognized and only the correct isomer results from the enzyme-catalysed reaction. If the enzymes of protein synthesis picked up the 'wrong' isomer of an amino acid in the course of synthesizing a protein molecule, for example, that molecule would fail to adopt the correct, biologically active, conformation.

Because high yields are vital, error-correcting mechanisms have evolved

No chemical reaction can give a yield of 100%, whether it be enzyme-catalysed or not. When it comes to duplicating information in cell division, which occurs by enzyme-catalysed chemical processes, even a small fraction of one percent of error is unacceptable.

The macromolecule in cells that carries information from one generation to the next is the double-stranded molecule of DNA (Fig. 1.6). When a cell divides the double strand has to be duplicated, and the duplication process is catalysed by enzymes. Special mechanisms have evolved to guarantee extremely high fidelity in copying an existing DNA sequence. The enzymic

Fig. 1.6 Molecular model of DNA. Courtesy of M.J. Derham, Department of Chemistry, Manchester Polytechnic, Manchester, UK.

processes are in themselves highly accurate, but in addition there are subsequent 'proofreading' or error-correcting mechanisms to purge any mistakes that may have crept in. The result is that in the duplication of DNA, the error rate (that is, the incidence of a base being placed incorrectly in the sequence) is less than 1 in 10^8. (Assuming that there are about 500 words to the page in this book and that on average each word has six letters, this is equivalent to printing a book of about 34 000 pages without a single misprint!)

Following the replication of DNA, the information contained in its sequence of bases is transcribed into **mRNA** and this information is then used to direct the assembly of chains of amino acids in the correct sequence to make protein molecules. The level of precision needed for RNA replication and for protein synthesis is somewhat less than that needed for DNA because these molecules are not carrying the genetic message from generation to generation. Also, RNA and protein molecules occur in the cell in multiple copies and so having a few incorrect ones matters less. RNA biosynthesis has an error rate of about 1 in 10^5, and protein synthesis of about 1 in 10^2–10^3. One 'bad' protein molecule in a thousand or so is not likely to constitute too serious a problem to the cell in most cases. In contrast, one 'false' mRNA molecule may give rise to a number of 'bad' protein molecules in the course of its life, and can be tolerated less readily. Nevertheless, the requirement for high fidelity demands that energy be spent to purge the errors. This is presumably worthwhile in an evolutionary sense. It comes down to a balance between using energy as efficiently as possible (in order to compete in a competitive environment) and achieving acceptable levels of error-free molecules enabling life and reproduction to continue successfully.

See *Molecular Biology and Biotechnology* Chapter 1

Exercise 5

What are the possible consequences of constructing an enzyme protein with one 'wrong' amino acid in it?

1.4 Making small molecules

The molecules most characteristic of cells are large and complicated. However, they are all constructed by linking together monomer units in various permutations (Table 1.3). These monomer units have to be obtained or manufactured in order for life to continue. Animals, and a number of other organisms, obtain their monomers from their food. They take in proteins and carbohydrates and other nutrients and, typically, degrade them to monomer units. In addition, they are capable of manufacturing some of their monomer units from simpler precursors, or in some cases, of transforming some types of monomers into others. This is all part of the metabolic scheme of things. To give one example, like all other organisms, humans require 20 different amino acids to make the proteins they require. However, they need only about nine of these in a preformed state in the diet: the rest may be manufactured, some almost 'from scratch', others by small transformations. Tyrosine, for example, is obtained by hydroxylating phenylalanine (Fig. 1.7), and a number of amino acids, including glutamate, are connected with intermediates of the tricarboxylic acid cycle (Fig. 1.7). The pathways are often

☐ Because animals require about half of the 20 amino acids to be supplied in their diet, the proteins they eat must contain a reasonable selection of these essential amino acids. However, some plant proteins are deficient in certain amino acids, notably lysine and tryptophan, and thus a diet consisting of these plants as the *sole* source of protein, will lead to malnutrition. The **prolamines** are a group of proteins from cereals containing up to 15% of proline and 30–45% glutamic acid, but only low concentrations of essential amino acids. **Zein** (from maize) lacks tryptophan and lysine and **hordein** (from barley) lacks lysine.

Table 1.3 *Monomer units and their corresponding macromolecules*

Monomer units	Macromolecule	Examples of subsequent modification (post-polymerization)
20 Amino acids	Proteins	Hydroxylation, phosphorylation glycosylation, acylation
10–20 Types of monosaccharide	Polysaccharides	Sulphation
4 Nucleotides (base–sugar–phosphate)	Polynucleotides (DNA, RNA)	Methylation, formation of 'unusual bases'

---

mRNA: *the common abbreviation for messenger RNA, which forms the intermediary between DNA and protein synthesis. It carries the information specifying how to join up amino acids in a specific sequence to form a particular polypeptide.*

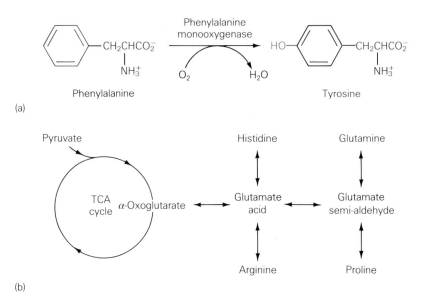

(a)

(b)

Fig. 1.7 (a) Tyrosine may be manufactured from phenylalanine by the addition of a hydroxyl group. If there is ample phenylalanine in the diet, tyrosine is not an essential amino acid. Curiously, although the reaction shown is biosynthetic and, indeed, requires NADPH, it also represents the first step in the catabolism of dietary phenylalanine (see *Energy in Biological Systems*, Chapter 7). (b) A number of the amino acids may be obtained from or metabolized *via* the tricarboxylic acid cycle, depending upon metabolic need. Glutamate is the key intermediate. It is difficult to say whether a given pathway is anabolic or catabolic. Perhaps it is better to think of this group of metabolism (and the intermediates) as a 'pool' rather than as members of a 'pathway'.

□ Although heterotrophic organisms take in macromolecules (proteins, nucleic acids, polysaccharides) in their diets, these macromolecules are not used as such, but are hydrolysed in the gut to their monomeric units (amino acids, nucleosides, monosaccharides) before they can be absorbed. Subsequently they may be oxidized to supply energy and their carbon skeleton rearranged to provide different building blocks, or they may be reassembled to form the type of macromolecules characteristic of that particular organism.

rather tortuous, forming a network of reactions called **intermediary metabolism**. Part of the reason is, as already explained, that enzymes catalyse only small chemical transformations of molecules and a number of enzymes need to act in sequence to achieve a major change of structure.

Another reason for complicated pathways is that metabolic routes are controlled. Typically, degradative or catabolic pathways are distinct from biosynthetic or anabolic ones. This enables them to be controlled individually according to a cell's needs and nutritional status at a given time. Control of the complicated network of metabolic pathways is important because, by regulation, the cell is able to integrate its many biochemical routes so as to avoid waste of energy and raw materials and minimize the production of toxic wastes.

Fixing carbon and nitrogen

Green plants and photosynthetic bacteria synthesize *everything* from CO_2, H_2O and NH_3, using the energy of light to drive the manufacturing processes. The process of turning CO_2 into organic compounds is called **carbon *fixation*** and this is the role of photosynthesis. Most of the organic compounds that form the monomers for macromolecule production contain nitrogen in addition to carbon, hydrogen and oxygen. This nitrogen has to come from somewhere if the **biomass** on the earth is to increase, and the source is atmospheric nitrogen. Nitrogen also, therefore, has to be fixed (Fig. 1.8). However, green plants cannot carry out this process themselves, and only certain prokaryotes (on which we all ultimately depend) can do it.

The processes of carbon and nitrogen fixation are described in detail in Chapters 2 and 5. They are both energy-requiring processes and hence require a supply of ATP. In addition, they involve a change of oxidation state, CO_2 to $(CH_2O)_n$, and N_2 to NO_3^- or NH_4^+, and consequently reducing

□ **The Haber process** is the industrial way of fixing atmospheric nitrogen: $N_2 + 3H_2 \rightleftharpoons 2NH_3$. Chemically speaking, it employs moderate temperatures (500 °C) and high pressures (300 000 kPa, 300 atm), both of which can be looked upon as inputs of energy. In addition, energy will have already been used in producing the hydrogen used in the process. A catalyst is required too (see Box 5.1).

Fixation: we talk about 'fixing' CO_2 or N_2. The term comes from the French. The idea is that these free gaseous substances become 'fixed' or incorporated into more solid organic compounds.

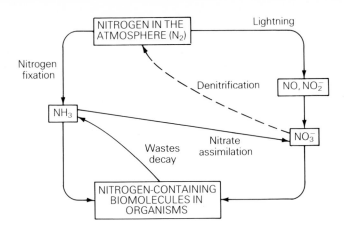

Fig. 1.8 The nitrogen cycle. Only certain prokaryotic organisms can fix nitrogen. Electric storms fix a small amount, and industrial processes such as the Haber process also fix atmospheric nitrogen to make fertilizer, although the energy cost is considerable.

Exercise 6

Explain why nicotinamide adenine dinucleotide phosphate is so called (see also Fig. 1.9).

equivalents are required in addition to ATP. In the majority of cases this reducing power is provided by nicotinamide adenine dinucleotide phosphate ($NADP^+$, Fig. 1.9). The reduced form of this coenzyme (NADPH) is produced either using light energy, or sometimes, by oxidizing other compounds.

Making monomers

All organisms are closely related in terms of the chemistry of their constituents and even of the isomers they use (see 'Optical purity', Section 1.3). The metabolic pathways by which the various compounds are transformed are also closely similar. From this point of view, the requirement of an animal is for protein in the diet: this protein will contain the 20 L-amino acids. It does not matter very much whether the animal is vegetarian or carnivorous, whether it eats chicken or beef or single cell protein: the amino acid content of the protein is usually the same.

Aside from the fundamental processes of carbon and nitrogen fixation, many cells can manufacture the majority of the monomer units they need for the construction of macromolecules. Presumably, cells that require certain of their monomers in the preformed state, as is the case with animal cells and some of the amino acids, have lost their ability to carry out the synthesis. Perhaps they have become adapted to a life-style in which these monomers have always been obtainable in their food, so that it became inefficient to continue to manufacture the enzymes for doing the job.

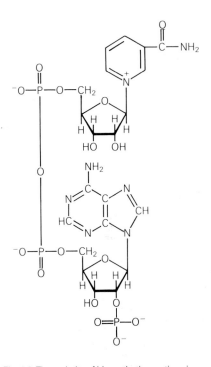

Fig. 1.9 The majority of biosynthetic reactions in which a reduction takes place require the participation of reduced nicotinamide adenine dinucleotide phosphate, $NADP^+$. Most oxidative catabolic processes generate NADH (or sometimes $FADH_2$) instead. $NADP^+$ differs from NAD^+ solely in the presence of an extra phosphate group. The other properties of the compounds, including the oxidation–reduction potentials, are identical. However, the presence of the additional phosphate allows the compound to be recognized by enzymes different from those which deal with NAD^+.

1.5 Building macromolecules

Organisms depend on macromolecules to carry out the major life processes of enzyme catalysis, provision of structural materials, energy storage and information handling (Table 1.3). Although all cells have proteins, nucleic acids and polysaccharides, built according to the same chemical plan, the *particular* types of macromolecules are highly specific to individual species. Human proteins differ from chimpanzee proteins, for example. Enzymes that carry out the same function in different organisms may diverge quite widely in structure, even though the reactions catalysed may be identical.

For a given type of cell or organism to build its own type of macromolecule demands **information**. When the genetic message is carried on from

Exercise 7

How could a human protein be distinguished from a chimpanzee protein with the same function?

Box 1.3
Chemical polymers

A variety of artificial polymers of immense industrial and commercial importance are formed by condensation reactions. In 1909 Baekeland produced the first synthetic plastic, called Bakelite, by combining formaldehyde and phenol under heat and pressure:

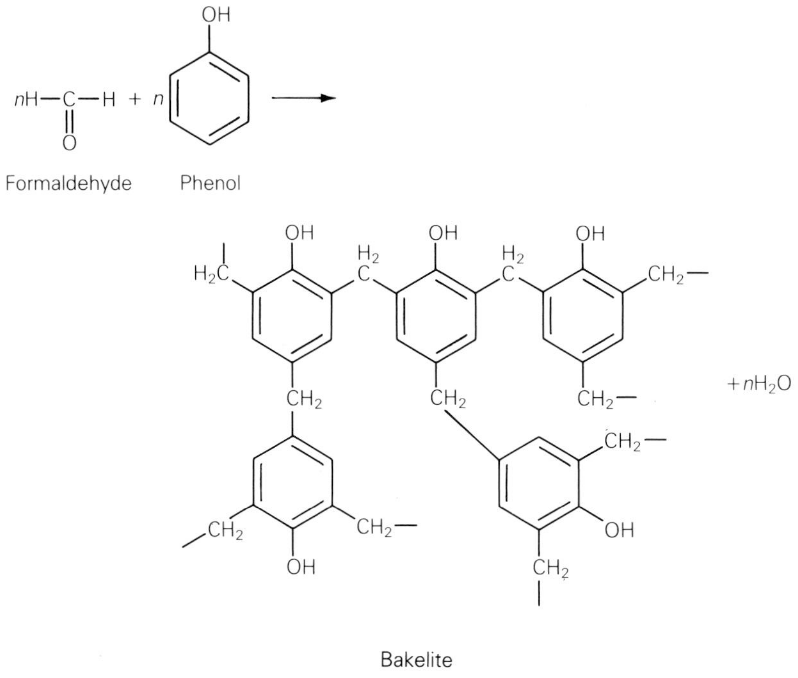

Formaldehyde   Phenol

$+nH_2O$

Bakelite

The polyester, *Dacron*, is produced by a condensation of terephthalic acid and ethylene glycol:

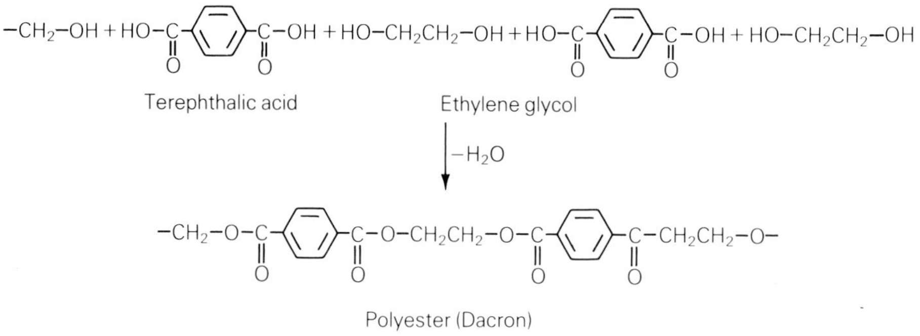

Terephthalic acid   Ethylene glycol

$-H_2O$

Polyester (Dacron)

Dacron is widely used in clothing, carpets and cords. When produced as thin sheets, called *Mylar*, it is used to manufacture recording and computer tapes. It has even been used to make synthetic blood vessels.

generation to generation at cell division, it is this message that bears the information for joining up amino acids into a protein sequence characteristic of one cell type rather than another. The macromolecules, DNA and RNA, are the medium by which information is stored and handled. Therefore, when any macromolecule is synthesized, there is an informational input.

─── Exercise 8 ───

Suggest three ways in which humans store information in everyday life.

Some kinds of macromolecules are maintained with high precision, some are not

Nucleic acids and proteins are maintained as far as possible with absolute precision. They could not carry out their functions if this were not the case. In contrast, for other kinds of macromolecules, the control over structure is less precise or important. *Glycogen* (Section 4.3) is a polymer of glucose used for energy storage in mammalian and bacterial cells. It does not matter very much exactly how many glucose units or precisely how many branch points there are in the structure for the molecule to function properly (Fig. 1.10). Indeed, the function of the molecule is to grow larger or smaller by the gain or loss of glucose residues, depending upon the nutritional status and the energy demands of the cell or organism. Therefore, the control over the biosynthesis of glycogen, in structural terms, is not very strict. It is simply necessary to produce a structure that the appropriate enzymes can act upon. More importantly, it is the **biosynthesis** and **degradation** processes of glycogen that need to be controlled very precisely, so that the energy storage material is either laid down or mobilized, according to metabolic demands.

Requirements for macromolecular biosynthesis

It is possible to specify the various requirements for the construction of a given macromolecule. In general, although the monomer units are different, amino acids for proteins, nucleotides for nucleic acids, the other requirements are common. Thus, there is a common requirement for information so that monomer units are linked together in the appropriate sequence. There is also to be a requirement for energy because more complicated structures are being built up from simpler ones. Finally, there will also, often, be a requirement for some special structures, such as ribosomes, to ensure that the

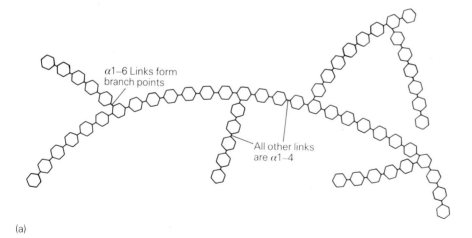

(a)

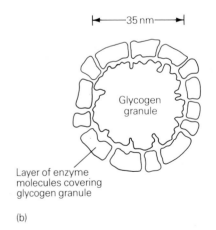

(b)

Fig. 1.10 Glycogen is a branched-chain polysaccharide made entirely of glucose units. Because of the branching, a giant, roughly spherical molecule results which can gain or lose glucose residues from its surface depending upon metabolic needs. (a) Part of glycogen structure (schematic). (b) It is likely that the surface of the glycogen is covered with the enzymes responsible for synthesis and breakdown. Which of these processes occurs at a particular time depends on control of the enzymes by feedback and by hormonal influences.

Glycogen: a branched polymer consisting only of D-glucose molecules.

See *Cell Biology*, Chapter 1

process proceeds in the correct way. Proteins cannot be built in free solution. They are assembled in the ribosomes and emerge in their mature conformations only after processing in the Golgi apparatus (see Box 1.5). In the case of storage polysaccharides such as glycogen, it seems likely that an enzyme complex is required to build them, and also that glycogen molecules are not made *de novo*, but either grow or shrink depending upon metabolic needs. This does, of course, beg the question of where new glycogen molecules come from when a cell divides!

Condensation reactions

The biosynthesis of macromolecules requires energy. The various ways in which this energy is actually used will be described briefly to emphasize the point that although proteins, nucleic acids and polysaccharides are very different in their structures and properties, they are all built by means of condensation reactions (Fig. 1.3). Such molecules are invariably drawn from a homologous series or small group of similar molecules. Proteins are composed of amino acid residues of general formula $RCHNH_2 CO_2H$. When a water molecule is removed between the amino and carboxyl groups of two such molecules, a peptide bond is formed (Fig. 1.11). This process may continue with the formation of a long, unbranched chain of amino acid residues called a polypeptide.

Nucleic acids consist of chains of nucleotides (nitrogenous base–pentose sugar–phosphate). The backbone of the molecule is a sugar–phosphate chain (Fig. 1.12). The chain is built by a series of condensations between the hydroxyl group on the sugar and a phosphate group (Fig. 1.13).

Polysaccharides are formed by repeated condensations between the hydroxyl groups of sugar molecules. The resulting bonds are called **glycosidic linkages** (Fig. 1.14).

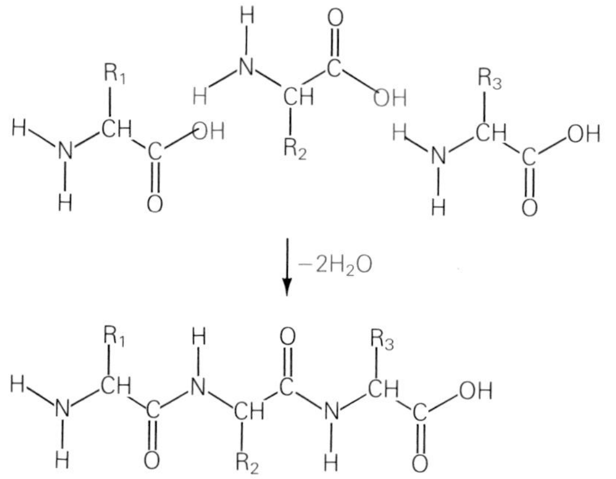

- A nucleoside consists of a base and a sugar (adenine + ribose → adenosine), and a nucleotide is a base–sugar–phosphate compound (adenine + ribose + phosphate → AMP) or base–sugar–polyphosphate (ATP). ATP is also, therefore, a nucleoside triphosphate. The major nucleoside triphosphates are ATP, CTP, GTP and UTP, and all have similar, high, free energies for the hydrolysis of their pyrophosphates.

- Adding sugar units to a molecule seems always to involve nucleoside-sugars. Examples include adding a mono-saccharide to another monosaccharide to make a disaccharide such as sucrose, or adding a monosaccharide to a polypeptide to make a glycoprotein (adding glucose and galactose to collagen). UDP-Glc was discovered by Leloir in the 1950s as an intermediate in the conversion of galactose 1-phosphate into glucose 1-phosphate. For his discovery Leloir was awarded a Nobel Prize in 1970.

- In a glycosidic link, the keto or aldo carbon atom of one of the sugars is always involved. The other one may be any of the carbons of the second sugar.

Fig. 1.11 Condensation between amino acid molecules yields a peptide (see Fig. 1.3).

Exercise 9

Describe chemically what a glycosidic link is.

Supplying the energy to drive condensations

It is possible to think of ATP, or another nucleoside triphosphate, being employed as a dehydrating agent in the condensation reactions that yield polypeptides, nucleic acids and polysaccharides. The **hydrolysis** of ATP to ADP and P_i 'consumes' a molecule of water and this hydrolysis is energetically favoured. However, dehydration is not something that can easily go on in the bulk water of a cell, because there is 'too much' water. The

De novo: *means synthesized 'from scratch' rather than being formed from other organic compounds. However, its meaning is often broader. One could speak of* de novo *synthesis of cholesterol from acetate units, for example.*

Reference Kornberg, A. (1968) The Synthesis of DNA, *Scientific American*, **219(4)**, 64–8. A rather old but readable account by the leading figure in this field. For another account by Kornberg, see his recent (1989) book *For the Love of Enzymes*, Harvard University Press, USA.

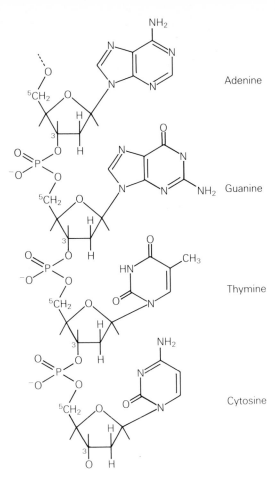

Fig. 1.12 DNA is a very long chain molecule consisting of a repeating sugar (deoxyribose)–phosphate backbone with any one of four nitrogenous bases attached to each deoxyribose.

Adenine

Guanine

Thymine

Cytosine

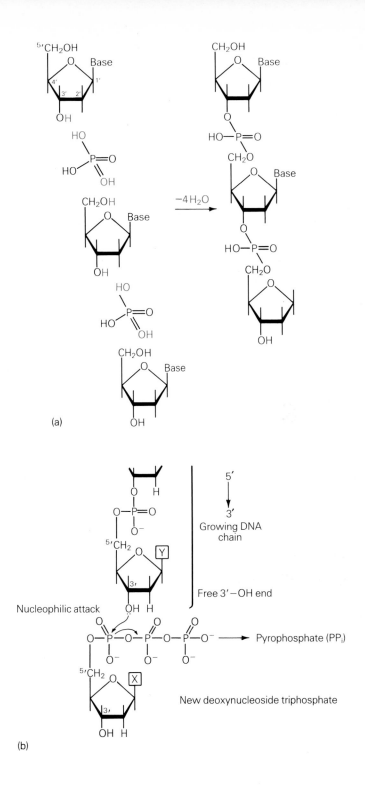

(a)

(b)

Fig. 1.13 DNA (and RNA) are formed by joining nucleotides in condensation reactions: (a) shows formally how the water molecules are removed. (b) The actual reaction is more complicated because the condensation is energetically unfavourable. In order to make it proceed, a nucleoside triphosphate (which supplies the new nucleotide unit) is hydrolysed with the release of pyrophosphate. The chain grows only in the 5′ → 3′ direction.

Fig. 1.14 The production of di- and polysaccharides formally involves the removal of a water molecule (condensation) between two monosaccharide units. In practice this reaction is energetically unfavourable and it needs to be driven (see Fig. 1.16).

□ The four bases required for DNA are attached to *deoxyribose*. Hence, the precursor nucleoside triphosphates are dATP, dCTP, dGTP, and dTTP. The prefix 'd' indicates 'deoxy'. dTTP is often written simply as TTP because the form with ordinary ribose does not occur naturally.

—————— *Exercise 10* ——————

State the Law of Mass Action.

action of ATP has to be targeted to the specific elements of water (OH,H) in the monomer units that it is necessary to remove.

The chapters in this book describe the variety of ways that this targeting is achieved. Broadly, it can be said that the process happens by the ATP (or **NTP**) chemically combining with one of the monomer units in question to form an 'activated' species. This then combines, in a second reaction, with the second monomer unit. In the case of nucleic acid biosynthesis, for example, the units that actually add on to the growing nucleic acid chain are nucleoside triphosphates (Fig. 1.13). The combination leaves behind a nucleoside monophosphate – the nucleic acid chain has increased by one unit; and a double phosphate (pyrophosphate, PP_i) unit departs. Typically, this pyrophosphate is immediately hydrolysed to give two orthophosphates, an energetically favourable reaction, 'pulling' the overall reaction according to the Law of Mass Action (Fig. 1.15).

In polysaccharide biosynthesis, initially a sugar–nucleotide diphosphate compound is formed in an energetically favourable reaction. This then reacts with the growing polysaccharide, releasing a nucleotide diphosphate unit (Fig. 1.16).

In the case of proteins, the sequence is more complex. An active complex is formed from the amino acid and ATP and this leads to the formation of an

Pyrophosphate   2 mol Orthophosphate

Fig. 1.15 In many of the biosynthetic reactions shown in this chapter, a pyrophosphate is produced. The subsequent hydrolysis of this pyrophosphate to two molecules of inorganic orthophosphate (catalysed by pyrophosphatase) is energetically favourable, thus 'pulling' the biosynthetic reaction. Note that in physiological conditions all these phosphates would be ionized.

Box 1.4
Branching in
macromolecules

In the biosynthesis of nucleic acids (a), protein (b) and polysaccharides (c), there are opportunities for branching or cross-linking in the polymers formed. In practice this never seems to occur in nucleic acids, occurs occasionally in proteins, and is common among the polysaccharides. The degree of branching seems to be linked to the polymers' functions. For example, nucleic acids are linear information stores; some structural proteins are cross-linked for strength; polysaccharides such as glycogen are highly branched to produce a spherical molecule that can rapidly grow or shrink from many points on its surface (see Fig. 1.10).

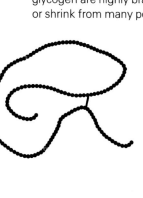

(a)

(b)

(c)

(a) DNA does not branch and (b) proteins are sometimes cross-linked. However, polysaccharides such as glycogen (c) are multi-branched structures.

NTP: *the abbreviation for 'nucleoside triphosphate'; similarly, dNTP means deoxynucleotide triphosphate. These abbreviations are used when the exact nucleoside triphosphate participating in a reaction (e.g. ATP, UTP) is either unknown or may be any of four.*

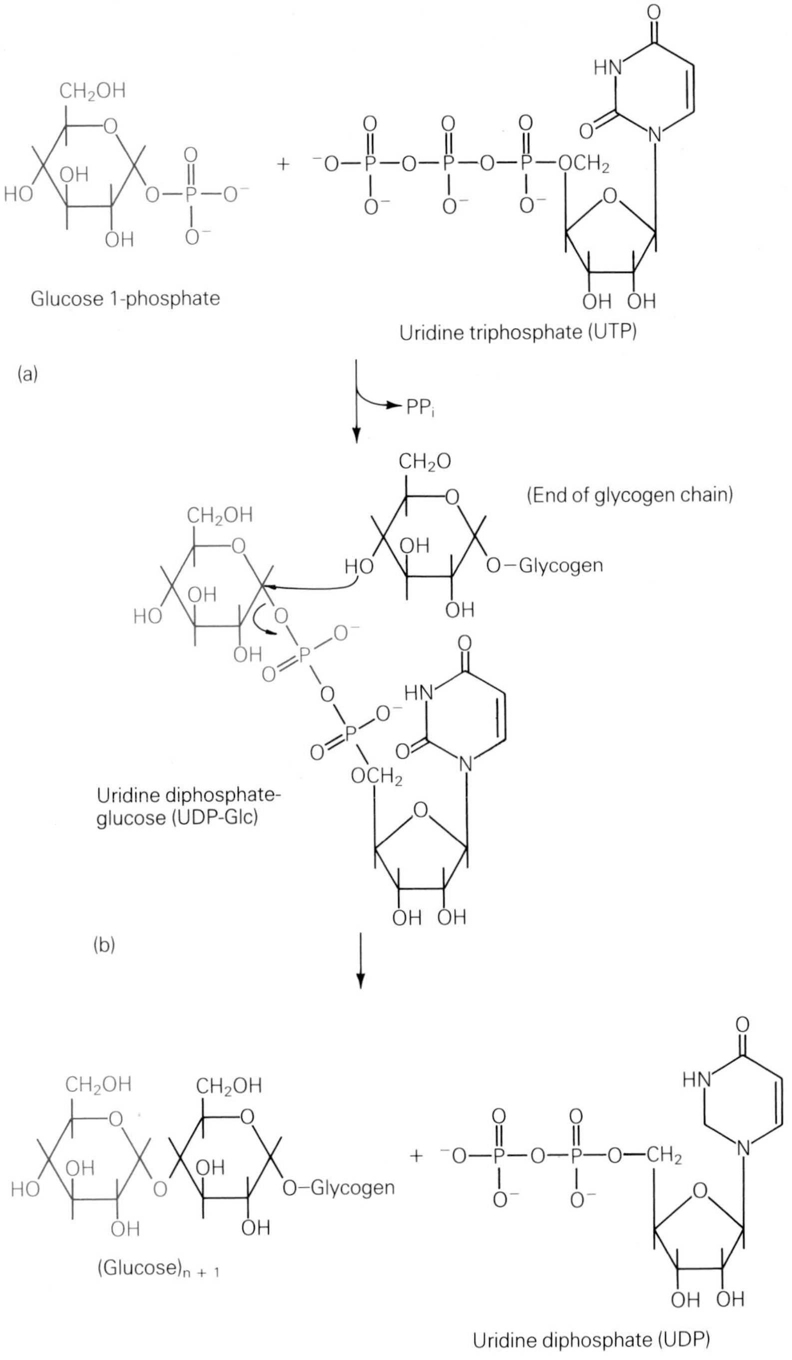

Fig. 1.16 The addition of a glucose residue to the end of a glycogen chain is made energetically favourable by means of the intermediary compound, uridine diphosphate-glucose. Compare the overall reaction shown in Fig. 1.14. The two reactions here are:

(a) Glucose 1-phosphate + UTP → UDP-glucose + PP$_i$
(b) UDP-glucose + (glucose)$_n$ → UDP + (glucose)$_{n+1}$
               glycogen   glycogen

Note that the pyrophosphate (PP$_i$) will be hydrolysed in two molecules of inorganic phosphate.

□ Gramicidin, an antibiotic from *Bacillus brevis*, is a cyclic decapeptide with a sequence of five amino acid residues repeated in the ring: -[D-Phe-L-Pro-L-Val-L-Orn-L-Leu]-$_2$. It is synthesized by an enzyme complex of M_r 280 000. Note that ornithine (Orn) is not a typical component of proteins and is not specified by the genetic code. It occurs in the urea cycle (*Energy in Biological Systems*, Chapter 7) and as a precursor of polyamines. The activated amino acids involved in gramicidin biosynthesis are amino acyl adenylates, and the amino acid residues are transformed to thiol groups of 4'-phosphopantatheine covalently joined to the enzyme. Neither the genetic code, nor mRNA, is involved in the biosynthetic sequence.

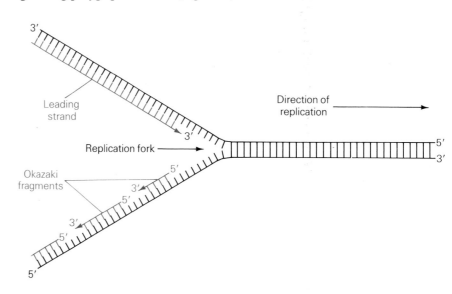

Amino acid

ATP

PP_i — pyrophosphate → $2P_i$

H_2O

mixed anhydride of carboxylic acid and adenylic acid

transfer RNA (tRNA)

amino acyl tRNA

Ribosome

one amino acid residue added to growing polypeptide

Fig. 1.17 The process of protein synthesis is far from the simple condensation shown in Fig. 1.11, although the reactions shown here achieve this. As with the other biosynthetic reactions, the condensation is energetically unfavourable, and energy needs to be used to drive it. In the case of adding an amino acid to a growing polypeptide, ATP and GTP are hydrolysed.

active aminoacyl-***tRNA*** complex. This can then add its amino acid on to the growing polypeptide chain (Fig. 1.17).

Original parent molecule

First-generation daughter molecules

Second-generation daughter molecules

Fig. 1.19 When DNA replicates, the new strands formed are indeed entirely new, and the double helix resulting from the events shown in Fig. 1.18 consists of one 'old' and one 'new' DNA strand. The process is therefore said to be '*semi*conservative'.

Leading strand

Direction of replication

Replication fork

Okazaki fragments

Fig. 1.18 The replication of the DNA double helix takes place at a replication fork where the two strands have separated. Each of the strands acts as a template and new deoxynucleotide triphosphates pair with existing bases in the template to be 'zipped up' to form a new DNA strand. The enzyme of DNA synthesis (DNA polymerase) will work only in the 5' → 3' direction. Therefore, along one strand (the 'lagging strand'), short sections a few hundred bases long are formed (Okazaki fragments), which are joined up later.

tRNA: *the abbreviation for transfer RNA, a family of small RNA molecules that participate in protein synthesis, bringing amino acids to the ribosome–mRNA complex for addition to the growing polypeptide.*

Semiconservative: *in making a new DNA double helix it would be possible to envisage a replica of the molecule being formed. One helix would be two 'old' strands and the other, two 'new' strands. In fact, the new molecules each consist of one new and one old strand, so that half of each molecule is* **conserved** *in the replication process.*

To sum up, all polymerization processes proceed by broadly similar types of condensation reactions, which usually involve at least two steps. These processes operate by fully understandable chemistry. There is sometimes a tendency to think of mysterious, half-understood processes when molecules are said to be 'activated'; this is not the case. For example, it is not feasible to use ATP as a dehydrating reagent in aqueous solution. Therefore, sequences of reactions have evolved which allow the overall process to be achieved *via* intermediates or on the surface of an enzyme. Thus, it is likely that the tortuous way in which the pathways and enzymes operate is also the result of the chemical processes that have to go on in order to achieve the required biosyntheses.

Information

The ultimate repository of a cell's information is the DNA in the chromosome of the nucleus. To produce new informational macromolecules, DNA or RNA are duplicated together with the information contained within their structures. In general, one molecule (usually, but not always, DNA) acts as a template while a new molecule is built on it (Fig. 1.18). The new molecule is not a *copy* of the old one but its *complement*: a photographic negative is a useful analogy. In fact, when DNA is duplicated by a process of **semiconservative** replication, two new double helixes of DNA are produced which are identical to the parent double helix (Fig. 1.19).

See *Molecular Biology and Biotechnology*, Chapter 5

When proteins are produced, the information is not duplicated but **translated**. The information is taken from the repository and the instructions acted upon to produce a sequence of amino acids to make a particular protein. The idea of translation comes from the notion that the 'language' of nucleic acids (a sequence of bases along a molecule), is translated into the 'language' of proteins (a sequence of amino acids in a polypeptide).

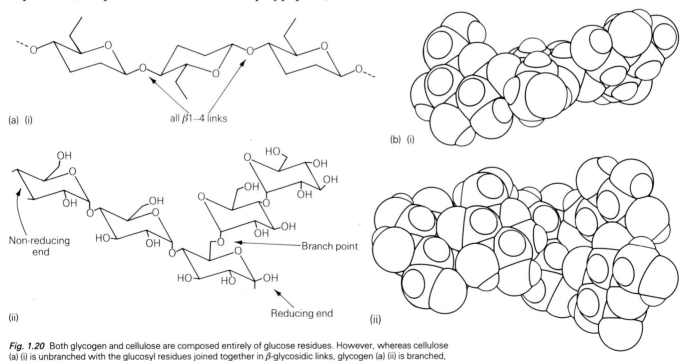

Fig. 1.20 Both glycogen and cellulose are composed entirely of glucose residues. However, whereas cellulose (a) (i) is unbranched with the glucosyl residues joined together in β-glycosidic links, glycogen (a) (ii) is branched, and has α1–4 and α1–6 links. The structures are quite distinct and are only recognized by the appropriate biosynthetic enzymes. (b), (i) and (ii), Computer-generated models of (a) (i) and (ii) respectively, which emphasize the difference in overall conformation between the two structures. Courtesy M.J. Derham, Department of Chemistry, Manchester Polytechnic, Manchester, UK.

Reference Clark, B.F.C. (1984) *The Genetic Code and Protein Biosynthesis* 2nd edn, Outline Studies in Biology, No. 83. Edward Arnold, London. An easily accessible account of protein synthesis. It includes thoughts on levels of molecular organization and their relationship to biosynthetic requirements.

Exercise 11

What is cellulose and where is it found?

The situation with polysaccharides is a little more subtle. The information blueprint is in the enzymes that catalyse polysaccharide biosynthesis. One set of enzymes will produce glycogen from glucose but another set will produce **cellulose** (Fig. 1.20). Enzymes that build glycogen will not add glucose units to a growing cellulose molecule because the shape and structure of the molecule do not match the enzyme's requirements.

1.6 Self-assembly

Not only does a cell have to synthesize its component molecules and macromolecules, it also has to have the means of assembling them to form the various structures such as membranes, cell walls, ribosomes, and mitochondria. Most of the molecules and macromolecules recognize each other's specific shapes and come together to form these structures spontaneously (**self assembly**), at least once the molecules are in the right locations. Commonly, little energy is required: the folding of the various macromolecules is largely spontaneous (and is presumably built into their information content), and once folded correctly, complementary regions recognize each other and associate together. Little is known about the detailed mechanisms involved in these processes. In some cases other molecules are involved. For example, when it is necessary to get proteins through certain membranes to their final locations, molecular 'chaperones' may be required.

Membranes

Many biosynthetic processes in cells either take place on membranes or occur within membrane-bound cavities. Thus, proteins destined for export from the cell are synthesized on the endoplasmic reticulum, and many proteins are post-translationally modified by the addition of oligosaccharide units, a process which goes on in the endoplasmic reticulum and Golgi apparatus. Much of the substance of any cell is membrane. Membranes are assemblies of lipid bilayers, with protein molecules of various types 'inserted' into them (Fig. 1.21). Unlike other cell structures, membranes, or at least their lipid foundation, are not built from macromolecules but from smallish units that self-assemble into layers.

The processes that govern the biosynthesis of membrane lipids are quite well understood. Long-chain fatty acids and other types of molecules such as glycerol and organic bases are required, and these are combined in enzyme-catalysed reactions to give the phospholipids and glycolipids (Chapter 8) characteristic of a particular type of membrane. Even the long-chain fatty acids are characteristic of a particular type of cell. Each type of cell has its own enzymes that control which types of fatty acids and lipids are synthesized.

A few types of long-chain fatty acid are essential in the diets of mammals. These are the so-called **essential fatty acids** (Table 1.4), but their role is more likely to be in the formation of prostaglandins and other substances rather than in membrane formation (Section 8.3). In general, therefore, cells have

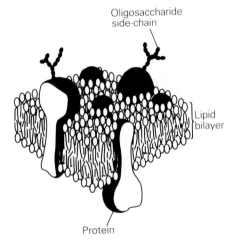

Oligosaccharide side-chain

Lipid bilayer

Protein

Fig. 1.21 Biological membranes are formed from a lipid bilayer into which proteins are inserted. Some proteins pass all the way through the lipid bilayer while others are confined to one or other of the faces. Many of the proteins exposed on the outer surface of the plasma membrane are glycosylated with oligosaccharide chains attached to them. Membranes are typically asymmetrical and their lipid composition reflects their origin. Membranes are supramolecular self-assembling structures.

Exercise 12

What is an oligosaccharide?

Table 1.4 Essential fatty acids

Name	Structure	Abbreviation
Linoleic	$CH_3(CH_2)_4CH\!=\!CHCH_2CH\!=\!CH(CH_2)_7CO_2H$	$18:2^{\Delta 9,\,12}$
Linolenic*	$CH_3CH_2CH\!=\!CHCH_2CH\!=\!CHCH_2CH\!=\!CH(CH_2)_7CO_2H$	$18:3^{\Delta 9,\,12,\,15}$
Arachidonic	$CH_3(CH_2)_4(CH\!=\!CHCH_2)_3CH\!=\!CH(CH_2)_3CO_2H$	$20:4^{\Delta 5,\,8,\,11,\,14}$

* Linolenic acid makes up 10–20% of the total fatty acids in triglycerides and phosphoglycerides in mammals.

Cellulose: *a long, unbranched polymer consisting only of D-glucose units linked β1–4.*

Self-assembly: *furniture that you have to assemble from the parts yourself, is known as self-assembly furniture. Self-assembling macromolecular complexes come together themselves (without any human assistance!) because their charges, shapes and hydrophobicities are complementary.*

Box 1.5
Biosynthesis and membranes

Some parts of the processes of biosynthesis are intimately connected with the membrane systems of the cell. For example, proteins that are destined for export from the cell are synthesized on ribosomes attached to the endoplasmic reticulum (rough endoplasmic reticulum). The newly synthesized polypeptides find their way into the cavities of the endoplasmic reticulum (a). Later they may be exported from the cell when those cavities fuse with the plasma membrane.

Many proteins are modified by the addition of other chemical entities such as carbohydrate units, however. This formation of glycoproteins takes place in another system of membranous cavities called the Golgi apparatus (b). In the Golgi apparatus, the addition of sugar units is a highly organized sequential process, each step being dependent on the previous reaction in the series. Furthermore, the different parts of the Golgi membrane have different compositions. The *cis* face, which is thought to develop from the endoplasmic reticulum, has a composition closely similar to that of the endoplasmic reticulum. In contrast, the *trans* face has a composition similar to that of the plasma membrane with which it may eventually fuse, releasing the fully processed protein from the cell (c).

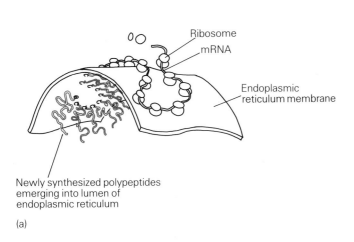

Ribosome
mRNA
Endoplasmic reticulum membrane
Newly synthesized polypeptides emerging into lumen of endoplasmic reticulum

(a)

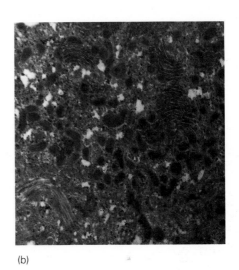

(b)

(a) Synthesis of proteins for export from the cytoplasm takes place on the rough endoplasmic reticulum.
(b) Electron micrography of the membrane systems of the Golgi apparatus.
(c) Glycosylation is initiated in the lumen of the endoplasmic reticulum and continues as the newly-formed polypeptides continue their journey through the Golgi vesicles until they are exported to the exterior by exocytosis.

Endoplasmic reticulum
cis Face
trans Face
Plasma membrane
Protein synthesis on endoplasmic reticulum (see Fig. a)
Golgi apparatus (see Fig. b)
Vesicles/granules
Exocytosis

(c)

Reference De Duve, C. (1984) *A Guided Tour of the Living Cell*, Scientific American Books, New York. A beautifully illustrated overview of cellular processes.

Reference Landry, S.J. and Gierasch, L.M. (1991) Recognition of nascent polypeptides for targeting and folding, *Trends in Biochemical Sciences*, **16**, 159–63. Up-to-date coverage of the recognition and role of molecular chaperones in protein folding.

the ability either to manufacture the types of membrane phospholipid they require or to 'redesign' dietary lipids to their own requirements.

Although much is known about the basic biosynthetic mechanisms for manufacturing lipids, relatively little is known about how the molecules find each other and self-assemble in the correct way. Cell membranes are never built from scratch, however. Membranes are always formed from existing membranes, and indeed, membranous organelles such as mitochondria appear to be formed by division of existing mitochondria (*Cell Biology*, Chapter 5).

Box 1.6
Fossil fuels

The carbon compounds of organisms are not always completely broken down to carbon dioxide and water following death. Coal, oil and natural gas, the fossil fuels, all originate from the biosynthetic processes which resulted in the fixation of carbon. However, fossil fuels only formed if the organic matter was buried before micro-organisms oxidized it. Geological forces then acted on the buried matter resulting in compaction, heating and the promotion of chemical reactions. This resulted in a loss of oxygen and other elements to leave, in the case of oil and natural gas, a material composed almost solely of hydrocarbons, and, with regard to coal, one almost pure carbon. Most geologically-stored organic matter occurs in a dilute form in sedimentary rocks and is found concentrated in commercially exploitable amounts only in certain conditions.

Crude oil originated from dead plankton (microscopic marine plants and animals) and plant debris, swept to sea by rivers, which sank and became mixed with mud and sand in the sediments of inland seas and coastal marine basins. Coal is fossilized plant material dating principally from the Carboniferous period (285–345 million years ago). At that time vegetation was dominated by vast areas of conifers, ferns and similar plants. The death and partial decay of these plants in large swamps formed thick beds of peat from which present day coal fields are derived.

(a)

(b)

Fossils of typical plants of the Carboniferous period from which present day coal fields are derived.
(a) Partial frond of the fern *Neuropteris rectineruis* (Manchester Museum, specimen LL. 2575).
(b) Portion of frond of the clubmoss *Lepidostrobus* sp (Manchester Museum specimen LL. 7999).
Both specimens courtesy Dr J.R. Nudds.

Reference Alberts, B. *et al.* (1989) *Molecular Biology of the Cell* 2nd edn, Garland, New York, pp. 451–8.

Box 1.7
Inorganic biological materials

Biosynthetic activities may involve the deposition of inorganic material, e.g. the impregnation of calcium salts into bone tissue or the formation of the silica skeletons of the unicellular or colonial algae called *diatoms*.

Photomicrograph of skeletons of a number of different diatoms. Courtesy G. Oppermans, Buxton Micrarium, UK.

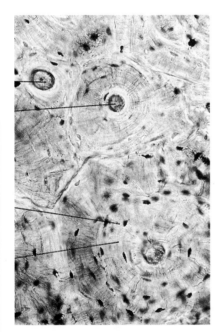

Photomicrograph of hard bone.

Bone does not have many industrial uses, although following sterilization it is used as a fertilizer, 'bonemeal'. However, diatomaceous earth, derived from the cell walls of diatoms, has many uses.

Diatoms have a complex wall, composed of two similar but unequal sized portions which fit together like a box with an overlapping lid. The cell walls are impregnated with silica producing intricate patterns. The walls of dead cells resist degradation and sink forming sedimentary deposits ('diatomaceous earth'), which vary in consistency from soft earth to compact rock. These deposits may be extensive covering several square miles and be hundreds of feet thick. Diatomaceous earth was used in clarifying sugars and syrups, but it has been adapted for many industrial filtration applications, including processing oils, solvents, alcoholic and non-alcoholic beverages, antibiotics and a variety of chemicals. Its second major use is as a filler or extender in paper, paints, bricks, tiles, plastics, soaps and detergents. Other uses include heat and sound insulation, as a carrier of herbicides and fungicides and as a fine abrasive in silver polishes. Under the name *kieselguhr* (derived from the German words *kiesel* and *guhr* meaning *flint* and *sediment* respectively), it has been used as the inert ingredient of dynamite.

1.7 Overview

Cells must be capable of making all the substances of which they are composed or of obtaining them from the environment. In fact, they do both. Making new cellular substance is called biosynthesis.

Making new cellular compounds requires energy which comes from the sun in the case of photosynthetic organisms or by oxidatively degrading organic compounds in the case of all others. The intermediary or energy currency used to drive the energy-requiring biosynthetic reactions is ATP. Some organisms can make all of their cellular compounds starting from CO_2, H_2O and a nitrogen source. Only certain microorganisms can fix nitrogen from the air and all other organisms are dependent on this process. Many of the metabolic transformations that go on in cells are modifications of small organic molecules. These are constantly being re-tailored to suit the needs of each particular cell.

Although metabolic pathway charts appear to be very complicated, and probably a good half of the reactions shown may be classed as biosynthetic ones, when they are looked at in detail a great number of common points emerge. Not only are the majority of the pathways similar or even identical in widely different cell types and organisms, but also the reactions themselves are based on quite a small number of enzymic transformations. Especially in the biosynthesis of biological macromolecules, a few common types of monomer unit are linked together in condensation reactions, employing the energy released by the hydrolysis of nucleoside triphosphates. By means of a few simple reactions a great deal is achieved in beautifully interlocked and controlled routes to achieve the biological aims of producing new cellular material.

1. In darkness, plants still need a supply of energy to survive, and they obtain this (like all other organisms) by oxidizing organic compounds, such as stored starch, made during the hours of sunlight.

2. A currency may have no intrinsic value of its own (a banknote, for example) but is recognized as an intermediary enabling two 'processes' to be linked. I work to obtain (valueless) banknotes but these may be used to buy food and goods.

3. Concentrated sulphuric acid is one and metallic sodium is another. It would not be a good idea to employ either of these in biological systems.

4. CO_2 needs to be converted to $(CH_2O)_n$, or carbohydrate.

5. If the wrong amino acid were in the active site, the enzyme might have no activity. At other places, the substitution may cause the protein to assume the wrong shape, preventing the active site from forming *its* correct conformation. At yet other sites, the substitution might have no effect at all.

6. A nucleotide is a base–sugar–phosphate unit, and $NADP^+$ consists of two of these linked through the phosphates. The bases are adenine and nicotinamide. One of the sugars has an extra phosphate (see Fig. 1.9).

7. The proteins could be sequenced and the amino acid sequences compared. Alternatively tryptic peptide maps may be prepared and may differ. Another possibility is that an antibody to the protein from one species may fail to react, or react to a lesser extent, with the protein from the other species.

8. In books, on magnetic disks or tapes, as bar codes, or photographs.

9. A glycoside link is an acetal or a ketal formed between an alcohol and an aldehyde or ketone.

Ring formation in sugars:

Hemiacetal

Acetal

This may now condense *via* an alcohol group of a second sugar, R'–OH, to form a full acetal. (Similarly for hemiketal and ketal).

Write this out for yourself using two glucose molecules and linking carbons 1 and 4.

10. The Law of Mass Action states that equilibria are drawn in the direction that removes a constraint, so that the equilibrium is re-established. For the reaction:

$$RCO_2H + H_2N–R' \rightleftharpoons RCONHR' + H_2O$$

removing water will cause the reaction to proceed from left to right to re-establish the equilibrium.

11. Cellulose is a β1–4 linked glucose polymer that forms the major bulk of plant cell walls.

12. An oligosaccharide is a chain, often branched, of a few sugar units. A 'few' in this case may indicate up to about 20 or so monosaccharide residues. Anything larger than this is a polysaccharide.

QUESTIONS

FILL IN THE BLANKS

1. *Biosynthesis* is synonymous with _____ which describes the metabolic processes whereby organisms build up _____ molecules from simpler ones. Some organisms can build all their cellular material from _____ , water and a source of _____ , but on the whole animal cells cannot do this. The macromolecules of the cell are built up from _____ units in an _____ requiring process which usually involves _____ or another nucleoside triphosphate. This process is a series of _____ reactions to produce long _____ molecules.

Choose from: complex, nitrogen, energy, condensation, anabolism, CO_2, monomer, ATP, chain.

2. In general, the majority of the energy released by living cells comes from _____ processes such as the catabolism of _____ and _____ obtained from or stored food materials. Plant cells differ in that they obtain energy from _____ , but they still store _____ and _____ for use in periods of darkness. The formation of the _____ building blocks of cellular macromolecules takes place either *de novo*, and requires _____ power: alternatively these units are obtained from food materials. The coupling together of monomers to form _____ is neither oxidative nor reductive but is described as a _____ reaction requiring the removal of _____ .

Choose from: fats (2 occurrences), sunlight, monomer, macromolecules, water, oxidative, carbohydrates (2 occurrences), reducing, condensation.

MULTIPLE-CHOICE QUESTIONS

3. Which of the following is/are macromolecules?

A. Phospholipids
B. Cholesterol
C. Hyaluronic acid
D. Glycogen
E. Transfer RNA

4. Which of the following combinations of compounds may be used to support some form of biosynthesis?

A. FAD, ATP, NADPH, GTP
B. GTP, ATP, NAD^+, UDP-Glc
C. GTP, ADP, dGTP, dCTP
D. GTP, ATP, UDP-Glc, $NADP^+$
E. GTP, UDP-Glc, amino acyl-tRNA, NADPH

5. Which of the following macromolecules has a branched structure?

A. Cellulose
B. DNA
C. Glycogen
D. Insulin
E. Rubber

6. Which of the following biosynthesis is/are dependent on the participation of membrane systems?

A. Glycogen from glucose phosphate
B. DNA from deoxyribonucleotide triphosphates
C. Glycoproteins
D. Collagen
E. Phospholipids

SHORT-ANSWER QUESTIONS

7. Give two examples of reactions that would be described as condensations (use biological molecules as the starting materials). What is the reverse of condensation?

8. List some of the advantages of using enzymes to catalyse reactions in living systems.

9. What are the main features that distinguish the reactions for making small molecules from those in which these monomers are joined to make macromolecules.

10. For each of the biological macromolecules, proteins, polysaccharides and nucleic acids, say what the monomer units are and in each case identify the origin of the information that ensures the monomers are joined in the correct order.

11. In terms of the biosynthesis of macromolecules, specify what is meant by an 'activated' molecule and briefly explain why such molecules are required.

12. What is the molarity of water in a cell that consists of 70% water?

ESSAY QUESTIONS

13. Write a short essay on Macromolecule Biosynthesis: the Common Features of the Process.

14. Write an essay on the Storage of Information.

15. Write an essay to explain *in simple terms* why ATP is required in biosynthetic reactions and give *one* detailed example of how it participates in such reactions.

16. Try to think of the problems that might be encountered if a cell tries to assemble macromolecules from monomers outside the cell.

17. What features are necessary in order for a structure (e.g. a membrane, a micro-tubule, a virus) to be self-assembling.

2
Photosynthesis

Objectives

After you have read this chapter, you should be able to:

☐ describe how CO_2 is fixed into C_3 to C_6 organic acids by the enzymes of the Calvin cycle;

☐ explain how photosynthetic cells have evolved different biochemical strategies in order to maximize CO_2 fixation or to cope with adverse environmental conditions;

☐ discuss how these biochemical processes are regulated at both enzymic and developmental levels.

2.1 Introduction

Light energy is used by both prokaryotic and eukaryotic cells to generate metabolic energy to drive biosynthesis. The process, photosynthesis, occurs only in cells that contain chlorophyll and then only in certain organelles, chromatophores or chloroplasts (Fig. 2.1). The immediate products of the absorption of light energy in photosynthesis are ATP and NADPH. These compounds are then used in the process of carbon assimilation, whereby atmospheric CO_2 is incorporated into organic compounds. The overall

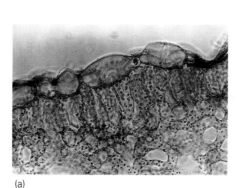
(a)

(b)

See *Cell Biology*, Chapter 5

Fig. 2.1 (a) Photomicrograph of a vertical section through a leaf of *Ranunculus* sp. (×185) showing numerous chloroplasts in the cells. Courtesy of M.J. Hoult, Department of Biological Sciences, Manchester Polytechnic, Manchester, UK. (b) Electron micrograph of chloroplast from a typical C_3 plant. The chloroplast contains membranes (thylakoids) in which the electron carriers necessary for light-driven ATP and NADPH production residue. It also contains a soluble phase, the stroma, which contains all of the enzymes and cofactors for CO_2 fixation.

References Foyer, C.H. (1984) *Photosynthesis*, Cell Biology Monographs, John Wiley, London; Halliwell, B. (1984) *Chloroplast Metabolism, The Structure and Function of Chloroplasts in Green Cells*. Oxford University Press, Oxford, Gregory, R.P.F. (1989) *Biochemistry of Photosynthesis*, 3rd edn. John Wiley, New York. These are all very readable and relatively inexpensive books on photosynthesis which include sections on CO_2 fixation.
Reference Leech, R. (1991) Harvesting the sun. *Biological Sciences Review*, **3(5)**, 13–16. An easily read overview of the development of chloroplasts.

process of carbon assimilation, which is dealt with in this chapter, requires the formation of carbon–carbon bonds and the reduction of carbon from its oxidation state in CO_2 to that in carbohydrate, $(CH_2O)_n$. This is why reducing power in the form of the reduced coenzyme, NADPH, is required as well as ATP.

See *Energy in Biological Systems*, Chapter 3

Ultimately, of course, all of the organic compounds in a green plant or a prokaryote are formed as a result of this process. Furthermore, practically all life on earth is dependent on photosynthesis because all organisms other than photosynthetic ones require preformed organic molecules (carbohydrates, fats, amino acids) in order to live. During periods of darkness, even photosynthetic organisms must survive and carry on their metabolism using organic compounds which they have manufactured and stored during light periods. At such times their metabolism is similar to that of all the heterotrophic organisms on earth.

The 'light reactions' of photosynthesis, whereby ATP and 'reducing power' in the form of NADPH are generated, have been described in *Energy in Biological Systems*, Chapter 3. This chapter is concerned with the **fixation of CO_2** and its transformation into organic compounds. One of the products of photosynthesis is glucose, empirical formula $C_6H_{12}O_6$. It is well established that such a molecule or its precursors are not produced by joining together six carbon atoms from CO_2. Rather, there is a complicated sequence of reactions, commencing with one CO_2 molecule carboxylating an acceptor, C_5, molecule, **ribulose 1,5-bisphosphate** (Fig. 2.2a). The details of the enzyme-catalysed reaction by which this **carboxylation** occurs are now well understood.

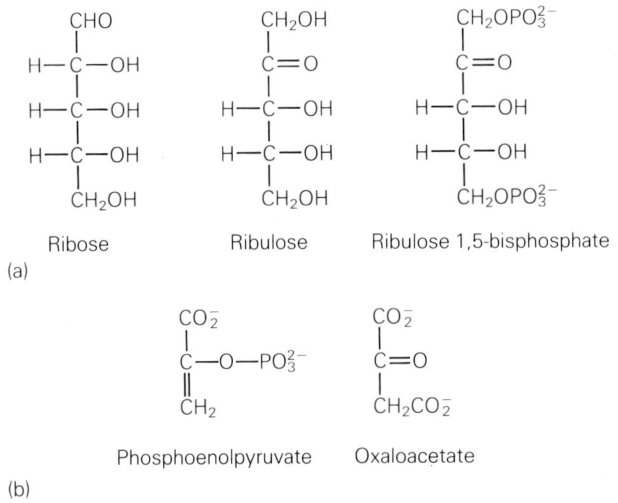

Exercise 1

Explain where the hydrogen atoms come from when CO_2 is reduced to $(CH_2O)_n$ during photosynthesis.

Fig. 2.2 (a) Ribulose 1,5-bisphosphate is the immediate acceptor for CO_2 in C_3 photosynthesis. Ribulose is the ketopentose corresponding to ribose (an aldopentose). (b) In C_4 plants the immediate acceptor of CO_2 in phosphoenolpyruvate resulting in the production of oxaloacetate. However, this 'fixed' CO_2 is later released to be combined with ribulose 1,5-bisphosphate, whereafter the steps in the process of carbon assimilation are the same as in C_3 plants.

2.2 Carbon fixation

All photosynthetic carbon fixation proceeds ultimately via the carboxylation of ribulose 1,5-bisphosphate. In the so-called 'C_3 plants' this compound is the initial or immediate CO_2 acceptor. In 'C_4 plants' it is phosphoenolpyruvate; this reaction yields a C_4 compound oxaloacetate (Fig. 2.2b). Subsequently these reactions are effectively reversed in another part of the plant to yield CO_2 which combines with ribulose 1,5-bisphosphate. The reasons for

Exercise 2

What is the difference between ribose and ribulose?

Fixation of carbon: *incorporating the carbon from atmospheric CO_2 into organic compounds.*
Carboxylation: *the addition of a CO_2 unit to an organic compound.*

Table 2.1 *Types of carbon dioxide fixation*

Type of CO_2 fixation	CO_2 acceptor	First fixation product	Examples of plant
C_3	Ribulose 1,5-bisphosphate	Phosphoglycerate	Spinach, wheat
C_4	Phosphoenolpyruvate	Oxaloacetate	Maize, sugar cane
CAM	Phosphoenolpyruvate	Oxaloacetate	Succulents, pineapple

☐ The distinction between C_3 and C_4 plants may not be as clear cut as previously supposed. Recently, certain plant species have been shown to be able to develop C_3- or C_4-fixing pathways depending on environmental conditions.

------ *Exercise 3* ------

List some ways in which bacterial photosynthesis differs from green plant photosynthesis.

☐ Calvin used the newly available isotopically labelled compounds in his work. ^{14}C had been prepared in 1940 by Ruben and Kamen, but the isotope did not become available for scientific work until after World War II ended. Radioisotopes are much easier to detect than non-radioactive isotopes (such as ^{13}C) which require a mass spectrometer. Calvin used the ability of ^{14}C to blacken photographic film as the detection method.

carrying out carbon fixation this way will be discussed in Section 2.5. They represent one way in which plants have evolved to cope with high light intensities and low CO_2 concentrations (see Table 2.1). Yet other plants are called **crassulacean acid metabolism** or **CAM** plants (Section 2.6). These also trap CO_2 to produce oxaloacetate initially, but they do it in a different way from C_4 plants. This phenomenon was originally found to be characteristic of plants of the family *Crassulacaea*, but is actually much more widespread (Table 2.1).

In the case of photosynthetic bacteria, carbon fixation occurs essentially by a C_3 mechanism. However, the enzyme responsible for catalysing the carboxylation of ribulose 1,5-bisphosphate in prokaryotic organisms is much simpler in structure than that found in eukaryotic plant cells (Section 2.4).

2.3 Initial events in CO_2 fixation in C_3 plants

The **mesophyll cells** of C_3 plants (Fig. 2.1) such as spinach or wheat characteristically possess a number of chloroplasts which have a double outer membrane, called the **envelope**, and a highly organized internal membrane structure known as the thylakoid membranes. These membranes contain all the necessary proteins and pigments for light-driven electron transport. Surrounding the thylakoid membranes, but confined by the envelope, is a protein-rich matrix which contains the enzymes and cofactors responsible for CO_2 fixation. The first step in this process is the combination of CO_2 with ribulose 1,5-bisphosphate to form an unstable C_6 molecule which immediately breaks down to yield two molecules of the C_3 compound, glycerate 3-phosphate, usually called 3-phosphoglycerate.

Calvin's experiments

The details of the initial steps of CO_2 fixation were discovered in 1946 by Melvin Calvin who incubated the microalga *Chlorella pyrenoidosa* in light in a solution containing radioactive $^{14}CO_2$. The incubations were carried out for time periods from a few seconds to several minutes. The algae were then rapidly killed and the radioactive cell contents extracted in boiling ethanol. Any radioactive compounds present in the extract were separated by two-dimensional paper chromatography and then located by **autoradiography**: The developed, dried chromatography sheet is placed against a sheet of X-ray film in the dark. The β-particles given off by any radioactive compounds on the sheet cause blackening of areas of emulsion, thus revealing their positions on the chromatogram. The labelled compounds may be identified chemically by comparing their behaviour in chromatography (in other words, the position to which they move on the paper) with samples of compounds of known structure.

When the alga was exposed to $^{14}CO_2$ for only a few seconds, Calvin found that only the C_3 sugar 3-phosphoglycerate was formed. Longer exposure (a minute or so) gave rise to many ^{14}C-labelled compounds. This experiment is summarized in Figure 2.3. Calvin's first (and not unreasonable) explanation

Mesophyll cells: *cells of the internal parenchyma of a leaf: they are the major photosynthetic cells.*

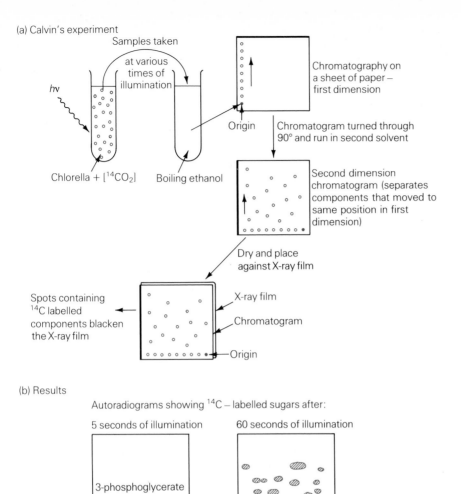

(a) Calvin's experiment

Samples taken at various times of illumination

hv

Chromatography on a sheet of paper — first dimension

Origin

Chromatogram turned through 90° and run in second solvent

Chlorella + [$^{14}CO_2$]   Boiling ethanol

Second dimension chromatogram (separates components that moved to same position in first dimension)

Dry and place against X-ray film

Spots containing ^{14}C labelled components blacken the X-ray film

X-ray film

Chromatogram

Origin

(b) Results

Autoradiograms showing ^{14}C — labelled sugars after:

5 seconds of illumination   60 seconds of illumination

3-phosphoglycerate

Exposed X-ray film

3-phosphoglycerate

Fig. 2.3 Calvin's experiment demonstrating that 3-phosphoglycerate was the first product of carbon fixation. This experiment was one of the earliest uses of radioactive carbon $^{14}CO_2$ to study plant metabolism. For his work on carbon fixation, Calvin was awarded the Nobel Prize.

that there was a C_2 acceptor molecule for the CO_2 was subsequently shown to be incorrect. In fact the immediate acceptor for the $^{14}CO_2$ turned out to be ribulose 1,5-bisphosphate, resulting in the formation of the unstable 2-carboxy-3-keto-D-arabinitol-1,5-bisphosphate. This appears only transiently and immediately breaks down to produce two 3-phosphoglycerate molecules. The initial carboxylation reaction is catalysed by 1,5-bisphosphate carboxylase. Notice that only one of the C_3 products resulting from this initial reaction carries the $^{14}CO_2$ (Fig. 2.4). The overall initial reaction, from the carboxylation of ribulose 1,5-bisphosphate to the generation of the two 3-phosphoglycerate molecules, is highly exergonic and has a free energy change, $\Delta G^{0'}$, of approximately 50.8 kJ mol^{-1}.

The fate of 3-phosphoglycerate

Following the assimilation of CO_2 into 3-phosphoglycerate, two processes occur. The first is the conversion of some of the 3-phosphoglycerate molecules

Reference Gibbs, M. and Latzko, M. (eds) (1979) *Encylopedia of Plant Physiology – Photosynthesis*, vol. II, New Series, Springer-Verlag, Heidelberg. This is a reference text, but contains many useful articles on aspects of CO_2 fixation e.g., Lilley, R. McC. and Walker, D.A., pp 41–52, on reconstituting the enzymes of CO_2 fixation *in vitro*.

Overall sequence:

$$C_5 + C_1 \longrightarrow C_6 \xrightarrow{\;H_2O\;} 2 \times C_3$$

Detailed reactions:

$$
\begin{array}{c}
\mathrm{CH_2O\,\textcircled{P}} \\
| \\
\mathrm{C{=}O} \\
| \\
\mathrm{H{-}C{-}OH} \\
| \\
\mathrm{H{-}C{-}OH} \\
| \\
\mathrm{CH_2O\,\textcircled{P}}
\end{array}
$$
Ribulose 1,5-bisphosphate

$+CO_2 \downarrow$

$$
\begin{array}{c}
\mathrm{CH_2O\,\textcircled{P}} \\
| \\
\mathrm{^-O_2C{-}C{-}OH} \\
| \\
\mathrm{C{=}O} \\
| \\
\mathrm{H{-}C{-}OH} \\
| \\
\mathrm{CH_2O\,\textcircled{P}}
\end{array}
$$
2'-carboxy-3-keto
D-arabinitol 1,5-bisphosphate

$+ H_2O \downarrow$

$$
\left[
\begin{array}{c}
\mathrm{CH_2O\,\textcircled{P}} \\
| \\
\mathrm{HO{-}C{-}CO_2^-} \\
| \\
\mathrm{HO{-}C{-}OH} \\
| \\
\mathrm{H{-}C{-}OH} \\
| \\
\mathrm{CH_2O\,\textcircled{P}}
\end{array}
\right]
$$
Hydrated intermediate

$\downarrow$

$$
\begin{array}{c}
\mathrm{CO_2^-} \\
| \\
\mathrm{H{-}C{-}OH} \\
| \\
\mathrm{CH_2O\,\textcircled{P}}
\end{array}
\quad + \quad
\begin{array}{c}
\mathrm{CH_2O\,\textcircled{P}} \\
| \\
\mathrm{HO{-}C{-}CO_2^-} \\
| \\
\mathrm{H}
\end{array}
$$

3-phosphoglycerate   3-phosphoglycerate
(Glycerate 3-phosphate)   (Glycerate 3-phosphate)

Fig. 2.4 Initial fixation of CO_2 to produce 3-phosphoglycerate. Note that only one of the two 3-phosphoglycerate molecules contains newly-fixed CO_2.

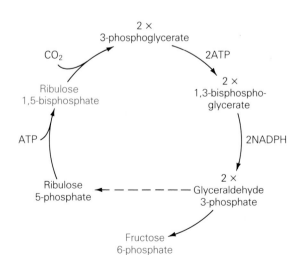

Fig. 2.5 The Calvin cycle. The cycle generates hexose phosphate (fructose 6-phosphate) which can yield other carbohydrates for use or storage, and regenerates ribulose 1,5-bisphosphate so that more CO_2 may be fixed.

into the hexose, fructose 6-phosphate. This can then be converted into a variety of oligo- and polysaccharides for storage. The second process is the regeneration of ribulose 1,5-bisphosphate, the initial acceptor of CO_2, from 3-phosphoglycerate. Although these metabolic pathways will be described separately they are in reality linked since some fructose 6-phosphate is itself needed for producing more ribulose 1,5-bisphosphate. The whole sequence is referred to as the Calvin cycle (Fig. 2.5).

PRODUCTION OF FRUCTOSE 6-PHOSPHATE. The set of reaction pathways by which fructose 6-phosphate is generated from 3-phosphoglycerate are largely similar to those of ***gluconeogenesis*** (Chapter 3) and are shown in Figure 2.6. Most of these steps represent rearrangements of the molecules involved, but note that one step is a reduction. This is the step that produces

☐ In order to connect the metabolic pathways in the chloroplast stroma and cytosol, photosynthetically fixed carbon must be exported from the chloroplast. This is achieved by the triose phosphate/ phosphate transporter which is present in the chloroplast envelope.

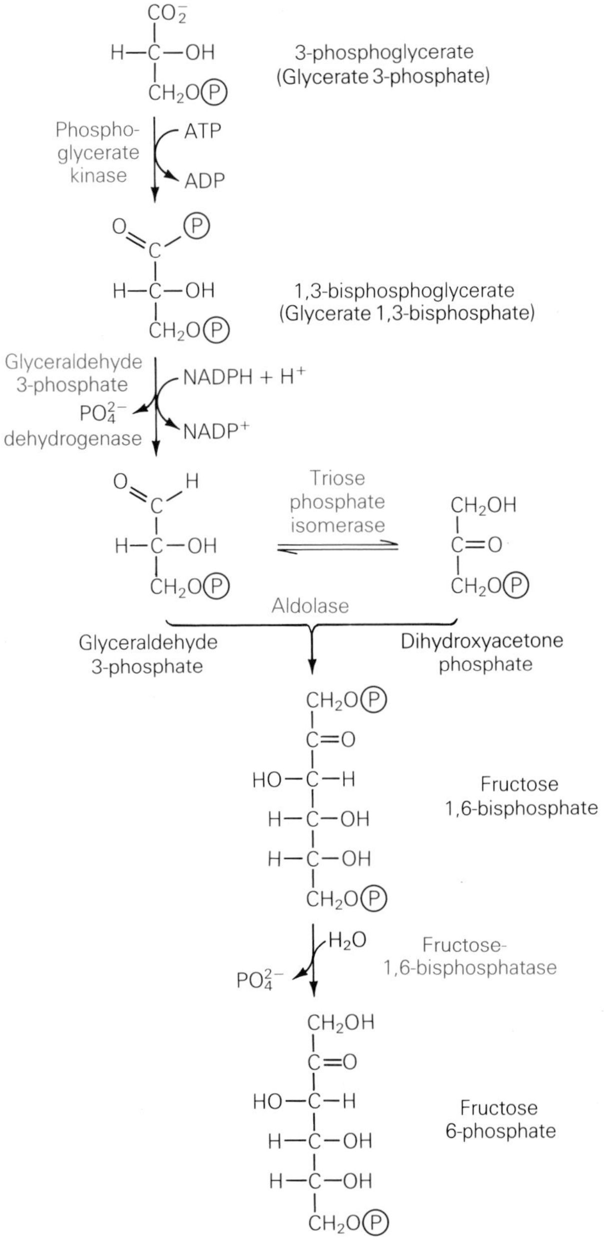

Fig. 2.6 Reduction of 3-phosphoglycerate and generation of hexose (fructose 6-phosphate). The latter may then give rise to other hexoses such as glucose 6-phosphate which may be used for the biosynthesis of starch or cellulose.

Gluconeogenesis: literally, the generation of new glucose. The term refers to the set of enzyme-catalysed reactions by which glucose is generated from most carbohydrate precursors.

Exercise 4

How might fructose 6-phosphate be converted into glucose 6-phosphate for starch biosynthesis?

glyceraldehyde 3-phosphate from 3-phosphoglycerate, a reduction of a carboxylic acid to an aldehyde. This is the only point in the sequence where a reduction occurs and its effect is to change the oxidation state of the newly incorporated carbon atom from that in CO_2 to that in $(CH_2O)_n$, or carbohydrate. The chloroplast enzyme **glyceraldehyde-3-phosphate dehydrogenase** is the enzyme responsible. Although it may be perceived as 'working backwards' to produce the glyceraldelyde derivative, it should be remembered that enzymes do nothing more than catalyse reactions in one direction or the other depending upon the conditions prevailing.

The cofactor for the reduction ('reducing power') is NADPH rather than NADH. For each CO_2 incorporated eventually into fructose 6-phosphate through the pathway described, one molecule of NADPH, and in addition two molecules of ATP, are required (Section 1.2).

☐ NADPH is, in general, employed in biosynthetic reactions in preference to NADH which tends to participate in catabolic reactions. These latter feed reducing equivalents into the electron transport chain for ATP production. Using two cofactors of identical oxidation–reduction potential enables the cell to control biosynthetic and catabolic processes separately. For example, different ratios of $NADPH:NADP^+$ and $NADH:NAD^+$ may be maintained within a cell.

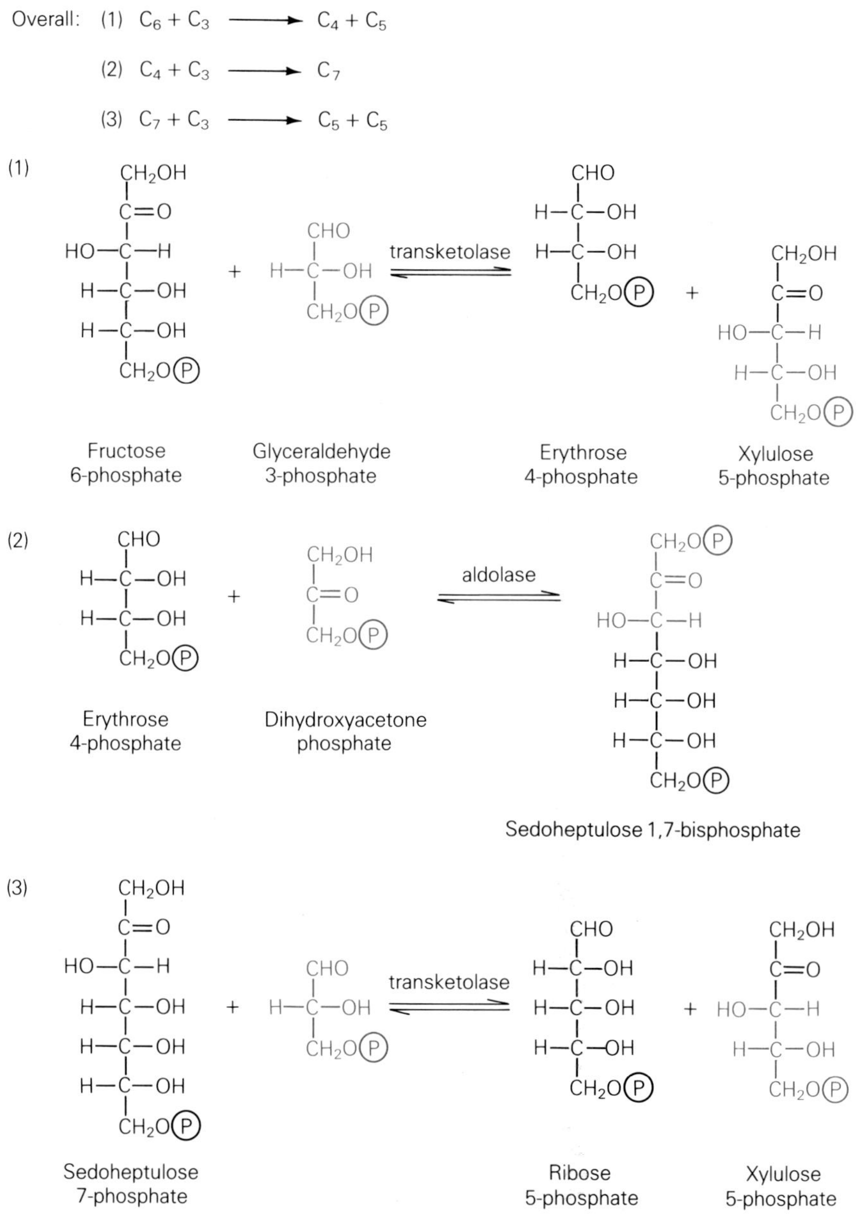

Overall: (1) $C_6 + C_3 \longrightarrow C_4 + C_5$

(2) $C_4 + C_3 \longrightarrow C_7$

(3) $C_7 + C_3 \longrightarrow C_5 + C_5$

Fig. 2.7 Transformations that regenerate ribulose 1,5-bisphosphate enabling more CO_2 to be fixed. Notice that dihydroxyacetone phosphate (reaction 2) may be produced by the action of triose phosphate isomerase on glyceraldehyde 3-phosphate (see Fig. 2.6). A phosphatase is required to produce sedoheptulose 7-phosphate (reaction 3) from sedoheptulose 1,7-phosphate (reaction 2).

REGENERATION OF RIBULOSE 1,5-BISPHOSPHATE is vital to provide a continuous supply of CO_2-acceptor. The process can be considered as a rearrangement of C_3 and C_6 sugars to give the C_5 ribulose 1,5-bisphosphate. Although a number of enzymes are involved in this process, the key ones are **transketolase** and **aldolase**, which catalyse the reactions shown in Figure 2.7. Transketolase, which requires thiamine pyrophosphate as its cofactor, transfers the 2-carbon unit ($-C-CH_2OH$) from the ketoses fructose 6-phosphate and sedoheptulose 7-phosphate to the aldose, glyceraldehyde 3-phosphate (reactions (1) and (3) in Figure 2.7). Aldolase catalyses the reaction involving dihydroxyacetone phosphate but is much less specific as regards the aldehyde that is the other substrate participating in the reaction (which is an aldol condensation). In reaction (2), the 4-carbon sugar erythrose 4-phosphate is utilized for this purpose. The net result of reactions (1–3) is to produce three 5-carbon sugars from one C_6 and three C_3 sugars. In addition to transketolase and aldolase the action of four other enzymes is needed to make the whole cycle, often called the Calvin cycle, operate. These are, respectively: phosphopentose isomerase, which converts ribose 5-phosphate to ribulose 5-phosphate; phosphopentose epimerase, responsible for converting xylulose 5-phosphate into ribulose 5-phosphate; a phosphatase, which removes one phosphate group from sedoheptulose 1,7-bisphosphate; and finally, phosphoribulokinase, catalysing the production of the initial acceptor for CO_2, ribulose 1,5-bisphosphate from ribulose 5-phosphate in an ATP-requiring reaction (Fig. 2.8).

OVERALL STOICHIOMETRY. In order to produce one molecule of a hexose, e.g. fructose 6-phosphate, from newly incorporated CO_2, the Calvin cycle (commencing with the carboxylation of ribulose 1,5-bisphosphate and ending with the regeneration of this CO_2 acceptor) has to operate six times. The reactions leading to the complete synthesis of one molecule of hexose are shown in Figure 2.9. It should be emphasized that out of this complex scheme of biochemical transformations, only one of the three fructose 6-phosphate

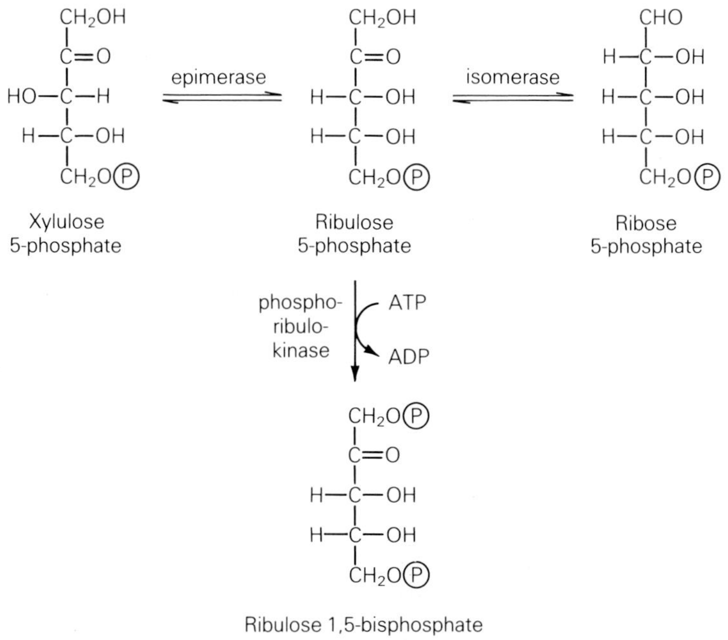

Exercise 5

Try to write out in detail the changes that take place in the conversion of 3-phosphoglycerate into glyceraldehyde 3-phosphate.

☐ The Calvin cycle enzymes are usually considered to operate as independent soluble entities. However, when rather gentle isolation procedures are used, multienzyme complexes which contain enzymes for several consecutive reactions can be isolated. It is thought that these complexes are rather loosely held together and would be disrupted by the harsh isolation methods normally used. **Reference** Contero *et al.* (1988) *European Journal of Biochemistry* **173**, 437–43.

Fig. 2.8 The final transformations that lead to the production of ribulose 1,5-bisphosphate, the CO_2 acceptor.

Exercise 6

What is the difference between xylulose and ribulose?

Transketolase: *an enzyme that catalyses the transfer of a C_2 unit from a ketose sugar to an aldose sugar.*
Aldolase: *an enzyme that catalyses the transfer of a C_3 unit from a ketose sugar to an aldose sugar.*

(a)	6	Ribulose 1,5-bisphosphate + $6 CO_2$ + $6 H_2O \rightarrow$ 12 3-phosphoglycerate
(b)	12	3-phosphoglycerate + 12 ATP $\rightarrow$ 12 1,3 Diphosphoglycerate + 12 ADP
(c)	12	1,3-Diphosphoglycerate + 12 NADPH $\rightarrow$ 12 Glyceraldehyde 3-phosphate + $NADP^+$ + 12 P_i
(d)	5	Glyceraldehyde 3-phosphate $\rightarrow$ 5 Dihydroxyacetone phosphate
(e)	3	Glyceraldehyde 3-phosphate + 3 Dihydroxyacetone phosphate $\rightarrow$ 3 Fructose 1,6-bisphosphate
(f)	3	Fructose 1,6-bisphosphate $\rightarrow$ 3 Fructose 6-phosphate + 3 P_i
(g)	2	Fructose 6-phosphate + 2 Glyceraldehyde 3-phosphate $\rightarrow$ 2 Erythrose 4-phosphate + 2 Xylulose 5-phosphate
(h)	2	Erythrose 4-phosphate + 2 Dihydroxyacetone phosphate $\rightarrow$ 2 Seduheptulose 1,7-bisphosphate
(i)	2	Sedoheptulose 1,7-bisphosphate + $2 H_2O \rightarrow$ 2 Sedoheptulose 7-phosphate + 2 P_i
(j)	2	Sedoheptulose 7-phosphate + 2 Glyceraldehyde 3-phosphate $\rightarrow$ 2 Ribose 5-phosphate + 2 Xylulose 5-phosphate
(k)	2	Ribose 5-phosphate $\rightarrow$ 2 Ribulose 5-phosphate
(l)	4	Xylulose 5-phosphate $\rightarrow$ 4 Ribulose 5-phosphate
(m)	6	Ribulose 5-phosphate + 6 ATP $\rightarrow$ 6 Ribulose 1,5-bisphosphate + 6 ADP

Fig. 2.9 Enzyme-catalysed reactions involved in the overall scheme of CO_2 fixation in photosynthesis and the required stoichiometry.

molecules generated at step (f) is retained. The other two fructose 6-phosphate molecules are recycled in order to regenerate ribulose 1,5-bisphosphate. The overall sum of the reactions is:

$$6 \text{ Ribulose 1,5-bisphosphate} + 6 CO_2 + 11 H_2O + 18 \text{ ATP} + 12 \text{ NADPH} \rightarrow$$
$$6 \text{ Ribulose 1,5-bisphosphate} + \text{Fructose 6-phosphate} +$$
$$18 \text{ ADP} + 17 P_i + 12 NADP^+$$

2.4 Ribulose-1,5-bisphosphate carboxylase ('Rubisco')

Ribulose-1,5-bisophosphate carboxylase (usually referred to as Rubisco) may form more than a sixth of the protein of the chloroplast and can reach concentrations of up to 25% (mass/volume) in the chloroplast stroma. The stroma must therefore be much more like a gel than a dilute aqueous solution of protein. Rubisco can rightly be held to be the 'most abundant protein in nature' and is, indeed, widely distributed in many classes of organism from photosynthetic, carbon-fixing bacteria to higher plants.

The reason for the abundance is not fully understood, but it may in part be due to the relatively low catalytic rate of the enzyme. The carboxylation of only three or four molecules of ribulose 1,5-bisphosphate can be catalysed per second per enzyme molecule. In comparison a 'typical' enzyme may turn over 1000 or more substrate molecules in the same time.

Rubisco from all plant sources consists of large, catalytic subunits (M_r 55 000) and small regulatory subunits (M_r 16 000). The catalytic subunits are synthesized within the chloroplast and the gene that codes for them is located on the chloroplast chromosome. In contrast, the regulatory subunit is coded by a gene located in the nucleus. Rubisco from photosynthetic bacteria is relatively simple in structure consisting either of a single catalytic and a regulatory subunit, or even of just dimers of the large catalytic subunit, that is, no regulatory subunit is present. The Rubisco from higher plants is much larger and more complex. The enzyme comprises eight large and eight small subunits (Fig. 2.10) giving a total M_r of 500–600 000.

The synthesis of such large amounts of this enzyme and its assembly into a fully functional carboxylase complex within the chloroplast obviously pose some problems which have been solved in a rather elegant way. After synthesis of large subunits within the chloroplast stroma and the bringing

□ Chloroplasts, like mitochondria, have their own DNA and their own protein-synthesizing machinery. Bringing the small subunits, manufactured in the cytosol, into the chloroplast stroma requires that they pass through the chloroplast membranes. There are special proteins called chaperonins that enable this transfer and subsequent assembly to occur. **Reference** Hemmingson, S.M., Woolford, C., Van der Vies, S.M. *et al.* (1988) Homologous plant and bacterial proteins chaperone oligomeric protein assembly. *Nature* **333**, 330–4.

Reference Knaff, D.B. (1989) Structure and regulation of ribulose-1,5-bisphosphate carboxylase-oxygenase. *Trends in Biochemical Sciences* **14**, 159–60.

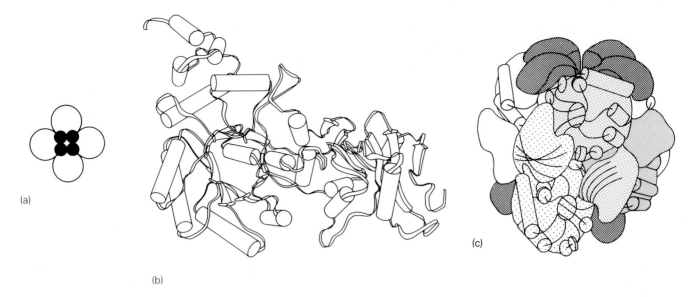

(a)

(b)

(c)

Fig. 2.10 (a) Sketch to show arrangement of eight large and eight small subunits in Rubisco. (b) Schematic diagram of one subunit of *Rhodospirillum rubrum* Rubisco. This bacterial Rubisco is a dimer of large subunits and differs considerably from green plant Rubisco. (Redrawn from Schneider, G., Lindqvist, Y., Brändeń, C-I. and Lorimer, G. (1986) *EMBO Journals* **5**, 3409–15). (c) Sketch to show the structure of tobacco Rubisco which has the subunit structure L_8S_8. (Redrawn from Voet, D. and Voet, J.G. (1990) *Biochemistry*, John Wiley & Sons, New York.)

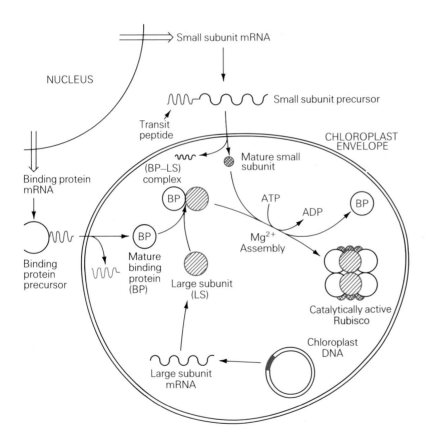

Fig. 2.11 Synthesis and assembly of Rubisco. To produce the catalytically active enzyme within the chloroplast stroma requires the synthesis of small and large carboxylase subunits in the cytosolic and the chloroplast stroma respectively.

in of the small subunits from the cytosol, the two types of subunits are then correctly assembled by a binding protein (BP), which is itself made in the cytosol and transferred into the chloroplast (Fig. 2.11). The assembly reaction requires the hydrolysis of ATP and the presence of Mg^{2+}.

Photorespiration

Whilst Rubisco normally accepts CO_2 in the carboxylation reaction (Fig. 2.4), it is *also* able to accept O_2 instead and function as an oxygenase. This process of **photorespiration** leads to the degradation of ribulose 1,5-bisphosphate to 3-phosphoglycerate and 2-phosphoglycollate (Fig. 2.12). The reaction involves the same active site on Rubisco as the carboxylation reaction, and because oxygen is always present the two reactions compete with each other. Interestingly, the selectivity displayed by Rubisco for CO_2 over O_2 is virtually the same in a wide range of species (Table 2.2).

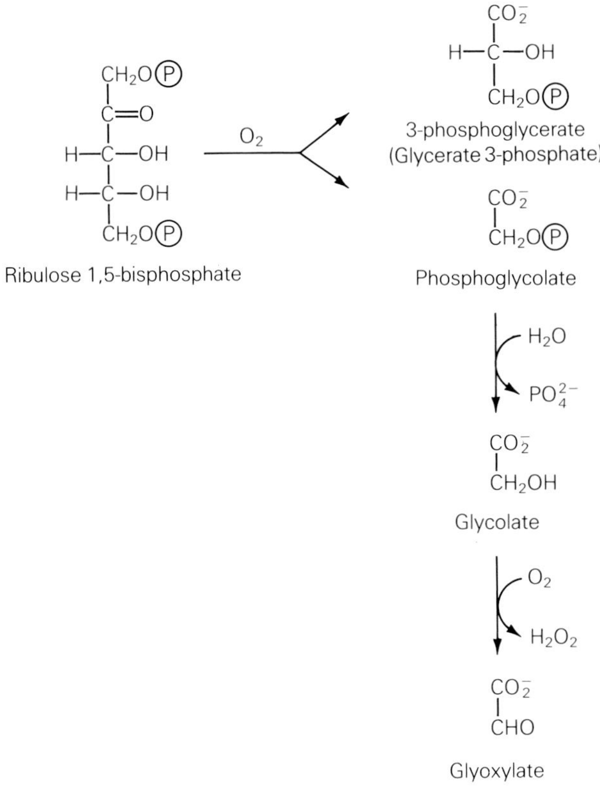

Fig. 2.12 Summary of photorespiration and glycolate metabolism. These processes require the action of enzymes in three organelles: the chloroplast (stroma), the peroxisome and the mitochondrion.

Box 2.1
Suppressing the oxygenase activity of Rubisco

Oxygenase activity can be suppressed by increasing the carbon dioxide concentration. Commercially, with crop plants that are cultivated in closed environments, the CO_2 concentration in the atmosphere is raised several-fold to increase the crop yield. Thus cultivation of greenhouse tomato plants under 1.0% CO_2 rather than the atomspheric concentration of around 0.035% CO_2 has been found to raise crop yield substantially.

Photorespiration: is generally regarded as 'wasteful'. It occurs when Rubisco picks up O_2 instead of CO_2 and produces glycolate.

Reference Woodrow, I.E. and Berry, J.A. (1988) Enzymic Regulation of Photosynthetic CO_2 Fixation in C3 Plants. *Annual Review of Plant Physiology* **39**, 533–94.

Table 2.2 *Species variation in CO$_2$/O$_2$ specificity ratio for Rubisco*

	Source of Rubisco	Specificity ratio
Green algae	*Chlamydomonas reinhardtii*	61
	Scenedesmus obliquus	62
C$_3$ Plants	*Nicotiana tabacum* (tobacco)	77
	Spinacea oleracea (spinach)	80
	Lolium perenne (rye grass)	80
C$_4$ Plants	*Zea mays* (maize)	78
	Sorghum bicolor (sorghum)	70

Following the splitting of ribulose 1,5-bisphosphate, the 2-phospho-glycollate is first converted to glycollate via the action of phosphoglycollate phosphatase. The glycollate is then transported to **peroxisomes** where the action of glycollate oxidase produces glyoxylate. A product of this reaction is H$_2$O$_2$ which is converted to H$_2$O and O$_2$ by catalase. Finally, in the peroxisome, glyoxylate can be transaminated to produce glycine using either serine or glutamate as amine donor.

$$\text{Glyoxylate + Serine} \xrightarrow[\text{aminotransferase}]{\text{Serine--glyoxylate}} \text{Glycine + Hydroxypyruvate}$$

$$\text{Glyoxylate + Glutamate} \xrightarrow[\text{aminotransferase}]{\text{Glutamate--glyoxylate}} \text{Glycine + } \alpha\text{-Oxoglutarate}$$

The glycine then enters the mitochondria where it may be oxidatively decarboxylated yielding NH$_3$ and NADH in addition to CO$_2$.

The remaining glycine carbon atom which is attached to the tetrahydrofolate cofactor (Section 6.6), is transferred to a second glycine by the enzyme serine transhydroxymethylase and results in the formation of a serine molecule.

$$\text{Glycine + Methylene tetrahydrofolate} \xrightarrow{\substack{\text{Serine} \\ \text{transhydroxymethylase}}} \text{Serine + Tetrahydrofolate}$$

Thus, unlike CO$_2$ fixation, the photorespiratory pathway also involves two other cell compartments, the peroxisome and mitochondrion, in addition to the chloroplast (Fig. 2.13).

Box 2.2
Building a better Rubisco

At the present time Rubisco has been crystallized and has been subjected to X-ray diffraction analysis. The structure of its large and small subunits are known to high resolution. This information, along with the new technique of protein engineering, is being used in a number of laboratories in attempts to produce a 'better' enzyme, that is, one that has a faster rate of turnover. A second and equally important problem being addressed is that of photorespiration in which Rubisco catalyses the energetically wasteful oxygenase reaction rather than the more useful carboxylation reaction. The aim of protein engineering (see *Molecular Biology and Biotechnology*, Box 9.6) in this case is to improve the selectivity of Rubisco for CO$_2$ rather than O$_2$ as a substrate.

Peroxisomes: *small organelles bounded by a single membrane and containing catalase and peroxidases.*

Reference Ogren, W.L. (1984) Photo-respiration: Pathways, Regulation and Modification. *Annual Review of Plant Physiology* **35**, 415–42. Comprehensive review; fairly intense reading but a very good source of references relating to photorespiration.

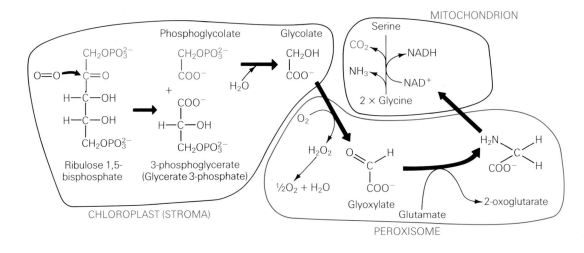

Fig. 2.13 Compartmentation of the steps of glycolate metabolism.

Overall, photorespiration is an energetically wasteful process since the decarboxylation of ribulose 1,5-bisphosphate occurs without any gain in metabolic energy. Indeed, ATP has been utilized in the formation of ribulose 1,5-bisphosphate from ribulose 5-phosphate. Looked at another way, it can be seen that considerably more energy is required to fix a given quantity of CO_2 when photorespiration is occurring than when it is not. In many temperate plant C_3 species, so-called because the first CO_2 fixation product is 3-phosphoglycerate, photorespiration can be responsible for the degradation of 20% or more of the fixed CO_2. Since photorespiration increases at a greater rate with temperature than does CO_2 fixation, it might be expected to provide a severe metabolic limitation for subtropical and tropical plant species. However, an alternative strategy for CO_2 fixation has evolved in these plants which minimizes photorespiration. In this strategy the first carboxylation product is not 3-phosphoglycerate but rather the C_4 acid, oxaloacetate. These species are termed C_4 plants.

Regulation of carbon fixation

Rubisco catalyses the rate-limiting step in carbon fixation. As might be expected, therefore, it is subject to control by a variety of factors, including the provision of sufficient NADPH and ATP by the light reactions, the pH of the stroma, the Mg^{2+} concentration, and the presence of an inhibitor (Fig. 2.14).

THE pH. Rubisco has a pH optimum that lies above that of the stroma in the dark. The light reactions, which cause H^+ ions to be pumped out of the stroma and into the thylakoid lumen, raise the pH of the stroma to about 8.5, and thereby activate the carboxylase.

Mg^{2+} CONCENTRATION. The acidification of the thylakoid lumen in the light reactions is accompanied by the release of Mg^{2+} into the stroma, resulting in an increase in the stromal Mg^{2+} concentration to about $10 \, \text{mmol dm}^{-3}$. Rubisco is stimulated by Mg^{2+} and consequently the rise in concentration of Mg^{2+} serves to activate the enzyme further.

NADPH produced in the light reactions allosterically activates Rubisco.

Reference Edwards, G.E. and Walker, D.A. (1983) *C3, C4, Mechanisms and Cellular and Environmental Regulation in Photosynthesis.* Blackwell Scientific, Oxford. Advanced level reference text, very useful for detailed aspects of C3 and C4 metabolism.

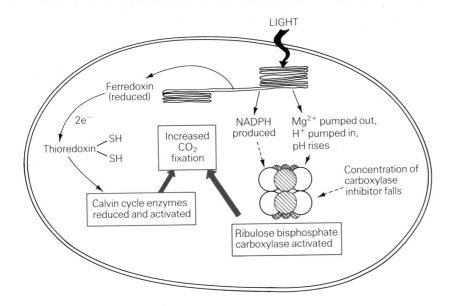

Fig. 2.14 Regulation of CO_2 fixation by light. The Calvin cycle is stimulated by light due to activation of Rubisco by NADPH, Mg^{2+}, increased stromal pH and lowered inhibitor concentration. Other Calvin cycle enzymes are activated by reduction of internal disulphide bonds to –SH groups.

INHIBITOR. Rubisco appears usually to be 'switched off' in the dark by a potent inhibitor bound to the enzyme. The inhibitor, 2-carboxy-D-arabinitol 1-phosphate (see Box 2.3) is a C_6 compound, similar in structure to the C_6 transition compound formed by carboxylation of ribulose 1,5-bisphosphate (Fig. 2.4). Illumination, in some unknown way causes a depletion of this inhibitor to occur, thus the greatest enzymic capacity for CO_2 fixation is found when the light is brightest, that is, at midday.

The enzymes that catalyse the remainder of the Calvin cycle reactions, and process newly fixed CO_2, in the form of 3-phosphoglycerate, to produce the intermediates of this cycle (including the regeneration of ribulose 1,5-bisphosphate), are also activated in the light. However, in this case, activation takes place mainly as a result of reduction of disulphide bonds in the enzymes such as glyceraldehyde-3-phosphate dehydrogenase, and a number of other Calvin cycle enzymes, by the protein **thioredoxin**. This small protein (M_r 12 000) is converted to the reduced form by the small iron–sulphur protein ferredoxin (see section 5.3), which is itself reduced by the photosynthetic electron transport chain:

In the light

Oxidized ferredoxin $\overset{e^-}{\rightleftharpoons}$ Reduced ferredoxin

$$\text{Reduced ferredoxin} + \text{Oxidized thioredoxin} \underset{S}{\overset{S}{|}} \longrightarrow \text{Oxidized ferredoxin} + \text{Reduced thioredoxin} \overset{SH}{\underset{SH}{}}$$

$$\text{Reduced thioredoxin} \overset{SH}{\underset{SH}{}} + \text{Oxidized enzyme} \underset{S}{\overset{S}{|}} \rightleftharpoons \text{Oxidized thioredoxin} \underset{S}{\overset{S}{|}} + \text{Reduced enzyme (ACTIVE)} \overset{SH}{\underset{SH}{}}$$

In the dark

$$\text{Reduced thioredoxin} \underset{S}{\overset{S}{|}} + \tfrac{1}{2}O_2 \rightleftharpoons \text{Oxidized thioredoxin} \underset{S}{\overset{S}{|}} + H_2O$$

$$\text{Reduced enzyme} \underset{S}{\overset{S}{|}} + \tfrac{1}{2}O_2 \rightleftharpoons \text{Oxidized enzyme (INACTIVE)} \underset{S}{\overset{S}{|}} + H_2O$$

Exercise 7

Suggest why the inhibitor 2-carboxy-D-arabinitol 1-phosphate is relatively stable whereas the intermediate formed during the fixation of CO_2 is unstable.

☐ Enzymes activated by thioredoxin include fructose-1,6-bisphosphatase, sedo-heptulose-1,7-bisphosphatase, phospho-ribulokinase and glyceraldehyde-3-phosphate dehydrogenase, and, in C_4 plants, malate dehydrogenase. Note that thioredoxin *deactivates* glucose 6-phosphate dehydrogenase which operates in the pentose phosphate pathway (Chapter 3).

Reference Cseke, C. and Buchanan, B. (1986) Regulation of the formation and utilization of photosynthate in leaves. *Biochimica et Biophysica Acta* **853**, 43–63. Excellent review article covering regulation of CO_2 fixation: fairly advanced but readable. The Bioenergetics Review Series of this journal often have up-to-date, well-written articles covering aspects of CO_2 assimilation.

Box 2.3
2-Carboxy-D-arabinitol
1-phosphate inhibits
Rubisco

The transient C_6 intermediate formed upon carboxylation of ribulose 1,5-bisphosphate (see Fig. 2.4) and the inhibitor, 2-carboxy D-arabinitol-1-phosphate (CA1P) are closely similar in structure.

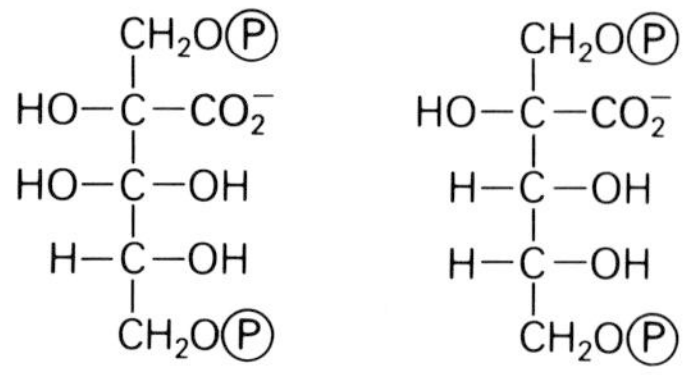

The inhibitor binds to Rubisco which has been carbamylated by attachment of a CO_2 molecule to a specific lysine residue in the protein (this is not the CO_2 which reacts with ribulose 1,5-bisphosphate). This modification is catalysed by an enzyme called 'Rubisco activase' and which serves to activate Rubisco. The figure shows the proposed interaction between the inhibitor CA1P, and the amino acid side-chains of Rubisco. The concentration of CA1P increases in the dark and decreases in the light.

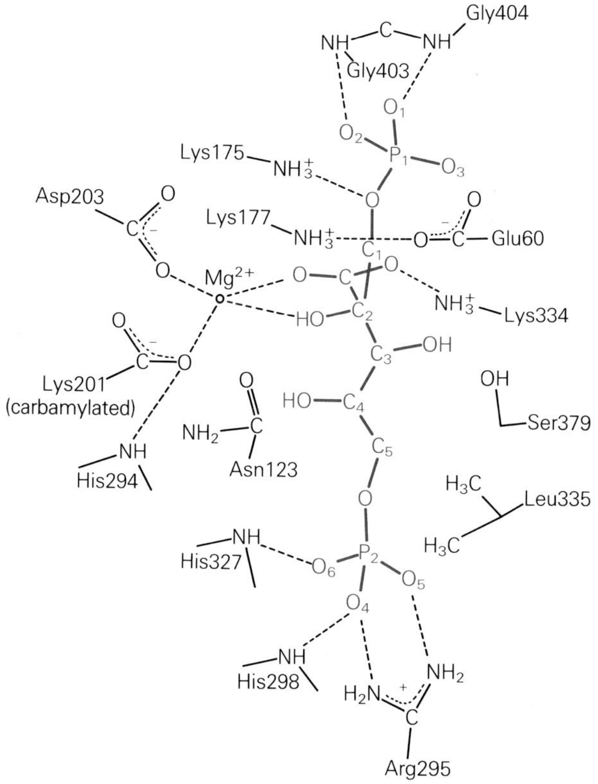

CA1P binding to Rubisco amino acid side-chains. From Anderson *et al.* (1989) *Nature* 337, 229–34.

Deactivation of the Calvin cycle enzymes takes place in the dark simply by oxidation, by oxygen, of the sulphydryl groups to yield the original disulphide bond.

Carbon fixation, like a large number of metabolic pathways, is closely controlled and is intimately linked to the availability of its substrates, that is ATP, NADPH, and, of course, CO_2. The regulation of the Calvin cycle enzymes is summarized in Figure 2.14. Whilst the so called dark reactions of photosynthesis can proceed in the absence of light *in vitro*, they are strongly light-stimulated *in vivo*.

2.5 The C₄ pathway

The leaves of C_4 plants possess a characteristic anatomy which is distinct from that of C_3 plants (Fig. 2.15). In the leaves of C_3 plants the mesophyll cells are distributed randomly throughout the leaf tissue and are the primary sites of CO_2 fixation into 3-phosphoglycerate as described earlier. In C_4 leaves, however, many of the mesophyll cells are arranged more closely around the **bundle sheath cells**, which are themselves more prominent than their counterparts in C_3 leaves. This arrangement is called ***Kranz anatomy***. The mesophyll cells in C_4 plants are also specialized, and fix CO_2 into the C_4 dicarboxylic acid, oxaloacetate (Fig. 2.16) in a reaction catalysed by phospho-enolpyruvate carboxylase, which is present in the cytosol and has higher affinity for CO_2 than Rubisco. In some species, the oxaloacetate is subsequently reduced to malate in the mesophyll chloroplasts and then transported to the bundle sheath chloroplast where it undergoes decarboxylation (by $NADP^+$-malic enzyme) to give NADPH, CO_2 and pyruvate. Finally, in the bundle sheath chloroplast, CO_2 is fixed as in C_3 plants by Rubisco, and enters the Calvin cycle. The pyruvate is returned to the mesophyll cell to be reconverted to phosphoenolpyruvate.

The net result of this cycle, shown in Figure 2.16, is that CO_2 is constantly transported into the bundle sheath cells. Moreover, since the rates of phosphoenolpyruvate carboxylation and subsequent malate decarboxylation are usually greater than the rate at which Rubisco is able to operate, the CO_2 concentration within the bundle sheath chloroplasts is maintained at a much high level than that in the atmosphere and suppresses oxygenase activity and therefore photorespiration. The C_4 cycle effectively functions as a CO_2 pump between the mesophyll and bundle sheath cells but costs metabolic energy to operate. For each round of the cycle, an ATP is hydrolysed to AMP and pyrophosphate:

$$\text{Pyruvate} + \text{ATP} + \text{Pi} \xrightarrow{\text{Pyruvate–phosphate dikinase}} \text{Phosphoenolpyruvate} + \text{AMP} + \text{PP}_i$$

☐ The C_4 grasses and cereals form an economically very important group of plant species, particularly in Third World countries. The C_4 group includes *Zea mays* (maize), *Sorghum bicolor* (sorghum), *Pennisetum americanum* (millet) and *Saccarum* species (sugar cane).

☐ The ^{14}C tracer technique was used in 1954 to discover the early products of C_4 photosynthesis in just the same way as used by Calvin for C_3 plants. The complete C_4 pathway, however, was not worked out until 1960 by two scientists in Australia, Hatch and Slack. The pathway is sometimes referred to as the Hatch–Slack pathway.

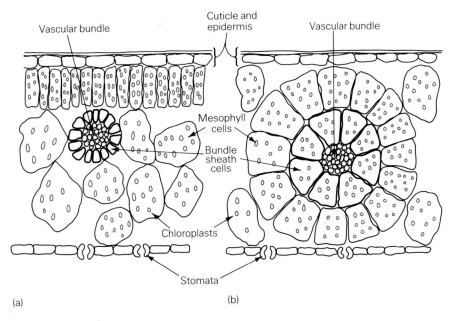

Vascular bundle   Cuticle and epidermis   Vascular bundle

Mesophyll cells

Bundle sheath cells

Chloroplasts

Stomata

(a)   (b)

See *Cell Biology*, Chapter 5

Fig. 2.15 Anatomy of typical C_3 (a) and C_4 (b) leaves. Mesophyll cells containing chloroplasts in both cases are involved in CO_2 fixation. However, in C_4 leaves fixation of CO_2 by Rubisco occurs in the bundle sheath cells whereas C_4 mesophyll cells are specialized to incorporate CO_2 initially into the dicarboxylic acid, oxaloacetate.

Bundle sheath cells: *large parenchyma cells, surrounding a vascular bundle in photosynthetic tissue, characteristic of C_4 plants.*

Kranz anatomy: *the arrangement of the bundle sheath cells in C_4 plants. Kranz is German for 'wreath', and refers to the ring-like arrangement of the photosynthetic cells around the vein.*

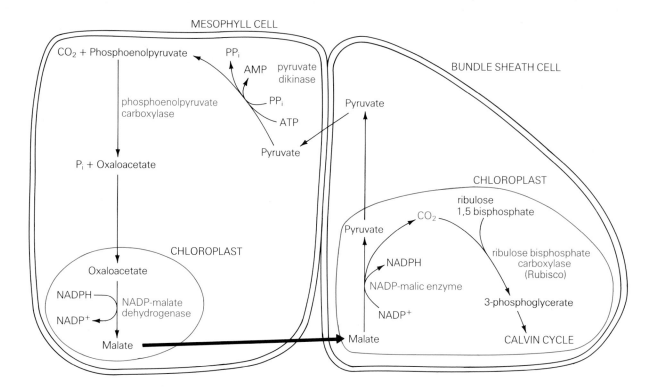

Fig. 2.16 The C_4 pathway of CO_2 fixation. Transport of malate from mesophyll to bundle sheath cells causes CO_2 to be carried between the two cells. For each molecule of CO_2 transported, a molecule of phosphoenolpyruvate must first be generated from pyruvate using the free energy from the hydrolysis of phosphoanhydride bonds.

Box 2.4
Alternative metabolic
routes in C_4 plants

The scheme described in the text represents the simplest C_4 carbon-fixing pathway. Plants that use this scheme of reactions are known as the $NADP^+$–malic enzyme ('NADPme') group of plants. Alternative schemes involve either the NAD^+–malic enzyme for decarboxylation of malate in the bundle sheath cells, or phospho-enolpyruvate carboxykinase which decarboxylates oxaloacetate to phosphoenol-pyruvate, also within the bundle sheath cell. These groups of C_4 species are known as the 'NADme' and 'PEPck' groups respectively. Figure 2.17 shows a simplified scheme of the flow of compounds in the three different C_4 type plants. The most notable differences between the NADPme group of plants and the NADme plus PEPck groups are that in the latter, aspartate is the C_4 compound transported into the bundle sheath whilst alanine, a C_3 compound, is transported back to the mesophyll cells following the decarboxylation reaction.

Since each CO_2 fixed by the C_4 route requires a phosphoenolpyruvate molecule as an acceptor, this reaction means that two extra phosphoanhydride bonds are used per CO_2 fixed. Thus the fixation of six CO_2 molecules to form a hexose requires 12 extra ATP molecules, so that 30 ATP molecules are needed overall (18 + 12) compared with the 18 needed for the production of a hexose by the C_3 scheme.

Reference Jones, C.A. (1985) *C4 Grasses and Cereals*, John Wiley, New York. Fairly advanced overview on the biology of C4 plants. However, the first few chapters give a very readable account of the biochemistry and ultrastructure of this group.

Regulation of C_4 enzymes

Some of the enzymes particular to the C_4 scheme of CO_2 fixation are regulated by reversible reduction using the thioredoxin system described earlier for C_3 enzyme regulation (Fig. 2.14). A key enzyme that is regulated in this way is the NADP–malate dehydrogenase (or NADP–malic) enzyme. However, another key C_4 enzyme, pyruvate–phosphate dikinase (PPDK), is regulated in a somewhat different way. During its catalytic cycle PPDK normally becomes phosphorylated at a histidyl residue in its active site. Once this active site histidyl has been modified, the enzyme can then by phosphorylated at a second residue (threonyl) by a regulatory enzyme, pyruvate–phosphate dikinase regulatory protein (PDRP), that catalyses the reversible phosphorylation of PPDK. It is this second phosphorylation which is unusual since ADP rather than ATP is the phosphate donor. Phosphorylation of the threonyl residue inactivates the PPDK. This scheme is shown in Figure 2.18. Activation of the PPDK occurs by dephosphorylation of the phosphothreonine and is maximal in the light. However, the precise mechanism by which activation and deactivation are regulated by light is unknown. It may be based on the relative levels of ATP and ADP in the cells, since the high ATP : ADP ratio found in the light would favour dephosphorylation of the phosphothreonine residue in PPDK due to relative lack of ADP to serve as a phosphate donor in the inactivation step.

Exercise 8

Summarize the chief differences between C_3, C_4 and CAM plants.

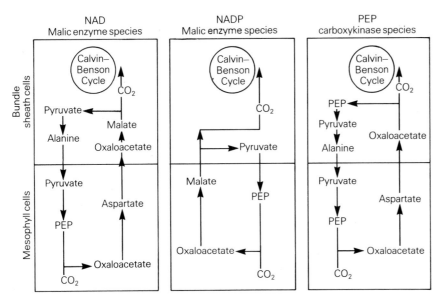

Fig. 2.17 Comparison of CO_2 fixation in the NADPme, NADme and PEPck subgroups of C_4 plants (reproduced from Hatch, M.D. (1976) CO_2 metabolism and plant productivity. in Burns, R.H. and Black, C.C. (eds) *The C4 pathway of photosynthesis: mechanism and function.* University Park Press, Baltimore.)

Reference Edwards, G.E., Nakomoto, H. Burnell, J.N. and Hatch, M.D. (1985) Pyruvate, P_i-dikinase and NADP–malate dehydrogenase in C_4 photosynthesis. *Annual Review of Plant Physiology* **36**, 255–86. A comprehensive review on two key enzymes in C_4 metabolism.

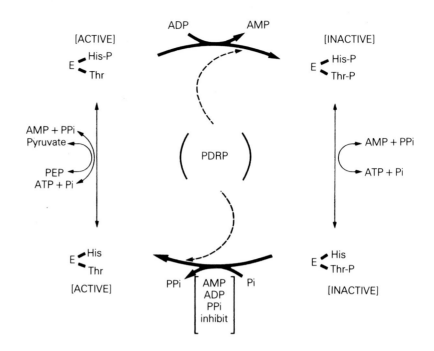

Fig. 2.18 Regulation of pyruvate-phosphate dikinase in C_4 plants by reversible protein phosphorylation. These reactions are believed to be responsible for the light/dark-mediated regulation of pyruvate, P_i dikinase in leaves. $E<^{His}_{Thr}$ represents pyruvate, P_i dikinase, $E<^{His-P}_{Thr}$ represents the enzyme phosphorylated on a catalytic site histidine by reacting with either ATP plus P_i or PEP (see section on the mechanism of reaction). $E<^{His-P}_{Thr-P}$ and $E<^{His}_{Thr-P}$ represent forms of the enzyme inactivated by phosphorylation of a threonine residue with or without phosphorylation on the catalytic histidine.

2.6 Crassulacean acid metabolism (CAM)

In the previous section, it was shown that C_4 plants employ the strategy of *spatially* separating the initial 'trapping' of CO_2 which occurs in mesophyll cells, and the fixation of CO_2 into 3-phosphoglycerate, which takes place in the bundle sheath cells. In CAM plants these events are separated not spatially by *in time*. Many of the enzymes that catalyse C_4 metabolism are also involved in CAM but are present in the same cell, rather than being present in specialized cells. However, the activities of these enzymes are maximally expressed at different times in the day.

Initially, CO_2 is fixed into oxaloacetate in a reaction catalysed by phospho-enolpyruvate carboxylase which occurs in the cytosol and shows maximal activity at night. At this time the stomata are fully open. The oxaloacetate is subsequently reduced to malate by $NADP^+$-malate dehydrogenase and immediately transported to large storage vacuoles where it remains until daytime. During the initial period when light becomes available, carbon fixation occurs using atmospheric CO_2 which is incorporated both into malate by the action of phosphoenolpyruvate carboxylase and also into 3-phospho-glycerate by the action of Rubisco. However, as the light period continues and the light intensity and temperature increase, the stomata close in order to prevent excessive water loss. Now, malate is mobilized from the vacuoles and

Reference Ting, I.P. (1985) Crassulacean acid metabolism. *Annual Review of Plant Physiology* **36**, 595–622. A comprehensive review on CAM metabolism.

Some species of CAM plants, including pineapple, are able to shift between CAM and C_3 metabolism in response to environmental conditions. For example, when pineapples are irrigated and water conservation is of less importance, C_3 fixation of CO_2 predominates. Under drought conditions the pineapple plant switches to CAM metabolism.

Pineapples belong to the family *Bromeliaceae* (bromeliads) which are typically terrestrial, often epiphytic monocots with a reduced stem and a rosette of fleshy, water-storing leaves. CAM is found in succulent plants of the *Cactaceae* as well as the *Crassulaceae*.

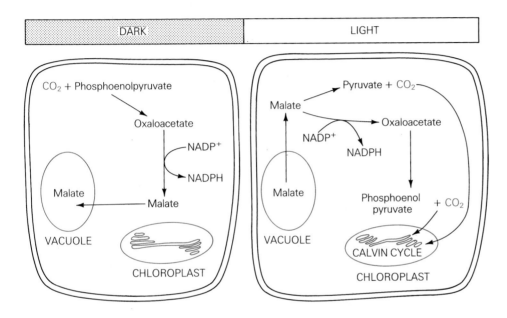

Fig. 2.19 CAM metabolism. In the dark, the net flow of fixed carbon is into the vacuole. In the light, decarboxylation of malate (or reduction to oxaloacetate) yields CO_2 which enters the chloroplast and is fixed by Calvin cycle enzymes.

Appropriately 1% of atmospheric CO_2 is present as $^{13}CO_2$ (rather than the more abundant $^{12}CO_2$). Since Rubisco discriminates against $^{13}CO_2$ to a much greater extent than phosphoenolpyruvate carboxylase, carbohydrates present in C_4 or CAM plants are relatively enriched in ^{13}C as compared with carbohydrates from C_3 plants. The enrichment in ^{13}C is usually expressed as a ratio, $\delta^{13}C$ between ^{13}C and ^{12}C present in the plant, where

$$\delta\ ^{13}C = \frac{^{13}C/^{12}C \text{ in plant} - {}^{13}C/^{12}C \text{ standard} \times 1000}{^{13}C/^{12}C \text{ standard}}$$

The 'standard' is a standard CO_2 prepared from a particular fossil source.

On this basis atmospheric CO_2 has a $\delta\ ^{13}C = -7‰$ (7 parts per thousand more $^{12}C_2$ than the fossil source), whilst C_3 plants have $\delta\ ^{13}C$ values of from -24 to $-34‰$. In C_4 plants and CAM plants $\delta^{13}C$ may be from -10 to -23% and -11 to -33% respectively. Thus a high (less negative) $\delta^{13}C$ value is a good indicator of C_4 or crassulacean acid metabolism.

the activity of phosphoenolpyruvate carboxylase falls to a low level. Rapid decarboxylation of the malate provides a high internal CO_2 concentration, perhaps as high as 1% compared with an atmospheric concentration of 0.035%. The CO_2 is then fixed by Rubisco and enters the Calvin cycle. A further consequence of the high CO_2 concentration produced is that, as in C_4 plants, photorespiration is effectively suppressed. Figure 2.19 summarizes the CAM cycle.

Most CAM species, as with C_4 plants, occur in tropical or arid climates; a commercially important CAM plant is *Ananus sativus*, the pineapple. It should also be noted, that like C_4 metabolism, CO_2 fixation by the CAM route also involves the use of 'extra' ATP. The latter must be used either in the regeneration of phosphoenol pyruvate from pyruvate or in the decarboxylation reaction catalysed by phosphoenolpyruvate carboxykinase (Fig. 2.16).

2.7 Overview

In all plant species, the fixation of CO_2 is at some point carried out by the carboxylation of ribulose 1,5-bisphosphate in a reaction catalysed by the enzyme ribulose-1,5-bisphosphate carboxylase or Rubisco. However, only in the C_3 group of plants is this the initial step in CO_2 fixation, with the C_3 molecule 3-phosphoglycerate being the first stable product. In the C_4 and CAM groups of plants CO_2 is fixed initially in a reaction catalysed by phosphoenolpyruvate carboxylase in which phosphoenolpyruvate carboxylase and CO_2 combine to produce the C_4 acid oxaloacetate. Subsequent events involve transport of oxaloacetate either to a special group of cells (the bundle sheath cells) in C_4 plants, or into the vacuole, in CAM plants. In C_4 plants the oxaloacetate is then converted to malate which is decarboxylated, releasing CO_2 within the bundle sheath cells for fixation by Rubisco. As the bundle sheath cells are surrounded by mesophyll cells which have performed the initial CO_2 fixation into oxaloacetate, this process amounts to the 'pumping' of CO_2 into the bundle sheath cells. In CAM plants similar biochemical events occur but the separation of carboxylation reactions is temporal rather than spatial. In these plants the initial CO_2 fixation and transport of oxaloacetate to the vacuole occurs at night. In the daytime this transport is reversed and decarboxylation of malate in the cytosol again releases CO_2 for fixation by Rubisco. Since, in CAM plants, the stomata are closed during the day to avoid water loss and open at night to take in CO_2, the CAM strategy also acts to concentrate CO_2.

In all of the plant groups, Rubisco can also use O_2 rather than CO_2 as a substrate. When this occurs, the process known as photorespiration takes place, resulting in the splitting of ribulose 1,5-bisphosphate without any useful fixation of CO_2. In C_4 and CAM plants, however, higher internal CO_2 levels are created and keep photorespiration at a relatively low level.

Virtually all of the enzymes involved in CO_2 fixation are strongly regulated, so that they express maximum activity at the time of greatest substrate availability. In the case of Rubisco, NADPH concentration, stromal pH and Mg^{2+} concentration, and the level of a specific inhibitor molecule (CA1P) are major regulatory factors. Other Calvin cycle enzymes are regulated by reversible reduction of disulphide bonds by the redox protein thioredoxin (which is maximally reduced in the light), or by reversible phosphorylation of threonine or serine residues. The net result of these regulatory strategies is that *in vivo* most of the Calvin cycle enzymes are strongly stimulated by light, although *in vitro* they will work in the dark, needing only the appropriate substrates. This is why the reactions catalysed by the Calvin cycle enzymes are usually referred to as 'the dark reactions'.

Answers to exercises

1. Immediately from NADPH and ultimately from H_2O.

2. Ribose is an aldose sugar or aldopentose, whereas ribulose is the corresponding ketose sugar ketopentose. They have identical empirical formulae. The addition of 'ul' indicates the keto sugar.

3. The Rubiscos are different, but also bacterial photosynthesis does not produce oxygen and the chlorophylls and other pigments involved are different. Bacteria have only one photosystem, not two (see also *Energy in Biological Systems*, Chapter 3).

4. By the action of phosphoglucomutase.

5. See Fig. 2.6.

6. They are epimers (see position of –OH on C-3, Fig. 2.8), but both are ketose sugars.

7. See Box 2.3 for the structures. Organic compounds with two hydroxyl groups on the same carbon atom are unstable.

8. In C_4 and CAM plants CO_2 is first fixed into the C_4 molecule oxaloacetate rather than reacting with ribulose 1, 5-bis-phosphate. The anatomy is different (Kranz anatomy). The effect of this initial step,

which costs energy, is that the CO_2 concentration can be increased so that the subsequent reaction with ribulose 1, 5-bisphosphate is more efficient. This enables the plant to photosynthesize at high light intensities but low CO_2 concentrations (with the stomata closed to prevent water loss). In C_4 plants the process of CO_2 trapping and CO_2 release to Rubisco are separated spatially. In CAM plants the two processes are separated temporally.

QUESTIONS

FILL IN THE BLANKS

1. In C_3 plants the first stable product of CO_2 fixation is _____ which is formed by carboxylation of the C_5 acceptor molecule _____ _____ by the enzyme _____ _____ , and subsequent hydrolysis of the transient C_6 intermediate. The _____ produced is phosphorylated to produce _____ and reduced by glyceraldehyde 3-phosphate dehydrogenase using _____ as electron donor to produce glyceraldehyde 3-phosphate. Some of the glyceraldehyde 3-phosphate is converted to the hexose _____ _____ , but most of it is recycled in a complex series of reactions to produce the C_5 sugar phosphate _____ _____ , which is then phosphorylated to regenerate the CO_2 acceptor _____ _____ . These reactions are collectively called the _____ _____ .

Choose from: 1,3-bisphosphoglycerate, Calvin cycle, fructose 6-phosphate, NADPH, 3-phosphoglycerate (2 occurrences), ribulose 5-phosphate, ribulose-1,5-bisphosphate (2 occurrences), ribulose-1,5-bisphosphate carboxylase.

2. The enzyme _____ _____ , which catalyses the first step in CO_2 fixation, comprises _____ _____ and _____ _____ subunits. The _____ subunits are synthesized in the _____ whilst the _____ subunits are synthesized in the chloroplast _____ . Assembly of the _____ and _____ subunits into an active oligomer requires the presence of a _____ _____ . The active _____ is subject to control by a number of factors. For example _____ binds and allosterically _____ . Also the _____ of the stroma _____ when the _____ membranes take up H^+ ions to a _____ that is optimum for _____ _____ activity. Other _____ _____ enzymes are activated by _____ of _____ bonds by the small redox protein _____ . This protein in turn receives its _____ from _____ . Since the _____ form of the latter protein accumulates in the light most of the _____ _____ enzymes are strongly _____ by light.

Choose from: activates, binding protein, Calvin cycle (2 occurrences), complex, cytosol, disulphide, electrons, ferredoxin, large (2 occurrences), large catalytic, NADPH, pH (2 occurrences), reduced, reduction, ribulose-1,5-bisphosphate carboxylase (2 occurrences), rises, small (2 occurrences), small regulatory, stimulated, stroma, thioredoxin, thylakoid.

3. C_4 and _____ plants differ from _____ species in that _____ and not the C_3 molecule _____ is the first stable CO_2 fixation product. This C_4 compound results from the carboxylation of _____ by the enzyme _____ _____. In C_4 plants this reaction takes place in the _____ cells, which surround the _____ _____ cells. The _____ is then reduced to _____ which is transported to the _____ _____ cells and decarboxylated. Fixation of the released CO_2 is then by the enzyme _____ _____ using _____ _____ as CO_2 acceptor. This scheme of CO_2 fixation means that CO_2 becomes _____ in the _____ _____ cells and the energetically _____ process of _____ is suppressed. The special organization of _____ and _____ _____ cells is called _____ anatomy.

Choose from: bundle sheath (3 occurrences), C_3, CAM, concentrated, Krantz, malate, mesophyll (2 occurrences), oxaloacetate, phosphoenolpyruvate, phosphoenolpyruvate carboxylase, 3-phosphoglycerate, photorespiration, ribulose 1,5-bisphosphate, ribulose-1,5-bisphosphate carboxylase, wasteful.

MULTIPLE-CHOICE QUESTIONS

4. Photorespiration takes place when the enzyme ribulose-1,5-bisphosphate carboxylase uses O_2 as a substrate rather than CO_2. Under these conditions, ribulose 1,5-bisphosphate is split into which of the following:

A. phosphoglycolate
B. glycolate
C. glyoxylate
D. phosphoglycerate
E. glyceraldehyde 3-phosphate

5. Ribulose-1,5-bisphosphate carboxylase is activated by which of the following ionic changes in the chloroplast stroma:

A. decreased Cl^- concentration
B. increased Mg^{2+} concentration
C. increased H^+ concentration
D. decreased H^+ concentration
E. decreased Na^+ concentration

6. Which of the following enzymes is involved in the initial fixation of CO_2 by the C_4 pathway?

A. phosphoenolpyruvate carboxylase
B. ribulose-1,5-bisphosphate carboxylase (Rubisco)
C. malate dehydrogenase
D. glycolate oxidase
E. phosphoribulokinase

SHORT-ANSWER QUESTIONS

7.   Compare CO_2 fixation in C_4 plants and CAM plants. Point out the similarities and major differences in CO_2 fixation between these types of plant.

8.   Make a table of the mechanisms used to regulate the Calvin cycle enzymes indicating the enzyme involved and how it is regulated.

9.   Describe the methods you might use to distinguish between C_3 and C_4 plants and indicate which ^{14}C-labelled compounds would be produced initially if leaves from the two types of plant were allowed to take up $^{14}CO_2$.

10.   Draw up an 'energy requirement' balance sheet for carbon fixation in a C_3 plant showing:

(a) the number of ATP molecules used in converting one fixed CO_2 into phospho-glyceraldehyde;
(b) the number of ATP molecules used (per CO_2 fixed) in regenerating ribulose 1,5-bisphosphate;
(c) the number of NADPH molecules used in (i) and (ii);
(d) the total number of ATP and NADPH molecules needed to 'fix' one hexose.

11.   Use the balance sheet constructed in Question 10 as the basis for drawing up a new balance sheet for carbon fixation in a C_4 plant.

(a) Include the ATP and NADPH needed in sections (a)–(c) in Question 10.
(b) Show the number of ATP 'high energy bonds' per CO_2 fixed in capturing and transporting CO_2 from mesophyll to bundle sheath cells.
(c) Determine the total number of 'high energy phosphate' bonds and molecules of NADPH that must be used to fix one hexose by the C_4 pathway.

3

Carbohydrates and gluconeogenesis

Objectives

After reading this chapter you should be able to:

☐ explain the necessity for glucose synthesis from non-carbohydrate precursors and to identify these precursors;

☐ describe the metabolic pathway for gluconeogenesis and explain why it is not a simple reversal of glycolysis;

☐ quantify the requirements for ATP and NADH for glucose synthesis from a variety of precursors;

☐ explain the roles of amino acids and glycerol as sources of glucose;

☐ appreciate why fatty acids cannot be converted to glucose in mammalian tissues;

☐ discuss factors which regulate the rate of glucose synthesis in liver.

3.1 Introduction

Glucose (Fig. 3.1) is an important nutrient for virtually all organisms, including mammals. Together with fatty acids, it serves as a major energy source for most vertebrate tissues. Some cell types cannot oxidize fatty acids and have an absolute requirement for glucose as an energy source. This is true of red blood cells and cells of the central nervous system, particularly the brain. Most of the glucose required is supplied in the diet as sucrose or polysaccharides, but under some circumstances, glucose is also synthesized from non-carbohydrate precursors. This process is called *gluconeogenesis*. Compounds which can be converted into glucose include lactate, amino acids and glycerol (Fig. 3.2). Gluconeogenesis occurs only in the liver and the kidney. Only gluconeogenesis in liver will be considered in this chapter; it is quantitatively much more important than that in kidney under most conditions (Table 3.1).

Gluconeogenesis is particularly important following anaerobic contraction of skeletal muscle when lactate produced is converted back to glucose in the

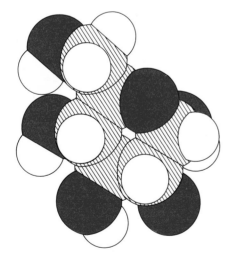

Fig. 3.1 Molecular model of α-D-glucose. Courtesy of C. Holloway, Biosym Technologies Ltd.

Table 3.1 *Sources of carbon for glucose synthesis in liver*

Precursor of glucose	Origin
Lactate	Glucose metabolism during muscle contraction
Amino acids, especially alanine	Diet
	Protein and amino acid metabolism in muscle and other tissues
	Glutamine metabolism in gut
Glycerol	Triglyceride breakdown in adipose tissue

Gluconeogenesis: *the synthesis of glucose from non-carbohydrate precursors. The term is derived from three Greek words which literally mean 'building new glucose'.*

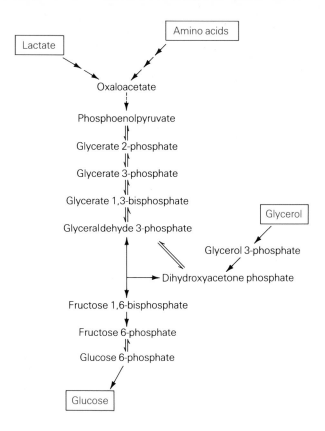

Fig. 3.2 Overall pathway of gluconeogenesis from lactate, amino acids and glycerol in liver.

— *Exercise 1* —

Gluconeogenesis from various substrates provides one source of glucose in short-term starvation. What is the other major source of plasma glucose under these conditions?

liver, and during fasting when amino acids and glycerol in particular are metabolized to produce glucose. Glucose formed in this way is released from the liver and helps to maintain the concentration of glucose in the plasma. Gluconeogenesis from amino acids is particularly significant in short-term starvation. During this condition, the requirement of the brain for glucose is satisfied in part by the breakdown of skeletal muscle to provide amino acids for metabolism to glucose in the liver. In prolonged starvation, brain metabolism is capable of adapting to use ketone bodies to replace some of its glucose requirement and this adaptation decreases the necessity for gluconeogenesis from amino acids derived from muscle breakdown.

3.2 Gluconeogenesis from lactate

Skeletal muscle uses glucose (or glucose 6-phosphate from glycogen breakdown) as an energy source for contraction. Glucose is metabolized in muscle via the glycolytic pathway to produce pyruvate. Some of the pyruvate is oxidized to CO_2 via the TCA cycle, and part of it reduced to lactate. When the muscle becomes anaerobic, or when the rate of pyruvate formation exceeds the rate at which it can be oxidized, such as during vigorous exercise, lactate formation increases. The formation of lactate from pyruvate serves to reoxidize the NADH which was reduced during glycolysis and is essential to allow anaerobic glucose metabolism to continue.

A proportion of the lactate released from skeletal muscle is oxidized in the heart, but the major part is taken up by the liver and converted to glucose (Fig. 3.2). The conversion of glucose to lactate in muscle followed by the reconversion of lactate to glucose in the liver is called the **Cori cycle** (Fig. 3.3).

☐ Vertebrate voluntary muscles can work aerobically or non-aerobically. Aerobically, much more energy can be obtained per mole of glucose. However, in the 'flight or fight' response, the large muscles need to contract strongly and rapidly. In this situation most of the blood is squeezed out of the muscle bed and there is no time between contractions for blood to flow. Consequently both oxygen and glucose supplies are cut off and energetic contraction can only be continued by using stored muscle glycogen anaerobically. This mode of action can only continue for a limited time. Eventually, blood must be allowed to flow again, to replenish the stores and to wash out the accumulated lactate.

☐ Lactate is negatively charged at pH 7 while glucose is neutral. The metabolism of glucose to lactate in glycolysis generates $2H^+$ per glucose, and would lead to acidosis in the absence of compensating factors. The conversion of lactate to glucose in the liver consumes two H^+ per glucose formed. Gluconeogenesis is thus important in the compensation of acidosis caused by anaerobic glucose metabolism. The conversion of neutral amino acids or of glycerol to glucose does not involve a net production or consumption of H^+ ions.

See *Energy in Biological Systems*

Reference Hers, H.G. and Hue, L. (1983) Gluconeogenesis and related aspects of glycolysis. *Annual Reviews of Biochemistry* **52**, 617–53. Comprehensive account, written at rather a high level, but nevertheless understandable.

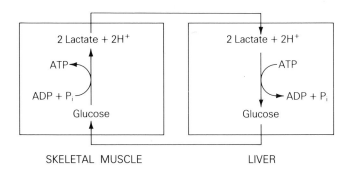

Fig. 3.3 The Cori cycle.

The metabolic pathway for converting lactate to glucose in the liver uses many of the enzymes of the glycolytic pathway. However, three of the reactions in glycolysis are effectively irreversible under physiological conditions and the gluconeogenesis pathway has different enzyme-catalysed reactions which bypass these three irreversible steps, thus allowing the overall resynthesis of glucose to occur (Table 3.2).

Lactate is first converted to pyruvate in the cytosol of the cell in a reaction catalysed by lactate dehydrogenase, with the reduction of NAD^+:

$$
\begin{array}{ccc}
\text{COO}^- & & \text{COO}^- \\
| & & | \\
\text{HO}-\text{C}-\text{H} + \text{NAD}^+ \rightleftharpoons & & \text{C}=\text{O} + \text{NADH} + \text{H}^+ \\
| & & | \\
\text{CH}_3 & & \text{CH}_3 \\
\text{L-Lactate} & & \text{Pyruvate}
\end{array}
$$

Table 3.2 *Different enzymes involved in gluconeogenesis and glycolysis in liver*

Gluconeogenesis	Glycolysis
Pyruvate carboxylase	Pyruvate kinase
Phosphoenolpyruvate carboxykinase	
Fructose-1,6-bisphosphatase	Phosphofructokinase
Glucose-6-phosphatase	Glucokinase

Reference Stryer, L. (1988) *Biochemistry*, 3rd edn, W.H. Freeman, New York. Contains a good, well-illustrated basic account of gluconeogenesis on pages 438–45.

Reference Lehninger, A.L. (1982) *Principles of Biochemistry*, Worth Publishers, New York. Contains a comprehensive account at this level. The energetic aspects are particularly well covered. The text also contains a brief account of gluconeogenesis in ruminants.

Conversion of cytosolic pyruvate to cytosolic phosphoenolpyruvate

In the glycolytic pathway, phosphoenolpyruvate is converted to pyruvate via the pyruvate kinase-catalysed reaction:

$$\text{Phosphoenolpyruvate} + \text{ADP} \rightarrow \text{Pyruvate} + \text{ATP}$$

which has a high negative standard free energy change ($\Delta G^{0'}$ $-31.5\,\text{kJ mol}^{-1}$) and is hence effectively irreversible. This reaction cannot therefore, be involved in the gluconeogenesis pathway. The reactions catalysed by pyruvate carboxylase and phosphoenolpyruvate carboxykinase bypass this irreversible reaction and achieve the conversion of pyruvate to phosphoenolpyruvate at the expense of the hydrolysis of one molecule of ATP and one molecule of GTP, that is, two pyriphosphate bonds.

Pyruvate is converted to phosphoenolpyruvate by the action of two enzymes working in succession: pyruvate carboxylase, which occurs only in the mitochondrial matrix, and phosphoenolpyruvate carboxykinase which is found in the liver cytosol in most species. The conversion of pyruvate to phosphoenolpyruvate is achieved as follows. Pyruvate needs first to be transported into mitochondria via a specific carrier system. In the mitochondrial matrix pyruvate reacts with bicarbonate and ATP to produce oxaloacetate in an energy-requiring reaction catalysed by **pyruvate carboxylase**:

COO⁻ | C=O + HCO₃⁻ + ATP ⟶ O=C(COO⁻)(CH₂—COO⁻) + ADP + Pᵢ | CH₃
Pyruvate   Oxaloacetate

This is a **carboxylation** reaction and pyruvate carboxylase, like other carboxylases, requires **biotin** as a coenzyme. The function of biotin is to bind CO_2 in an activated complex and transfer it to the pyruvate.

Pyruvate carboxylase has a specific and absolute requirement for acetyl CoA as an allosteric activator. The rate of pyruvate carboxylation within the mitochondrial matrix is highly sensitive to changes in the concentration of acetyl CoA in the physiological range. The oxaloacetate produced in this reaction is then effectively transferred to the cytosol by a shuttle system. The mitochondrial membrane is impermeable to oxaloacetate. Oxaloacetate reacts with glutamate to form α-oxoglutarate and aspartate in a reaction catalysed by mitochondrial glutamate oxaloacetate transaminase. α-Oxoglutarate and aspartate leave the mitochondria via specific transport systems and form glutamate plus oxaloacetate in the cytosol in a reaction catalysed by the cytosolic isoenzyme of glutamate aspartate transaminase (Fig. 3.4).

The cytosolic enzyme **phosphoenolpyruvate carboxykinase** catalyses the decarboxylation and phosphorylation of oxaloacetate to produce phosphoenolpyruvate plus CO_2. This reaction also involves the conversion of GTP to GDP:

$$\text{Oxaloacetate} + \text{GTP} \rightarrow \text{Phosphoenolpyruvate} + CO_2 + \text{GDP}$$

The sum of these reactions is:

$$\text{Pyruvate} + \text{ATP} + \text{HCO}_3^- \rightarrow \text{Oxaloacetate} + \text{ADP} + \text{P}_\text{i}$$

$$\text{Oxaloacetate} + \text{GTP} \rightarrow \text{Phosphoenolpyruvate} + \text{GDP} + CO_2$$

$$\text{Pyruvate} + \text{ATP} + \text{GTP} \rightarrow \text{Phosphoenolpyruvate} + \text{ADP} + \text{GDP} + \text{P}_\text{i}$$

$$\Delta G^{0'} + 0.8 \text{ kJ mol}^{-1}$$

□ Biotin is one of the B group of vitamins. It functions as a carrier of metabolically active CO_2 in a number of enzymes in addition to pyruvate carboxylase.

Attached to a lysine side-chain of enzyme

Molecular model of carboxybiotin. The metabolically active CO_2 is shown in red.

Exercise 2

Biotin is a cofactor for several carboxylase reactions. Name two other carboxylase enzymes in liver which require biotin as a cofactor. Give the reactions involved and indicate their importance in metabolism.

□ The liver mitochondrial membrane contains several specific transport systems for metabolites (*Cell Biology*, Chapter 5). These include carriers for phosphate, adenine nucleotides, malate, citrate, α-oxoglutarate, pyruvate, glutamate and aspartate. The carriers mainly involved in gluconeogenesis are those for malate, aspartate and pyruvate. The mitochondrial membrane is not permeable to NADH nor to oxaloacetate.

□ Carboxykinase catalyses the reaction of GTP and oxaloacetate to give phosphoenol pyruvate, GDP and CO_2. The reaction does not involve biotin and CO_2 is not 'activated'.

Exercise 3

In which other metabolic pathway in liver and adipose tissue is pyruvate carboxylase involved?

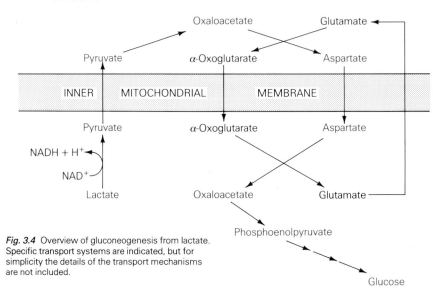

Fig. 3.4 Overview of gluconeogenesis from lactate. Specific transport systems are indicated, but for simplicity the details of the transport mechanisms are not included.

These sequential reactions achieve the conversion of pyruvate to phosphoenolpyruvate. The overall standard free energy change ($\Delta G^{0'}$) for this process is $+0.8$ kJ mol^{-1}. However, at the concentrations of metabolites present in the liver cell the free energy change for the conversion of pyruvate to phosphoenolpyruvate is negative and this reaction proceeds in the direction of phosphoenolpyruvate formation. It may be recalled that the standard free energy change for the removal of phosphate from phosphoenolpyruvate is about twice that for ATP hydrolysis to ADP.

Metabolism of cytosolic phosphoenolpyruvate to glucose

See *Energy in Biological Systems*, Chapter 5

Phosphoenolpyruvate is converted to fructose 1,6-bisphosphate by the reversal of the series of enzyme reactions of the glycolytic pathway:

1. Phosphoenolpyruvate + H$_2$O $\rightleftharpoons$ Glycerate 2-phosphate
2. Glycerate 2-phosphate $\rightleftharpoons$ Glycerate 3-phosphate
3. Glycerate 3-phosphate + ATP $\rightleftharpoons$ Glycerate 1,3-bisphosphate + ADP
4. Glycerate 1,3-bisphosphate + NADH + H$^+$ $\rightleftharpoons$ Glyceraldehyde 3-phosphate + NAD$^+$ + P$_i$
5. Glyceraldehyde 3-phosphate $\rightleftharpoons$ Dihydroxyacetone phosphate
6. Dihydroxyacetone phosphate + Glyceraldehyde 3-phosphate $\rightleftharpoons$ Fructose 1,6-bisphosphate

□ Organic compounds having two phosphate groups on different parts of the molecule are called *bis*phosphates, as in fructose 1,6-*bis*phosphate, which used to be called fructose 1,6-diphosphate. It is now taken that diphosphates have two phosphates that are joined together, as in adenosine diphosphate.

The reaction catalysed by phosphofructokinase in glycolysis has a $\Delta G^{0'}$ of -14.3 kJ mol^{-1} and is not reversible under physiological conditions. A different enzyme is involved in catalysing the conversion of fructose 1,6-bisphosphate to fructose 6-phosphate, namely fructose-1,6-bisphosphatase which catalyses the reaction:

Fructose 1,6-bisphosphate + H$_2$O $\rightarrow$ Fructose 6-phosphate + P$_i$

$$\Delta G^{0'} - 16.4 \text{ kJ mol}^{-1}$$

This reaction is much more likely to proceed because it involves a hydrolysis: ATP is not produced as a reversal of the phosphofructokinase reaction would be required. Fructose-1,6-bisphosphatase is inhibited by AMP and also by fructose 2,6-bisphosphate. The importance of this inhibition will be discussed later in relation to the control of the gluconeogenic pathway.

A number of inborn errors of gluconeogenesis have been recognized. The best understood is known as Type I glycogen storage disease or von Gierke's disease and is characterized by massive enlargement of the liver and kidneys. The condition results from a deficiency of the enzyme glucose-6-phosphatase. Subjects with this disease are unable to augment their blood glucose level by breaking down glycogen, or by gluconeogenesis and become hypoglycaemic on fasting. This hypoglycaemia is unresponsive to glucagon or adrenalin. Increased breakdown of glycogen leads to lactate production rather than glucose release.

Another rare disorder is due to a deficiency of the enzyme fructose-1,6-bisphosphatase. In this condition, glycogen breakdown is normal, but gluconeogenesis is impaired. Subjects with this disease suffer from hypoglycaemia on fasting after liver glycogen has been broken down. Another characteristic of this condition is lactic acidosis due to the inability of the subjects to metabolize lactate to glucose (see marginal note on p. 53).

Fructose 6-phosphate is converted to glucose 6-phosphate in a reversible reaction catalysed by phosphoglucoisomerase:

$$\text{Fructose 6-phosphate} \rightleftharpoons \text{Glucose 6-phosphate}$$

Glucose 6-phosphate is then hydrolysed to glucose in a reaction catalysed by glucose-6-phosphatase:

$$\text{Glucose 6-phosphate} + H_2O \rightleftharpoons \text{Glucose} + P_i \quad \Delta G^{0'} -12 \text{ kJ mol}^{-1}$$

As was the case with phosphofructokinase, the large negative standard free energy change of the ATP-dependent phosphorylation reaction catalysed by glucokinase or hexokinase ($\Delta G^{0'}$ -17 kJ mol^{-1}) means that the formation of glucose 6-phosphate is effectively irreversible. It cannot therefore be used for formation of glucose from glucose 6-phosphate, hence the need for the glucose-6-phosphatase-catalysed reaction. The glucose formed in the glucose-6-phosphatase-catalysed reaction leaves the liver cell via a passive transport system located in the cell membrane and enters the plasma. Alternatively, glucose 6-phosphate may be converted into liver glycogen.

□ If glycogen is being synthesized in liver, there is no need to remove phosphate from the glucose 6-phosphate because this is the starting material for glycogen synthesis. However, if it is *muscle* glycogen that is being replenished, it is necessary for the liver to make free glucose which enters the blood and passes to the muscles where it is rephosphorylated prior to glycogen synthesis. Sugar phosphates do not pass through cell membranes and are not transported in the blood stream.

Requirements for gluconeogenesis from lactate

The overall reaction for the conversion of two molecules of lactate into one molecule of glucose is:

$$2 \text{ Lactate} + 4 \text{ ATP} + 2 \text{ GTP} \rightarrow \text{Glucose} + 4 \text{ ADP} + 2 \text{ GDP} + 6 P_i$$

$$\Delta G^{0'} -38 \text{ kJ mol}^{-1}$$

The overall process as written is energetically favourable. However, the formation of one molecule of glucose is only achieved at the expense of the hydrolysis of *four* molecules of ATP plus *two* molecules of GTP. Two molecules of ATP are required for the conversion of two molecules of pyruvate to two molecules of oxaloacetate in the pyruvate carboxylase-catalysed reaction, while two molecules of GTP are needed for the conversion of two molecules of oxaloacetate to two molecules of phosphoenolpyruvate in the phosphoenolpyruvate carboxykinase-catalysed reaction. A further two ATP molecules are used in the reaction catalysed by phosphoglycerate kinase in which two molecules of glycerate 3-phosphate are converted to glycerate 1,3-bisphosphate. The ATP requirement of gluconeogenesis is met by the oxidation of fatty acids in the liver. GTP can be synthesized from GDP at the expense of ATP:

$$\text{GDP} + \text{ATP} \rightleftharpoons \text{GTP} + \text{ADP}$$

Reference Stanbury, J.B., Wyngaarden, J.B., Frederickson, D.S., Goldstein, J.L. and Brown, M.S. (1983) *Metabolic Basis of Inherited Disease*, 5th edn, McGraw Hill, New York. Contains much interesting information on inherited diseases in this area of metabolism. For reference only.

The synthesis of glucose also requires two molecules of NADH for the triose phosphate dehydrogenase-catalysed reaction. However, with lactate as substrate, this NADH is provided by the lactate dehydrogenase reaction and an additional input of reducing power is not required.

3.3 Gluconeogenesis from amino acids

Amino acids are also major precursors of glucose in liver (Fig. 3.2). After hydrolysis of dietary protein in the gut, amino acids enter the portal circulation and a proportion of these amino acids is taken up by the liver (Fig. 3.5). Amino acids are also produced by protein breakdown in skeletal muscle particularly during starvation or untreated *diabetes*. These amino acids are partially metabolized in the muscle. The amino groups released in muscle are transported in plasma mainly in the form of alanine and glutamine. Alanine produced by muscle is metabolized in the liver, while the glutamine is largely metabolized by the cells of the intestinal mucosa. A major

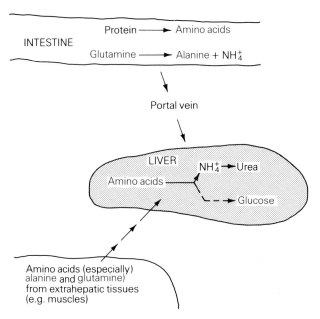

Fig. 3.5 The origins and transport of amino acids which are substrates for gluconeogenesis in liver.

end-product of glutamine metabolism in the gut is alanine, which enters the portal circulation and is metabolized in liver (Fig. 3.5). Alanine is quantitatively the major amino acid source for glucose formation in liver, although gluconeogenesis from other amino acids is also significant.

During amino acid metabolism in liver, urea is formed from amino acid nitrogen. The carbon skeletons of the amino acids are converted, via various complex metabolic pathways, to pyruvate, intermediates of the TCA cycle and acetyl CoA or acetoacetyl CoA. Since a metabolic pathway exists in liver for the conversion of oxaloacetate to glucose, then any amino acid which can be metabolized to oxaloacetate is a potential precursor of glucose. All amino acids except leucine can, therefore, be regarded as substrates for gluconeogenesis, although different amino acids are metabolized at widely different rates. Leucine cannot act as a precursor of glucose since it is

Diabetes: *a group of diseases characterized by a lack of insulin from the pancreas or an insensitivity to insulin (deficiency of receptors?). This results in a high and potentially fatal blood glucose level. The word* diabetes *comes from the Greek word for 'siphon' because untreated diabetic individuals excrete large volumes of urine.*

metabolized only to acetyl CoA and acetoacetyl CoA and these compounds cannot be metabolized to give a net production of glucose. Leucine is said to be **ketogenic**.

PATHWAY OF SYNTHESIS OF GLUCOSE FROM ALANINE. Alanine enters **hepatic** mitochondria and is transaminated with α-oxoglutarate to produce glutamate and pyruvate:

$$\text{Alanine} + \alpha\text{-Oxoglutarate} \rightleftharpoons \text{Glutamate} + \text{Pyruvate}$$

The pyruvate is carboxylated to form intramitochondrial oxaloacetate, which is indirectly transferred to the cytosol (see later). The pathway from cytosolic oxaloacetate to glucose is the same as that described above for lactate (Section 3.2).

In the case of glucose synthesis from alanine, the provision of cytosolic NADH which is required for the triosephosphate dehydrogenase-catalysed reaction must be obtained by transferring reducing power out of the mitochondria by the malate shuttle (Fig. 3.6). This is done by reducing intramitochondrial oxaloacetate to malate, a reaction catalysed by mitochondrial malate dehydrogenase. Malate leaves the mitochondria via a specific transport system and is reoxidized to oxaloacetate in the cytosol. This reaction involves the reduction of cytosolic NAD^+ and is catalysed by cytosolic malate dehydrogenase. Since two alanine molecules are required for the synthesis of one glucose molecule, this malate shuttle provides sufficient cytosolic NADH for glucose synthesis at the expense of intramitochondrial NADH.

□ Leucine is referred to as a ketogenic amino acid because its degradation produces acetoacetate, one of the so-called ketone bodies.

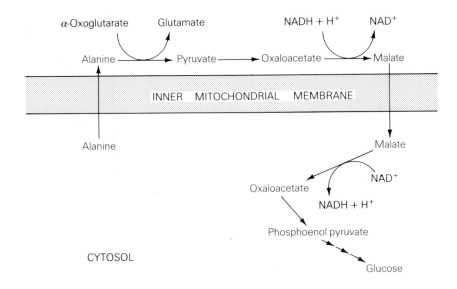

MITOCHONDRIAL MATRIX

Fig. 3.6 The malate shuttle for the efflux of oxaloacetate and reducing power from the mitochondria to the cytosol. Mechanisms of transport are not included.

THE CATABOLIC PATHWAYS OF OTHER AMINO ACIDS form oxaloacetate at some stage, either in the mitochondria or in the cytosol. This oxaloacetate is converted to glucose. As noted earlier, the conversion of oxaloacetate to glucose requires cytosolic NADH for the reaction catalysed by triose phosphate dehydrogenase. The degradative pathways of some amino acids involve reactions catalysed by cytosolic dehydrogenases which provide

Exercise 4

Gluconeogenesis requires a supply of NADH in the cytosol for the reactions catalysed by triose phosphate dehydrogenase. What is the source of this NADH if (a) lactate and (b) alanine is the initial substrate?

Hepatic: *pertaining to the liver (Greek* hepar, *the liver).*

the required NADH. For amino acids whose metabolism does not include cytosolic dehydrogenase reactions, reducing power is exported from the mitochondria by the malate shuttle as described earlier.

3.4 Gluconeogenesis from glycerol

Glycerol is also an important starting point for the synthesis of glucose, particularly during fasting (Fig. 3.2). Glycerol is produced by the hydrolysis of **triacylglycerols** (triglycerides) in adipose tissue, and reaches the liver via the circulation. Glycerol rapidly enters liver cells by diffusion.

Liver contains glycerol kinase which catalyses the reaction:

$$\text{Glycerol} + \text{ATP} \rightarrow \text{Glycerol 3-phosphate} + \text{ADP}$$

Glycerol phosphate dehydrogenase then catalyses the formation of dihydroxyacetone phosphate:

$$\text{Glycerol 3-phosphate} + \text{NAD}^+$$

$$\rightleftharpoons \text{Dihydroxyacetone phosphate} + \text{NADH} + \text{H}^+$$

Dihydroxyacetone phosphate is an intermediate in the pathway of glycolysis as well as of gluconeogenesis from lactate and amino acids, and may therefore be converted into glucose as described in Section 3.2.

Glucose synthesis from glycerol requires the hydrolysis of two molecules of ATP per molecule of glucose formed. Cytosolic NADH is not required and indeed, the oxidation of the glycerol 3-phosphate to dihydroxyacetone phosphate itself reduces cytosolic NAD^+ to NADH. Reoxidation of this NADH by the mitochondria produces three ATP molecules by oxidative phosphorylation per glycerol used. Thus the formation of glucose from glycerol can proceed with a net production of ATP.

3.5 Acetyl CoA cannot be converted to glucose in mammalian tissues

Mammalian tissues do not possess a metabolic pathway for the conversion of acetyl CoA to glucose. Acetyl CoA reacts with oxaloacetate to form citrate in the TCA cycle. The citrate is further metabolized to form oxaloacetate. However, this process does not result in a *net* synthesis of oxaloacetate, and therefore cannot lead to a net synthesis of glucose.

In the formation of citrate, two carbon atoms (acetyl group) are added on to the four carbon compound oxaloacetate to form the C_6 compound, citrate. In the conversion of citrate to oxaloacetate in the TCA cycle, two carbon atoms are lost as CO_2 (in the isocitrate and α-oxoglutarate dehydrogenase-catalysed reactions). Therefore additional oxaloacetate has not been formed. In contrast, the metabolism of lactate and amino acids results in a *net* production of oxaloacetate and hence of glucose. It follows that even-numbered fatty acids, which are metabolized to form acetyl CoA, cannot be starting points for the net synthesis of glucose in mammals, although the glycerol part of the triacylglycerol molecule can be used for glucose synthesis.

3.6 Regulation of gluconeogenesis

Gluconeogenesis is subject to both short-term and long-term regulation. As indicated, gluconeogenesis is particularly important after muscle contraction and during fasting. The major factors controlling the intrinsic rate of glucose

□ Triglycerides are hydrolysed to fatty acids and glycerol in adipose tissue both during exercise and during starvation. This process is controlled by hormones.

See *Energy in Biological Systems*, Chapter 6

--- Exercise 5 ---

Why is it not possible for mammalian tissues to achieve a net synthesis of glucose from acetyl CoA?

Triacylglycerols (TAG): *when glycerol is esterified with three molecules of long-chain fatty acid, the product is called a triacylglycerol. The former term was 'triglyceride'.*

Box 3.3
Gluconeogenesis from propionate is an important source of glucose for ruminants

Propionate ($CH_3CH_2COO^-$) can be metabolized via propionyl CoA to oxaloacetate and is therefore a potential substrate for gluconeogenesis. In non-ruminants, propionate itself is not a significant nutrient, although propionyl CoA is formed as an intermediate in the metabolism of methionine, valine, threonine and odd-numbered fatty acids, all of which are precursors of glucose.

No vertebrate is known to be able to produce cellulose-digesting enzymes. In ruminants, in which cellulose supplies a major portion of their energy and raw material, ingested cellulose is hydrolysed to glucose by bacteria in the rumen. This glucose is not released by the rumen bacteria, but is further metabolized via anaerobic reactions to produce the short-chain fatty acids acetate, propionate and butyrate. Ruminants therefore obtain little glucose from their digestive tracts, and the major nutrients entering the bloodstream are acetate, propionate and butyrate (*Energy in Biological Systems*, Chapter 6). Of these compounds, only propionate is a substrate for gluconeogenesis. Acetate and butyrate are used as energy sources and for fat synthesis. Since ruminants must obtain much of their glucose from the metabolism of propionate, the conversion of propionate to glucose is quantitatively an important pathway in ruminant liver.

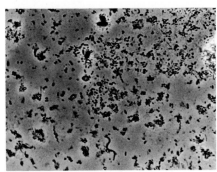

Photomicrograph showing bacteria from rumen of a sheep. Courtesy Dr K.T. Holland, Department of Microbiology, University of Leeds, UK.

synthesis are: (i) the activation of pyruvate carboxylase by acetyl CoA, the intramitochondrial concentration of which is elevated when fatty acid oxidation increases during fasting; (ii) inhibition of fructose-1,6-bisphosphatase by AMP; (iii) product inhibition of glucose-6-phosphatase by glucose and P_i. These regulatory mechanisms ensure that the pathways of gluconeogenesis and glycolysis are not simultaneously active. Thus, for example, an increase in the concentration of cytosolic AMP would stimulate phosphofructokinase and inhibit fructose-1,6-bisphosphatase, increasing the rate of glycolysis and inhibiting glucose synthesis respectively.

The hormone glucagon is secreted from the pancreas in response to a decreased blood glucose concentration (Fig. 3.7). Glucagon acts on liver cells to increase the rate of glucose secretion. One important effect of glucagon is to increase the rate of gluconeogenesis. When glucagon is added to perfused liver or to isolated liver cells, an increased rate of gluconeogenesis is observed within a few minutes. Since mechanisms involving increased protein

See *Cell Biology*, Chapter 8

Box 3.4
Studies using isolated hepatocytes

Much of our knowledge of the regulation of gluconeogenesis has been obtained since 1975 in experiments using isolated liver cells. Such cells can be prepared in large quantity by perfusion of the liver with the enzyme collagenase which disrupts the intercellular matrix releasing individual cells or hepatocytes. The isolated cells are sensitive to hormones and constitute a very useful experimental system for the study of liver metabolism.

Reference Denton, R.M. and Pogson, C.I. (1976) *Metabolic Regulation*, Outline Studies in Biology, Chapman and Hall, London. Although this book is now rather old, it contains an excellent account of methodological approaches to the study of control of metabolism, including gluconeogenesis, and is particularly useful for its description of tissue interrelationships in fat, carbohydrate and amino acid metabolism.

Box 3.5
Futile cycles and metabolic control

It is essential that biosynthetic and degradative pathways be separate at least to some extent, otherwise the direction of a pathway cannot be controlled. The general principle is that metabolism may be switched from the biosynthetic mode to the degradative mode, according to circumstance, by inhibiting one pathway whilst stimulating the other, or *vice versa*. Metabolic control has to work this way. Control by equilibrium (mass action) will not do. As an example, consider the possible fates of glucose 6-phosphate in liver. This might be used for glycogen synthesis, or for energy production via glycolysis, or be converted into ribose to be used in nucleic acid production, or be turned into free glucose to maintain the blood glucose level. These choices could not be made on the basis of mass action control of the equilibria of the various reaction pathways involved.

However, in having different pathways for biosynthetic and degradative routes, there is the chance of developing futile cycles. A case in point is in glycolysis and gluconeogenesis and the steps linking fructose 6-phosphate and fructose 1,6-bisphosphate (Fig.). The step is catalysed by phosphofructokinase in the glycolytic route, and by fructose-1,6-bisphosphatase in the gluconeogenic route. If these two steps operated at the same time, the net effect would be to convert ATP into ADP and inorganic phosphate. This cannot normally be allowed to happen, and it is observed that both phosphofructokinase and fructose-1,6-bisphosphatase are allosteric enzymes whose catalytic action are tightly regulated by the levels of a number of metabolites.

One instance is known where control is removed and the futile cycle is allowed to operate, and this is in bumble bees. In these insects, cycling of the two reactions appears to be used to generate heat (since the free energy released by what is effectively ATP hydrolysis is not conserved and appears as heat). This mechanism raises the temperature of the thorax of the insects and allows them to fly, even on a cold day. Honey bees are not able to do this.

Substrate cycling may also have an important role, in conjunction with adenylate cyclase, in amplifying small changes in the levels of AMP which may trigger other controls. (**Reference** Newsholme, E.A. and Start, C. (1973) *Regulation of Metabolism*, Wiley, London and New York).

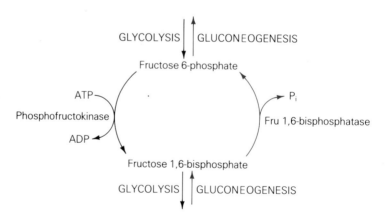

Exercise 6

Glucagon also stimulates glycogen breakdown and inhibits glycogen synthesis in liver by mechanisms involving protein phosphorylation. Describe these mechanisms.

synthesis occur over considerably longer time intervals than this, the short-term effect of glucagon is mediated by mechanisms which involve modification of the activity of pre-existing enzymes rather than by the synthesis of new enzyme molecules. Although a number of different mechanisms are probably involved, this effect is mediated in part by the phosphorylation, and consequent inhibition, of liver pyruvate kinase.

In gluconeogenesis, pyruvate is converted to phosphoenolpyruvate via the sequential reactions catalysed by pyruvate carboxylase and phosphoenol-

Reference Newsholme, E.A. and Leech, A.R. (1983) *Biochemistry for the Medical Sciences*, John Wiley, Chichester. Contains a comprehensive account of metabolism and its control, with emphasis on medically relevant aspects.

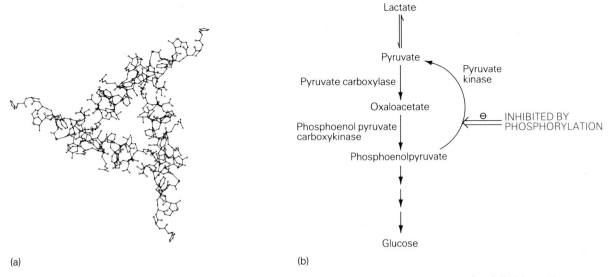

(a)   (b)

Fig. 3.7 (a) Structure of a glucagon trimer as determined by X-ray crystallography. Redrawn from Sasaki, K. *et al.* (1975) X-ray analysis of glucagon and its relation to receptor binding. *Nature*, **257**, 751–7. (b) Regulation of gluconeogenesis by glucagon via phosphorylation of pyruvate kinase.

See *Cell Biology*, Chapter 8

pyruvate carboxykinase (Fig. 3.7). Liver pyruvate kinase can catalyse the conversion of phosphoenolpyruvate back to pyruvate, thus establishing a substrate cycle, sometimes called a *futile cycle*. Liver pyruvate kinase is a substrate for cAMP-dependent protein kinase. Phosphorylation of the enzyme results in an increased K_m for phosphoenolpyruvate and hence an inhibition of the enzyme at physiological concentrations of substrate. Inhibition of pyruvate kinase results in a reduced rate of substrate cycling and an increase in the net conversion of pyruvate to glucose. Liver pyruvate kinase differs from the pyruvate kinases in other tissues which are not phosphorylated by cAMP-dependent protein kinase.

An additional mechanism by which glucagon stimulates gluconeogenesis has been elucidated relatively recently. In addition to the enzyme known as phosphofructokinase 1 (PFK 1) which catalyses the formation of fructose 1,6-bisphosphate in glycolysis, it has been found that a second enzyme catalyses the phosphorylation of fructose 6-phosphate. This enzyme, termed phosphofructokinase 2 (PFK 2) catalyses the phosphorylation of fructose 6-phosphate by ATP to form fructose 2,6-bisphosphate (Fig. 3.8). Fructose 2,6-bisphosphate, which is present in the cell at very low concentrations, is a potent activator of phosphofructokinase and an inhibitor of fructose-1,6-bisphosphatase (Fig. 3.9)

The enzyme which catalyses the formation of fructose 2,6-bisphosphate (PFK 2) possesses a second active site which has phosphatase activity and can catalyse the hydrolysis of fructose 2,6-bisphosphate back to fructose 6-phosphate. PFK 2 is phosphorylated by ATP in a reaction catalysed by cyclic AMP-dependent protein kinase, and this phosphorylation leads to an *inhibition* of the kinase activity of PFK 2 activity and an *increase* in its phosphatase activity (Fig. 3.10).

Thus, an increase in cyclic AMP-dependent protein kinase activity in response to glucagon causes an inhibition of the formation of fructose 2,6-bisphosphate, and a decrease in the rate of its hydrolysis. There is therefore a decrease in the cellular concentration of fructose 2,6-bisphosphate. PFK 1 activity is consequently reduced, leading to a reduction in the rate of

Fig. 3.8 Molecular model of fructose 2,6-bisphosphate.

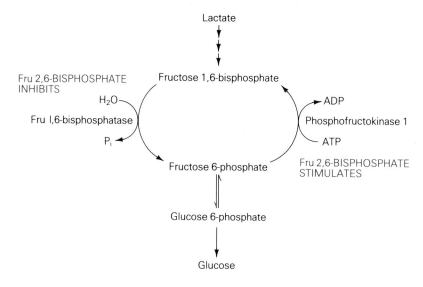

Fig. 3.9 Regulation of the fructose-1,6-bisphosphatase/phosphofructokinase 1 substrate cycle by fructose 2,6-bisphosphate.

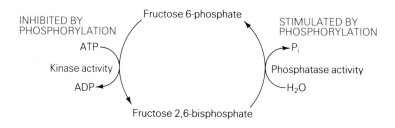

Fig. 3.10 Effect of phosphorylation of phosphofructokinase on its kinase and phosphatase activities, in the interconversion of fructose 6-phosphate and fructose 2,6-bisphosphate.

Exercise 7

Summarize the major mechanisms of short-term regulation of gluconeogenesis?

glycolysis. At the same time, fructose-1,6-bisphosphatase activity is increased leading to an increase in the rate of gluconeogenesis.

When the concentration of cyclic AMP in the cell falls, these changes are reversed. The concentration of fructose 2,6-bisphosphate increases, and both fructose-1,6-bisphosphatase activity and the rate of gluconeogenesis are reduced.

Long-term regulation of gluconeogenesis is achieved by alteration of the amounts of enzymes involved specifically in this pathway. In particular the synthesis of phosphoenolpyruvate carboxykinase and fructose-1,6-bisphosphatase increases markedly in liver cells stimulated by *glucocorticoid* hormones. In contrast, treatment of the liver with insulin decreases the rate of synthesis and the tissue content of enzymes specifically involved in the pathway of gluconeogenesis.

Glucocorticoids: *a category of naturally-occurring as well as synthetic steroids and steroid-like compounds that have effects on glucose metabolism. Example: hydrocortisone.*

Reference Martin, B.R. (1987) *Metabolic Control*, Blackwell Scientific, London. Contains an up-to-date account of the control of gluconeogenesis.

The disease, diabetes is characterized by an insufficiency in insulin production or of an insensitivity to insulin. In the juvenile onset type of diabetes, characterized by the absence of adequate levels of insulin, glucose uptake by muscle and adipose tissue is impaired and the resting blood glucose concentration is high. In untreated diabetes, the ratio of glucagon to insulin concentrations in the plasma is much higher than in normal individuals, and liver metabolism is influenced mainly by glucagon. In particular, gluconeogenesis in the liver is stimulated, and muscle protein is broken down to supply a source of carbon for glucose synthesis, although excess glucose is already present in the plasma. Diabetes also results in increased lipolysis, increased fatty acid oxidation and ketone body production, and increased glucose synthesis from glycerol.

3.7 Overview

Metabolic pathways exist in liver for the synthesis of glucose from non-carbohydrate precursors. The major precursors of glucose are lactate, amino acids and glycerol.

Glucose synthesis is not simply a reversal of glycolysis. The irreversible steps in glycolysis are bypassed by the reactions catalysed by the liver-specific enzymes pyruvate carboxylase, phosphoenolpyruvate carboxykinase, fructose-1,6-bisphosphatase and glucose-6-phosphatase. Both mitochondrial and cytosolic reactions are involved in glucose synthesis. The ATP and GTP required for glucose synthesis are provided by fatty acid oxidation.

The pathway of gluconeogenesis is subject to both short-term and long-term regulation by hormones. Gluconeogenesis is particularly important after muscle contraction, and during fasting when it helps to maintain the level of blood glucose.

Answers to exercises

1. Liver glycogen is the other major source of glucose. During fasting glycogen is broken down to provide glucose which is exported from the liver. See *Energy in Biological Systems*, Chapter 5.

2. (i) Acetyl CoA carboxylase

Acetyl CoA + HCO_3^- + ATP $\rightarrow$
$\qquad$ Malonyl CoA + ADP + P_i

The first reaction in the pathway of fatty acid synthesis from acetyl CoA (Chapter 8).

(ii) Propionyl CoA + HCO_3^- + ATP $\rightarrow$
$\qquad$ methylmalonyl CoA + ADP + P_i

This reaction is involved in the catabolism of threonine, methionine, valine and odd-numbered fatty acids (see Chapter 6).

3. Pyruvate carboxylase is involved in the pathway of fatty acid synthesis. See Chapter 8.

4. (a) From the lactate dehydrogenase-catalysed reaction in the cytosol; (b) from the export of intramitochondrial reducing power via the malate shuttle mechanism.

5. Metabolism of acetyl CoA to oxaloacetate in the TCA cycle does not lead to net oxaloacetate formation since two carbon atoms are lost as CO_2 via the isocitrate dehydrogenase and α-oxoglutarate-dehydrogenase catalysed reactions respectively.

6. Glucagon causes an increase in intracellular cyclic AMP concentrations and activates cAMP-dependent protein kinase. Protein kinase catalyses the phosphorylation and activation of phosphorylase kinase and the phosphorylation and inhibition of glycogen synthetase. Thus glucagon activates glycogen breakdown and inhibits glycogen synthesis in liver (*Energy in Biological Systems*, Chapter 5, and *Cell Biology*, Chapter 8).

7. The most important control is stimulation by the hormone glucagon. This is achieved mainly by cyclic AMP-dependent phosphorylation and inactivation of liver pyruvate kinase. A secondary mechanism of importance is the phosphorylation and inactivation of phosphofructokinase 2.

FILL IN THE BLANKS

1. The pathway of gluconeogenesis from lactate involves _____ reactions not encountered in glycolysis. These are catalysed by the enzymes _____ _____ .

_____ _____ , _____ _____ and _____ , respectively.

The conversion of two molecules of lacate to one molecule of glucose requires the hydrolysis of _____ molecules of ATP and _____ of GTP. The ATP is required in the reactions catalysed by _____ _____ and _____ _____ , while the GTP is required in the _____ _____ -catalysed reaction.

The three major non-carbohydrate sources of glucose production in liver are _____

, _____ _____ and _____ .

Choose from: amino acids, four (2 occurrences), fructose bisphosphatase, glucose-6-phosphatase, glycerol, lactate, phosphoenolpyruvate carboxykinase (2 occurrences), phosphoglycerate kinase, pyruvate carboxylase (2 occurrences), two.

MULTIPLE-CHOICE QUESTIONS

2. Which of the following can serve as precursors for net gluconeogenesis?

A. pyruvate
B. lactate
C. acetate
D. alanine
E. acetoacetate

3. Which of the following enzymes are operative during glycolysis but inoperative in gluconeogenesis?

A. triose phosphate isomerase
B. phosphofructokinase
C. aldolase
D. pyruvate kinase
E. glyceraldehyde-3-phosphate dehydrogenase

4. Which of the following are involved in gluconeogenesis but not in glycolysis?

A. $NAD^+/NADH$
B. biotin
C. ATP/ADP
D. carbon dioxide
E. GTP/GDP

5. State whether the following are True or False:

(a) When stored triglycerides are hydrolysed during starvation both the glycerol and fatty acid molecules produced are converted to glucose in the liver.
(b) The pathway of glucose synthesis from lactate is the exact reversal of the pathway of glycolysis.
(c) The production of glucose from neutral amino acids, but not glycerol, involves a net production of H^+.
(d) Leucine is a glucogenic amino acid because its metabolism produces acetoacetate.
(e) The synthesis of phosphoenolpyruvate and fructose-1,6-bisphosphatase in hepatocytes is stimulated by glucocorticoids.

6.   In what tissues and under what conditions is the synthesis of glucose from non-carbohydrate precursors important?

7.   In the pathway of glycolysis, glucose is converted to lactate. Why is gluconeogenesis from lactate not a simple reversal of this pathway?

8.   Discuss the permeability properties of the mitochondrial membrane in relation to glucose production from (a) lactate and (b) alanine.

9.   The conversion of the carbon skeleton of asparagine to glucose is a relatively simple process which does not require any mitochondrial reactions. Outline this metabolic pathway.

10.   The compound aminooxyacetate is a specific inhibitor of transaminase enzymes. In an experiment, isolated liver cells were incubated with either lactate or pyruvate and glucose synthesis was measured. Aminooxyacetate inhibited gluconeogenesis from lactate but had no effect on gluconeogenesis from pyruvate. Account for these observations.

11.   In an experiment, isolated liver cells were incubated with lactate in the absence of fatty acids. Under these conditions, some of the pyruvate derived from lactate is oxidized via the TCA cycle. Assuming that all the ATP necessary for glucose synthesis is obtained from pyruvate oxidation, what is the ratio of glucose molecules formed to lactate molecules consumed?

4

Polysaccharides

Objectives

After reading this chapter, you should be able to:

□ appreciate the energy requirements of processes for assembling monosaccharides into polysaccharides;

□ discuss the processes by which cells utilize a common sugar such as glucose to produce a variety of 'activated' precursors of monosaccharides;

□ describe the mechanisms by which activated precursors are linked in a precise sequence with appropriate inter-residue links;

□ appreciate the structural changes that may occur after polymerization;

□ explain how polysaccharides are assembled at specific locations both inside and outside the cell.

4.1 Introduction

The conformation adopted by a particular polysaccharide is unique and produces a surface that invests the molecule with a particular biological function. This conformation is principally dependent on the sequence of its constituent monosaccharide residues and in this respect it is like any other biopolymer. This chapter is concerned with the ways in which sugar residues are linked together in polysaccharides. However, in describing how cells produce polysaccharides it is also necessary to understand *where* polymerization occurs and what happens to the polymers before they take on a biological role.

The role of many polysaccharides is structural. As insoluble polymers they support and protect the fragile membrane bilayer surrounding the cytoplasm. Thus they must be located on the exterior face of the cell membrane, sometimes at a particular zone. To create an insoluble skeletal framework within the cell could be disastrous. Hence, cells accumulate precursor carbohydrate residues in the cytoplasm but produce polymers outside the cell.

The sequence of events starts with a monosaccharide and ends with a functional polysaccharide (Fig. 4.1). Not all of the events shown necessarily occur. Reserve polysaccharides like amylose are complete in structure and function at their point of synthesis. However, even simple polysaccharides like glycogen or amylopectin require structural alteration after the initial energy-dependent polymerization of glucose residues before they can fulfil their biological functions.

Reference Stoddart, R.W. (1984) *The Biosynthesis of Polysaccharides*. Croom Helm, London. A comprehensive survey of microbial, plant and mammalian polysaccharide biosynthesis in a little over 300 pages.

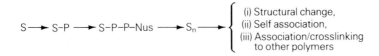

$$S \longrightarrow S\text{-}P \longrightarrow S\text{-}P\text{-}P\text{-}Nus \longrightarrow S_n \longrightarrow \begin{cases} \text{(i) Structural change,} \\ \text{(ii) Self association,} \\ \text{(iii) Association/crosslinking} \\ \qquad \text{to other polymers} \end{cases}$$

Fig. 4.1. Biosynthesis of a polysaccharide (S_n) from a monosaccharide (S). Intermediate stages are the formation of a sugar phosphate (SP) and an activated precursor, a nucleoside diphospho (pyrophosphoryl) sugar (S–P–P–Nus). Structural changes and/or polymer association after polymerization do not always occur.

The progression of events described in Fig. 4.1 has many parallels with the assembly of proteins from amino acids. However, there is one important difference at the beginning of the sequence, namely the step at which the activated precursor is added to the extending polymer. This requires not only a catalytic mechanism to produce the inter-residue bond, a peptide or glycosidic bond, but also some device for selecting a particular **residue** for addition: threonine rather than serine, or galactose rather than *N*-acetyl-glucosamine, for instance. For protein assembly this ordering of the primary structure is determined by the sequence of residues in another polymer, mRNA. No such template is available to enzymes catalysing the union of one sugar residue with another. The sequence of monosaccharide residues in a polysaccharide is determined by the specificity of the synthetase for the donated activated sugar and the accepting residue at the growing terminus of the polysaccharide chain. This difference in mode of assembly accounts for one of the features distinguishing proteins and polysaccharides, namely the variation in M_r that is found in any one type of polysaccharide in contrast to the very precise M_r characteristic of proteins. Indeed, the mechanism that terminates the extension of polysaccharides is poorly understood. The lack of a sequence coding template does not mean that carbohydrate polymers lack complexity or variety. The structures of oligosaccharides covalently bound in glycoproteins and glycolipids demonstrate the impressive ability of cells to assemble complex sequences of sugars.

4.2 Biosynthesis of a glycosidic bond

The incorporation of a sugar residue into a polysaccharide costs about 15–20 kJ mol^{-1} for each residue transferred. The activated precursors most commonly used are sugar derivatives of nucleoside diphosphate, namely, sugar nucleotides. The formation of a sugar nucleotide requires the cleavage of two phosphoanhydride bonds (Fig. 4.2). If the starting point is considered to be the free sugar entering the cell then phosphorylation by ATP accounts for the first of these. Alternatively, the hexose phosphate may be formed through gluconeogenesis or phosphorolysis of a reserve polysaccharide. However, the free energy liberated by removal of a single phosphoryl group from C-1 of the sugar residue is not sufficient to drive the formation of a glycosidic bond, and in a second step a diphospho or pyrophosphoryl group is generated in the form of a nucleoside diphosphate. The free energy change in this second reaction is small and, but for a subsequent reaction, only about half of the hexose 1-phosphate would be converted to the nucleoside diphosphate. This third reaction, the hydrolysis of pyrophosphate, one of the products of the previous reaction, ensures the virtually complete conversion to the sugar nucleotide (Section 1.5). The overall change in free energy for the second and third reactions is about -30 kJ mol^{-1}.

□ Groups at the active site of the synthetase catalyse the formation of an unstable oxonium ion in the sugar being transferred and release a stable nucleoside diphosphate. Nucleophilic attack by a specified hydroxylic oxygen of the accepting sugar residue leads to the formation of a glycosidic bond between the sugar residues.

Activated precursor, a sugar nucleoside diphosphate

Oxonium ion intermediate

Acceptor sugar residue

New glycosidic bond

□ In 1970 L.F. Leloir was awarded a Nobel prize for his pioneering work in identifying sugar nucleotides as precursors in the biosynthesis of polysaccharides. One year later Sutherland was similarly honoured for identifying the role of cyclic AMP. In 1947 C.F. and G.T. Cori received the prize for their work on the recycling of carbohydrates in the body.

□ A nucleotide is a monophosphate ester of a nucleoside. However, the activated sugar precursors that take part in biosynthesis are often described as sugar nucleotides despite the fact that the sugar and nucleoside are nearly always linked through a pyrophosphoryl group.

Residue: a convenient way of describing a monomer incorporated into a polymer.

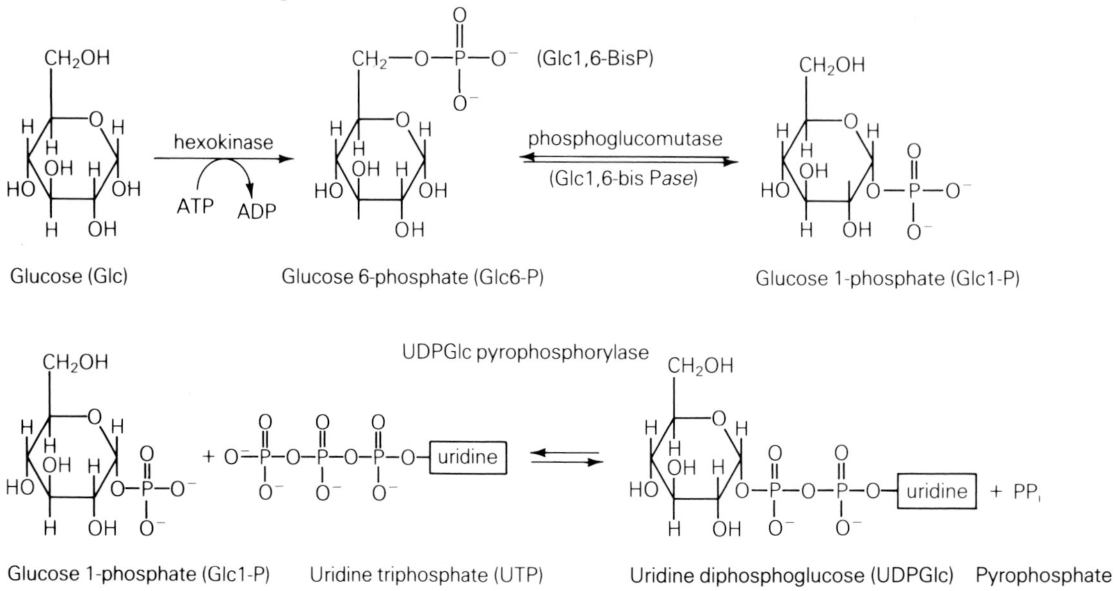

The chemical structures at the top show the conversion: Glucose (Glc) → hexokinase (ATP, ADP) → Glucose 6-phosphate (Glc6-P) (Glc1,6-BisP) → phosphoglucomutase (Glc1,6-bis Pase) → Glucose 1-phosphate (Glc1-P)

UDPGlc pyrophosphorylase

Glucose 1-phosphate (Glc1-P) + Uridine triphosphate (UTP) ⇌ Uridine diphosphoglucose (UDPGlc) + Pyrophosphate

pyrophosphatase

$$PPi + H_2O \longrightarrow 2P_i$$

Fig. 4.2. The synthesis of an activated precursor, uridine diphosphoglucose, from glucose. The formation of UDP-Glc from Glc 1-P is ensured by the subsequent hydrolysis of pyrophosphate (PP_i), which is a highly exergonic reaction.

□ The standard free energy of the reaction:

$$Glc\ 1\text{-}P + Glc_n \rightleftharpoons P_i + Glc_{n+1}$$

is about $-6\,kJ\,mol^{-1}$ and indicates that phosphorylase, the enzyme that catalyses the reaction, could be an agent of both phosphorolysis and synthesis. However, the ratios of P_i to Glc 1-P in most cells strongly suggests that synthetic reaction does not take place to any appreciable extent. More evidence against the role of phosphorylase as a synthetase is that the K_m for Glc 1-P of most enzymes ($1-50\ mmol\,dm^{-3}$ in plants) is far too high to allow binding. Glc 1-P is often present at a concentration of $100\ \mu mol\,dm^{-3}$ or less.

□ The hydroxyl group at C-1 of a pyranose hexose such as glucose, is not the same as those at carbons 2, 3, 4 or 6, but is part of a **hemiacetal** formed when the linear form of the structure closes to form a ring. Thus glucose 1-phosphate is not a simple phosphate ester in the way that glucose 6-phosphate is. The phosphate of glucose 1-phosphate could be thought of as a phosphoryl radical attached to the exocyclic oxygen at C-1.

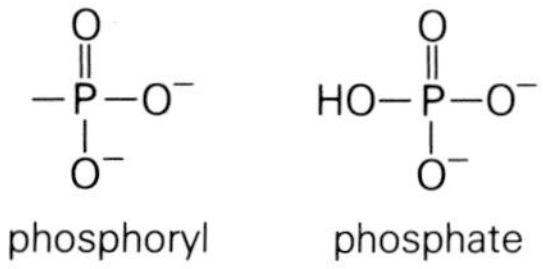

phosphoryl   phosphate

However, the formation of glucose 1-phosphate through phosphorylysis of glycogen proceeds by the formation of an oxonium ion and phosphate attack rather than involvement with a phosphoryl radical. This contrasts with the phosphorylation of the –OH at C-6 of glucose catalysed by hexokinase, where a phosphoryl radical is transferred from ATP to the oxygen of the –CH$_2$OH.

Sugar nucleotides may not be used directly

Sugar nucleotides are incorporated into polymers by one of three routes (Fig. 4.3). Which one occurs depends on the location of the poly- or oligosaccharide being assembled. The polysaccharide may be intracellular, extracellular or, particularly in plants, it may be produced in another population of cells far removed from the site of sugar nucleotide synthesis.

The extracellular assembly of polysaccharides is of vital importance to many cells although this requirement presents a problem. Characteristically, the lipid bilayer surrounding the cytoplasm is impermeable to charged molecules and translocation is selective and is mediated by transporter proteins. Compounds containing charged phosphate groups such as ATP are transported across the mitochondrial membrane, but neither they nor nucleotide sugars can penetrate the cytoplasmic membrane.

How then are extracellular structural polysaccharides assembled? There are two strategies. One is to produce an uncharged sugar such as the disaccharide sucrose that can be secreted by the cell. The other solution is to replace the

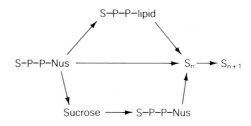

Fig. 4.3 Pathways of transferring a sugar (S) from an activated precursor nucleoside diphosphosugar (S–P–P–Nus) to extend the acceptor (S_n) molecule by one residue (S_{n+1}).

nucleoside with a lipophilic group that can ferry the diphosphosugar moiety through the lipid bilayer. A polyisoprenyl group containing anything from 55 to 120 carbon atoms assists the passage of the diphosphosugar. On arrival at the exterior face of the cytoplasmic membrane the lipid moiety remains embedded in the bilayer and polysaccharide assembly takes place at the site of translocation. This requirement alters the more usual pattern of addition: extension no longer occurs at the non-reducing terminus of the polymer but by insertion at the other end.

The structures of nucleoside and polyisoprenyl disphophosugars used in polysaccharide biosynthesis are shown in Fig. 4.4. Depending on the nature of the cell, prokaryote or eukaryote, there are variations in the number of isoprenyl units and in the structure of the α unit that is bound to the diphospho group. In prokaryotic cells, the most common polyisoprenyl moiety is undecaprenyl, a C_{55} unit, although C_{50} and C_{60} (10 and 12 isoprenyl units) also occur. As a group, the parent alcohols are sometimes referred to as bactoprenols; they are linked to both phosphoryl and pyrophosphoryl groups. In eukaryotic cells the length of the polyprenyl chain is longer; as a group the parent alcohols are called **dolichols**. Mammalian dolichyl chains comprise 17–21 isoprenyl units whereas those from plants and fungi comprise 14–24. Dolichyl moieties also differ from bactoprenols in containing a saturated C_5 unit linked to the pyrophosphoryl group.

☐ The biosynthesis of one of the peptide bonds in the cross-linking of glycan chains to form peptidoglycan during the assembly of the bacterial cell wall uses a similar strategy to that used to produce an α1–6-bond in the biosynthesis of glycogen and amylopectin to obtain the necessary free energy. The terminal peptide bond of the peptide attached to N-acetylmuramate is broken and another one immediately formed between the newly-released –COO⁻ and the –NH₃⁺ of the bridging peptide attached to the neighbouring glycan chain. D-Alanine is released.

☐ A third type of activated sugar carrying precursor has been proposed. Retinyl mannosyl phosphate a derivative of vitamin A, retinol, may play a part in the mannosylation of some mammalian glycoproteins, but its role is not clearly established.

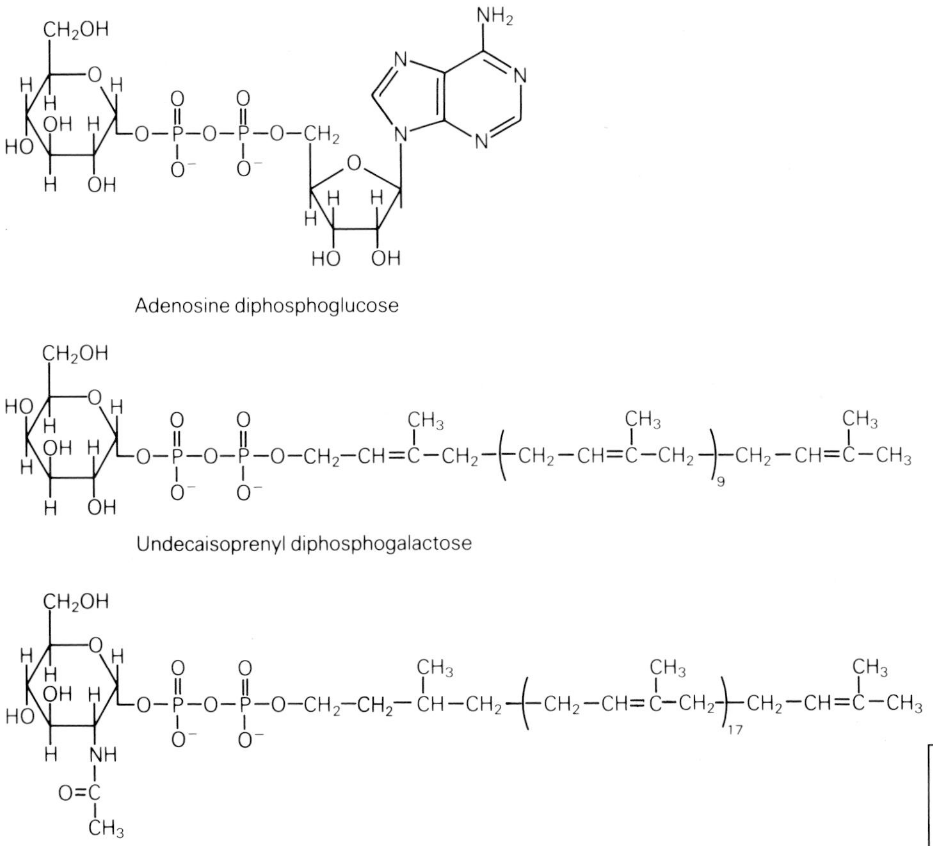

Adenosine diphosphoglucose

Undecaisoprenyl diphosphogalactose

Dolichyl diphospho-N-acetylglucosamine

Fig. 4.4 Examples of activated sugar precursors used in the biosynthesis of oligo- and polysaccharides.

— Exercise 1 —

A common way of assessing poly-saccharide synthetase activity is to measure the rate of transfer of radiolabelled monosaccharide from its activated precursor into polymer. How could such an experiment be carried out?

Structural alterations to sugar nucleotides before polysaccharide assembly

The type of monosaccharide residues required by the cell for incorporation into polymers are frequently not the same as those available to it in the environment or from endogenous cellular reserves. Monosaccharide residues are altered structurally by intracellular pathways according to the demands of the cell. With few exceptions the change from one sugar to another by epimerization of –OH groups, oxidation of –CH_2OH to –COO^- or introduction of an –NH_2 occurs before polymerization, in many cases after the formation of the sugar nucleotide. In contrast, the structural alterations of carbohydrates without change of configuration by sulphation, phosphorylation or methylation occur after the polymer has been formed. This post-assembly modification is a useful strategy since the structure, and therefore the conformation and properties of the polysaccharide, may be altered without depolymerization and reassembly, a process that would result in the loss of the free energy invested in making the glycosidic bonds.

The pre-polymerization structural modification of sugars is shown in Figs 4.5–4.7. Other metabolic pathways contribute to these alterations. Glutamine, a key intermediate in amino acid metabolism, is the donor of amino groups in the formation of glucosamine. Subsequent acetylation uses acetyl coenzyme as the donor. Sulphation requires 3'-phosphoadenosine-5'-phosphosulphate (PAPS) while oxidation requires NAD^+. More extensive alterations, changing one sugar to another, can occur before or after formation of the nucleotide.

A number of the metabolic pathways that produce sugar nucleotides start from fructose 6-phosphate (Fig. 4.5). The keto group of the open chain structure provides a site for structural alteration to mannose or for amination to glucosamine. N-acetylmannosamine is not formed at this point and is made from uridine diphospho-N-acetylglucosamine (UDP-GlcNAc).

The most widely used group of sugar nucleotides are those containing uracil and most are derived from uridine diphosphoglucose, UDP-Glc (Fig. 4.6). Oxidation followed by decarboxylation at C-6 of the glucose produces the nucleotides of glucuronate and xylose, respectively. Epimerization at C-4 of each of these nucleotides produces UDP-galacturonate and UDP-L-

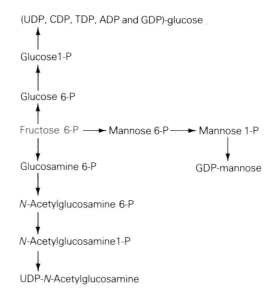

Fig. 4.5 Sugar nucleotides derived from fructose 6-phosphate.

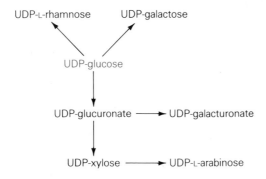

Fig. 4.6 Sugar nucleotides derived from uridine diphosphoglucose.

arabinose respectively. Oxidation, dehydration, reduction and epimerization of UDP-Glc yields UDP-L-rhamnose (UDP-Rha).

The second group of sugar nucleotides are those derived from UDP-GlcNAc, producing uridine diphospho-N-acetylgalactosamine (UDP-GalNAc), UDP-N-acetylmuramate and, by way of a number of modifications including a change of base, cytidine phosphoneuraminate (CMP-NeuNAc or CMP-sialate). N-Acetylmuramate is essential for assembly of the peptidoglycan in bacterial cell walls and sialate frequently terminates oligosaccharyl chains of mammalian glycoproteins and glycolipids. They do not appear to be formed in plants or fungi.

Thymidine diphosphosugars (TDP-sugars) appear to have the same role as the uridine derivatives.

Guanosine diphosphosugars (GDP-sugars) are ubiquitous (Fig. 4.7). The parent nucleotide GDP-mannose gives rise to GDP-mannuronate (GDP-ManUA), GDP-L-fucose (GDP-Fuc), GDP-D-rhamnose (GDP-Rha). GDP-colitose (3,6-dideoxy-L-galactose) is the activated donor for incorporation of one of the more hydrophobic sugars into the lipopolysaccharide coating of Gram-negative bacteria.

CDP is also the carrier for alcohols incorporated into the teichoic acids, components of cell wall surrounding Gram-positive bacteria. Cytidine diphosphosugars (CDP-sugars) appear to be exclusive to prokaryotic cells, especially Gram-negative ones where they provide the precursors of many of the 3,6-dideoxy sugars in the lipopolysaccharide coat such as paratose (3,6-dideoxy-D-glucose), abequose (3,6-dideoxy-D-galactose) and tyvelose (3,6-dideoxy-D-mannose).

Cytidine monophosphosugars (CMP-sugars) also function as donors. Both N-acetylneuraminate (sialate) and 3-deoxy-D-mannooctulosonate (KDO), a key sugar in the lipid A region of lipopolysaccharides, are donated from CMP derivatives.

Adenosine diphosphosugars (ADP-sugars) seem to be confined to prokaryotic cells and higher plants. For instance, ADP-glucose (ADP-Glc) acts as glucose carrier in the biosynthesis of the storage polysaccharides bacterial glycogen and plant starch.

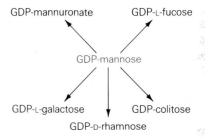

Fig. 4.7 Sugar nucleotides derived from guanosine diphosphomannose.

WHY SHOULD SUCH A VARIETY OF NUCLEOTIDES BE USED? There are only trivial differences in the free energy liberated when one base is used in place of another. It may have been a chance occurrence in the course of evolution that one nucleotide was selected rather than another. More likely, it may constitute an advantage to the cell in that metabolic control becomes possible. Using enzymes specific for the nucleotide moiety as well as for the glycosyl residue allows the sugar to be directed along different pathways and incorporated into different polymers.

Exercise 2

How could it be determined whether a polysaccharide synthetase is located in the cytosol, e.g. glycogen synthetase, or bound to a membrane, e.g. cellulose synthetase?

New bonds from old

Quite profound changes are made to the structures of some polysaccharides after polymerization. These changes are not in the constituent residues but rather in the glycosidic linkages between them. Strictly speaking, this is not biosynthesis in the sense of adding a residue, but rather the altering of bonds and consequently the overall changes in free energy are negligible. Thus, additional input of metabolic energy is not required. A further consequence of this is that such structural manipulations may occur on the periplasmic side of the cytoplasmic membrane, where ATP is not available.

Within the cell, the repositioning of short lengths of glucan chain to create the branched structures of the reserve polysaccharides glycogen and amylopectin is an essential step in producing a spherical polymer.

4.3 The biosynthesis of single sugar polysaccharides

Polysaccharides which contain only a single type of sugar, or **homoglycans**, are among the best known and most abundant of carbohydrates. This section will be limited to some of these well-characterized polysaccharides, namely glycogen, starch, cellulose and chitin.

Glycogen

The assembly of glycogen, the reserve polysaccharide of bacteria, fungi and mammals, involves a mechanism that produces a sequence of $\alpha1-4$ linked glucose residues with an occasional $\alpha1-6$ glucosidic bond forming a branch point. The formation of these two linkages illustrates the two methods of producing glycosidic bonds. The principal mechanism (formation of $\alpha1-4$ bonds) requires the participation of a pyrophosphoryl group. In the second mechanism (formation of $\alpha1-6$ bonds) existing $\alpha1-4$ bonds are changed to new configurations (Fig. 1.20). This reaction does not require an input of free energy from nucleoside triphosphate hydrolysis.

Glycogen biosynthesis also has a requirement that has occasionally to be met in the assembly of other polysaccharides, namely the participation of a **primer**. In this case it may be residual fragments of glycogen or a special glucoprotein that is at the core of the glycogen molecule. Glycogen cannot be built starting from single glucose molecules acting as acceptors.

THE PATHWAY OF GLYCOGEN SYNTHESIS is shown in Fig. 4.8. The initial step is the formation of the glucoprotein primer. Small lengths of glucan chain, probably no more than nine $\alpha1-4$ linked glucosyl residues long, are assembled at a number of points on protein backbone. The anchoring residues are probably tyrosyl, an unusual amino acid to be used for such purposes compared with typical glycoproteins. The enzyme involved is **glycogen initiator synthetase** and the glucose donor is UDP-Glc. Glucose residues continue to be donated from UDP-Glc but subsequent transfers are catalysed by a different enzyme, **glycogen synthetase**. During the extension of the $\alpha1-4$ glucan chains branch points are introduced by a transferase, **glucano $\alpha1-4 : \alpha1-6$ transglycosylase**. This enzyme usually repositions malto-heptaose units (chains of seven $\alpha1-4$-linked glucosyl residues). Following the introduction of branch points, extension of both arms of the branch continues with another round of branching occurring when the arms are sufficiently long. In this fashion, multibranched chains extend outward in every direction from the glucoprotein core until overcrowding of chain-end residues at the surface prevents their interaction with the synthetase and a glycogen molecule containing some 60 000 residues results.

□ The hydrolysis of released inorganic pyrophosphate drives the preceding step in the direction of synthesis in a number of enzyme-catalysed pathways. For example, the formation of acyl CoA from free fatty acid in the biosynthesis of lipids (Chapter 8) and the production of phosphodiesters in the biosynthesis of nucleic acids (*Molecular Biology and Biotechnology*, Chapters 1 and 2) and aminoacyl-AMP in protein synthesis.

See *Molecular Biology and Biotechnology*, Chapter 5

□ The breakdown of glycogen to glucose 1-phosphate also uses glucan transfer as part of the mechanism of glycogenolysis. Phosphorylase catalyses the phosphorylation of a glucosyl residue by inorganic phosphate to form glucose 1-phosphate. Thus phosphorylation is achieved without the use of ATP. The strategy of locally transferring lengths of glucan from a side-chain to a main chain, a break and make mechanism, ensures that as many as possible of the glucosyl residues are converted to glucose 1-phosphate by phosphorylase.

Reference Whelan, W.J. (1986) The initiation of glycogen synthesis, *BioEssays*, **5**, 136–46. Excellent essay by an acknowledged leader in the field.

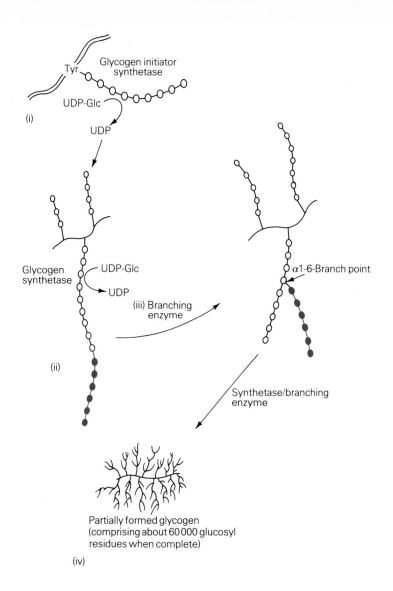

(i)

Tyr — Glycogen initiator synthetase

UDP-Glc

UDP

(ii)

Glycogen synthetase

UDP-Glc

UDP

(iii) Branching enzyme

α1-6-Branch point

Synthetase/branching enzyme

Partially formed glycogen
(comprising about 60 000 glucosyl
residues when complete)

(iv)

Fig. 4.8 The biosynthesis of glycogen involves (i) formation of the glucoprotein primer, (ii) synthesis of α1–4-linked glucan chains, (iii) branching of chains by forming an α1–6 branch point, and (iv) continued extension and branching to produce a more or less spherical glycogen molecule.

Box 4.1
Defects of polysaccharide synthesis cause human diseases

Defects in polysaccharide metabolism that cause disease are usually the result of impaired or absent hydrolases. An abnormally high accumulation of glycogen or glycosaminoglycans is often lethal. Occasionally a genetic defect affects the assembly of poly- or oligosaccharides. Andersen's disease (glycogen storage disease IV) is characterized by the absence of branching enzyme. Accumulation of long unbranched chains of α1–4-glucans in the liver causes cirrhosis and death.

I-Cell disease is a condition in which lysosomes fail to take up hydrolases destined for them. The defect is in the signal that targets these enzymes, namely the phosphorylation at C-6 of one or more of the mannosyl residues in the carbohydrate moiety of the glycoprotein hydrolase (*Cell Biology*, Chapter 1).

Incomplete oligosaccharide chains of cell surface glycoproteins and glycolipids are frequently found on malignant cells. Whether this is a cause or a consequence of malignancy is not clear.

THE CONTROL OF MAMMALIAN GLYCOGEN SYNTHESIS is inextricably linked to its catabolism and is governed by effector molecules such as glucose 6-phosphate (Glc 6-P), ATP and inorganic phosphate, and by hormones such as glucagon, insulin and adrenalin. Catabolism and biosynthesis are thus responsive to the energy state within the cell and to the requirements of the organism as a whole. Regulation of glycogen synthetase controls the rate of biosynthesis.

Glycogen synthetases exist in two interconvertible forms: *I* or *a* and *D* or *b*. The *I/D* nomenclature is derived from a property that distinguishes the two forms, namely their response to Glc 6-P. The activity of the *I* or *i*ndependent form is not affected by Glc 6-P whereas the *D* or *d*ependent form requires its presence for activity. The more recent *a* and *b* nomenclature attempts to clarify the terminology for glycogen synthetase and glycogen phosphorylase. In both cases the *b* form is dependent on an effector for activity whereas the *a* form is free of that constraint. The principal effector acting on phosphorylase *b* is AMP.

The role of hormones such as glucagon acting on liver cells, or adrenalin stimulating muscle cells, is both to liberate phosphorylase from intracellular effector control (by promoting the conversion of phosphorylase from the *b* to the *a* form) and to bring glycogen synthetase under the influence of Glc 6-P (by converting the synthetase from the *a* to the *b* form). Since the intracellular concentration levels of Glc 6-P are normally not great enough to stimulate the enzyme in the *b* form, the conversion switches off glycogen synthesis. Glucagon thus promotes phosphorolysis and stops glycogen synthesis (Fig. 4.9). The cell is thus prevented from synthesizing glycogen at a time when it is actively breaking it down, thus avoiding a wasteful cycle. As we shall see, the same co-ordination exists when the cell is building up glycogen reserves.

Interconversion between glycogen synthetase *a* and *b*, like that between the two forms of phosphorylase, is by phosphorylation/dephosphorylation of specific serine residues. Hormones that promote catabolic pathways initiate a cascade mechanism which commences with the activation of adenylate cyclase. The number of steps in the cascades controlling glycogen phosphorylase activation and glycogen synthetase inactivation however, are not the same although they share some enzymes (Fig. 4.9). The cyclic AMP-dependent kinase that activates phosphorylase kinase, the penultimate step in that cascade, acts directly on glycogen synthetase *a* to convert it to the *b* form.

Dephosphorylation of proteins is as important as phosphorylation in the control of metabolic pathways. Removal of the phosphoryl group by hydrolysis is catalysed by a number of phosphatases. Here again, enzymes are common to both inactivation and stimulation cascades. Protein phosphatase 1 catalyses the dephosphorylation of glycogen synthetase *b*, converting it to the active *a* form, and of glycogen phosphorylase *a*, to its inactive *b* form. Protein phosphatase 1 is controlled by an inhibitor which itself needs to be phosphorylated for activity. Phosphorylation of the inhibitor by cyclic AMP-dependent kinase ensures that phosphatase 1 is inactive. It is through this inhibitor that insulin is thought to promote the accumulation of glycogen. Insulin causes the dephosphorylation of the inhibitor inactivating it and so allowing the phosphatase to convert both glycogen synthetase to its active form and phosphorylase to the inert form.

The mechanisms of signal transduction across the cytoplasmic membrane, connecting the extracellular events of receptor-hormone interaction and the intracellular events generating second messenger are described in *Molecular Biology and Biotechnology*, Chapter 8.

☐ A common role of effectors is to couple the activity of enzymes to the energy requirements, or energy state, of the cell. AMP and ATP concentrations frequently modulate metabolic pathways in this way. Excess of glucose 6-phosphate probably means that glycolysis is inhibited because sufficient ATP has been accumulated and it would therefore be prudent of the cell to store excess glucose as glycogen. In these conditions glycogen synthetase *b* is active.

☐ Phosphorylation of glycogen synthetase occurs at multiple sites on different subunits of the enzyme and may well involve more than the three kinases described. The interaction between these sites is complex in that phosphorylation of one may assist subsequent phosphorylation or dephosphorylation. The multiplicity of kinases is matched by a multiplicity of phosphatases, of which at least three are known to exist. These too are synergistic in their action and are probably matched to specific phosphorylation sites.

─── *Exercise 3* ───

Draw the structures of the disaccharide termini at the reducing and non-reducing ends of a glycogen molecule. Is the reducing end capable of reducing?

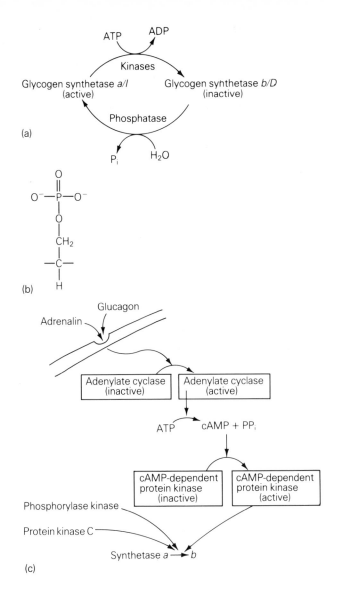

(a)

(b)

(c)

Fig. 4.9 Control of glycogen synthetase. (a) Phosphorylation of serine residues on the active *a*/*I* form converts the synthetase to the inactive *b*/*D* form dependent on Glc 6-P for activity. Phosphorylation is catalysed by at least three kinases. Dephosphorylation, catalysed by phosphatase and restores the synthetase to its active form. (b) phosphorylation of a serine residue. (c) The cascade initiated by the binding of hormones adrenalin (epinephrine) or glucagon to receptors on the cytoplasmic membrane. The result, an activated cAMP-dependent protein kinase, is one of three kinases that phosphorylate synthetase *a*/*I*.

BACTERIAL GLYCOGEN. In bacteria, both the substrate for glycogen synthetase and the control of glycogen synthesis are different from those in the mammalian cell. The sugar nucleotide involved is ADP-Glc. Unlike UDP-Glc in mammalian cells, which is used in a number of pathways, the compound in bacteria is produced solely for glycogen synthesis. Thus control of glycogen production in bacteria is by regulation of ADP-Glc pyrophosphorylase activity rather than at the synthetase step. In this way the unnecessary production of ADP-Glc and consequent waste of ATP is prevented.

Fig. 4.10 Starch grains in the endosperm of a seed viewed using polarized light. Courtesy M. Hoult, Department of Biological Sciences, Manchester Polytechnic, Manchester, UK.

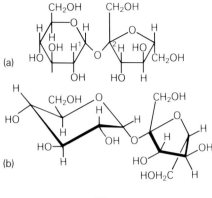

(a)

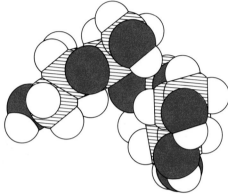

(b)

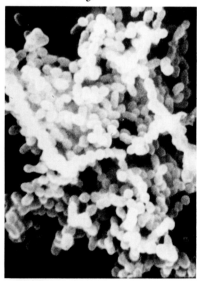

Fig. 4.11 The figure shows three ways of representing the structure of sucrose. Sucrose is formed by a (1 → 2) linkage between the carbonyl carbons in the separate sugars. Unlike maltose or lactose the disaccharide is therefore non-reducing. The structure is described as αD-fucopyranosyl-(1–2)-βD fructofuranoside. (c) Computer graphic model. Courtesy of C. Holloway, Biosyn Technologies Ltd.

Starch, amylose and amylopectin

Starch, the reserve polysaccharide of plants (Fig. 4.10) is a mixture of amylose and amylopectin, which are linear and branched glucans, respectively. The steps in the biosynthesis of these polysaccharides are much the same as those involved in glycogen biosynthesis. Thus, a primer is required to accept glucosyl residues and *de novo* assembly from glucose does not occur. The accepting molecule can be partially degraded starch. Alternatively, a small number of glucose residues are donated from UDP-Glc to a protein forming a glucoprotein core upon which the rest of the molecule is built. The principal nucleotide glucose donor in plants is ADP-Glc. However, in the formation of the more compact forms of amylopectin found in starch granules, UDP-Glc also participates. The introduction of $\alpha1$–6 glucosidic linkages producing branch points is by the transfer of existing lengths of $\alpha1$–4 glucan chains as occurs in the assembly of glycogen.

CONTROL OF STARCH SYNTHESIS lies at the point of ADP-Glc formation rather than in its use. The principal activator of ADP-Glc pyrophosphorylase is 3-phosphoglycerate and the main inhibitor is inorganic phosphate. The ratio of concentrations of these effectors is the key to the control of starch synthesis which is thus directly linked to CO_2 fixation.

SUCROSE (Fig. 4.11) is an intermediary in plant polysaccharide synthesis. The photosynthetic cells of higher plants are self-sufficient with respect to polysaccharide synthesis. Ribulose 1,5-*bis*phosphate or phosphoenolpyruvate act as CO_2 acceptors and the products are metabolized to hexose phosphate from which sugar nucleotides are produced. To provide non-photosynthetic cells of the plant with sugars, sucrose is secreted by photosynthetic cells, and travels through the vascular system of the plant to be absorbed, usually without hydrolysis, by the non-photosynthetic cells. Here

---

Box 4.2
Sucrose, glucans and dental decay

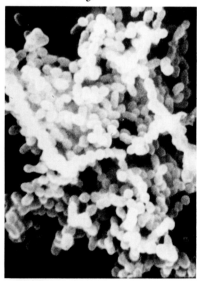

Whilst sucrose or household sugar, may satisfy those with a sweet tooth, it unfortunately also provides the component for synthesis of one of the major causes of tooth decay. Dental plaque is an insoluble matrix of polymers that first begins to form when glycoproteins present in saliva adhere to calcium phosphate on the tooth surface. Oral bacteria such as *Streptococcus mutans*, *sanguinis* and *salivarius* bind to this film (Fig.). Extracellular enzymes of *S. mutans* catalyse the polymerization of the glucose residues of sucrose into insoluble glucans that form an enveloping coat over the dental surface which in turn promotes adhesion of more protein and bacteria. This accumulation is called **plaque**. Acids secreted by the bacteria are hindered in their diffusion from the dental surface. Their prolonged presence is sufficient to promote the breakdown and demineralization of the teeth leading to caries or tooth decay.

Cells of *Streptococcus mutans*. Courtesy of Dr J. Verran, Department of Biological Sciences, Manchester Polytechnic, UK.

Reference ap-Rees, T. (1988) Hexose phosphate metabolism by nonphotosynthetic tissues of higher plants in *The Biochemistry of Plants*, vol. 14, pp. 1–33, Academic Press, New York. A detailed review of plant metabolism and the role of sucrose with respect to starch synthesis and breakdown.

it is converted to UDP-Glc and fructose. The sugar nucleotide is used for polysaccharide synthesis and fructose is phosphorylated to fructose 6-phosphate which is used in glycolysis or contributes to other metabolic demands of the cell. Two enzymes are capable of catalysing sucrose synthesis (Fig. 4.12) but are used in different ways. **Sucrose 6-phosphate synthetase** produces sucrose 6-phosphate in photosynthetic tissues. This reaction is reversible and the formation of the disaccharide is driven by the hydrolytic removal of the phosphoryl group catalysed by sucrose 6-phosphate phosphatase. In non-photosynthetic tissue the enzyme involved is **sucrose synthetase** which catalyses the production of UDP-Glc and fructose (that is, the reverse reaction) principally because of the high concentration of sucrose in the cells.

Why should sucrose be used for carbohydrate transport in place of glucose or fructose? There is some advantage in that the use of a dissaccharide generates about half the osmotic pressure than the equivalent mass of monosaccharide. The principal reason, however, is that the bond between the glucose and fructose residues has a standard free energy of hydrolysis approaching that of a pyrophosphoryl bond. It can thus be used to drive the reverse of an exergonic reaction. UDP acts as acceptor and is glucosylated by sucrose in the presence of sucrose synthetase to form UDP-Glc. The transport of sucrose around the plant is a strategy for transferring 'high energy bonds' from one cell to another.

□ A series of α1–6-galactose residues may be linked to the C-6 hydroxyl group of the glucosyl residue in sucrose. **Raffinose**, a trisaccharide; **stachyose**, a tetrasaccharide and **verbascose**, a pentasaccharide, are all derived by the transfer of a galactose residues from GDP-Gal to sucrose. The first transfer produces raffinose and is directly from the nucleotide. Subsequent additions are from the galactose donor, galactinol, *myo*-inosityl galactoside. Galactinol is widespread in the plant kingdom and is derived from UDP-Gal and *myo*inositol.

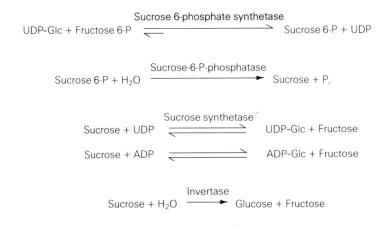

Sucrose 6-phosphate synthetase
UDP-Glc + Fructose 6-P $\rightleftharpoons$ Sucrose 6-P + UDP

Sucrose-6-P-phosphatase
Sucrose 6-P + H_2O $\longrightarrow$ Sucrose + P_i

Sucrose synthetase
Sucrose + UDP $\rightleftharpoons$ UDP-Glc + Fructose

Sucrose + ADP $\rightleftharpoons$ ADP-Glc + Fructose

Invertase
Sucrose + H_2O $\longrightarrow$ Glucose + Fructose

Fig. 4.12 Enzymes of sucrose metabolism.

galactose   sucrose   galactinol   galactose   *myo*-inositol

The oligosaccharides act as antifreeze agents in plants, counteracting the formation of damaging ice crystals in cells. **Melezitose**, a trisaccharide, consists of sucrose with an α-glucosyl residue at the C-3 hydroxyl of the fructose residue. It is present in the sweet exudations, or **manna**, produced by some plants in response to wounding by insects, although it is thought to be a product of the insect's rather than plant's metabolism. Some types of honey contain considerable quantities of the trisaccharide.

Cellulose

Cellulose, a β1–4 glucan, is the principal structural polysaccharide of the cell wall in higher plants and some algae and fungi (Fig. 4.13). It is also produced as an extracellular polymer by some bacteria such as *Acetobacter xylinum*. There is good experimental evidence that in the bacterial cell the nucleotide sugar precursor in cellulose synthesis is UDP-Glc, and the same precursor is most probably used in plants. The uncertainty arises because the synthesis of cellulose appears to be intimately bound up with the assembly of another structural polysaccharide, *callose*, a β1–3 glucan.

LOCATION AND ORGANIZATION OF CELLULOSE SYNTHESIS. Cellulose synthetase is a transmembrane protein that accepts nucleotide sugars at the cytoplasmic surface and lays down cellulose polymers on the external face (Fig. 4.14). A primer may be required.

Reference Pontis, H.G. (1978) On the scent of the riddle of sucrose, *Trends in Biochemical Sciences*, **3**, 137–9. A short review on the role of sucrose in moving carbohydrate between plant cells.

Reference Delmer, D.P. (1987) Cellulose biosynthesis, *Annual Review of Plant Physiology*, **38**, 259–90. An in-depth review of cellulose biosynthesis in both bacteria and plants.

□ An examination of the association of individual linear cellulose molecules reveals that alignment could occur in one of two patterns. The packing of adjacent strands could be in a parallel or antiparallel orientation. Polysaccharides have reducing and non-reducing ends and therefore like proteins and nucleic acids they have a polarity. In parallel arrangements polysaccharides have all the non-reducing ends at the same end of the microfibril. In antiparallel arrangements adjacent ends are non-reducing and reducing. The more stable of the two in cellulose is the antiparallel configuration but the complexity of achieving this association efficiently on the cell surface, when all polymers are being extended at the same end, is too great and the marginally less stable parallel association is formed.

Mercerization involves treating cellulose with strong alkali which allows the individual polymers to dissociate and subsequently reassemble in the antiparallel configuration.

□ The occurrence of β1–3-D-glucan as a structural polysaccharide is not confined to the higher plants. The cell walls of yeast and fungi contain β-linked glucans, a proportion of which are the linear β1–3-glucans which often associate into triple helices. The walls of some fungi such as *Saprolegnia* spp. contain both β1–3- and β1–4-glucans.

See *Cell Biology*, Chapter 6

Exercise 4

An isolated glucan synthetase capable of producing cellulose is suspected of catalysing the formation of a β1–3-glucan. How could this be confirmed?

□ Figure 4.15 illustrates the sequence of events that occurs during the budding and cellular division of the yeast *Saccharomyces cerevisiae*. When the expanding daughter cell reaches a size comparable to the parental cell, and has received a copy of the maternal DNA, then a barrier or primary septum of chitin is thrown across the waist, secondary septa follow and the cells separate leaving a bud scar on the mother cell.

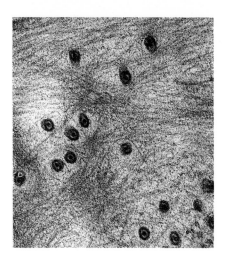

Fig. 4.13 Electron micrograph of a section through the wall of a plant cell. Note the fibres of cellulose ramifying throughout the wall.

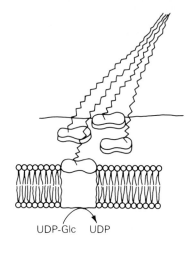

Fig. 4.14 Clusters of membrane-located cellulose synthetase molecules contribute to the ordered and rapid formation of cellulose microfibrils.

A study of freeze–fracture electron micrographs of membranes has revealed that extracellular cellulose microfibrils are attached to raised structures on the external surface of the cell. Individual cellulose molecules, once formed, bind together through hydrogen and hydrophobic bonds to form fibrillar conglomerates. The association of individual glucans would be assisted if they were produced adjacent to one another. The raised structures are thought to be groups of specially orientated cellulose synthetase molecules, synchronously assembling single glucan molecules which then form microfibrils on the external face of the cell. The geometry with which microfibrils are laid down is determined by microtubules, acting from within the cell possibly by dictating the positioning of the cellulose synthetase clusters.

SYNTHESIS OF CELLULOSE. It is now accepted that UDP-Glc is the substrate for cellulose synthetase. The difficulties encountered in establishing this have brought to light yet another mechanism by which cells regulate their metabolic pathways. The enzyme that catalyses the synthesis of cellulose was found also to catalyse the assembly of the β1–3 glucan, callose, the principal structural component of the plant cell plate. The proportion of each poly-saccharide synthesized from the common precursor and by the same enzyme may be varied by effectors such as GTP and ATP, Ca^{2+} and Mg^{2+}. The physiological significance of this is that it is part of the plant cell's response to wounding. Disturbance of the effector concentrations in damaged cells brings about the localized creation of callose bulkheads, with a subsequent isolation of the damaged area, and a cessation of cellulose synthesis.

Chitin

Chitin is a polymer that resembles cellulose both in the structure and con-formation of the single molecules and in the manner of its association into microfibrils. In yeast and fungi it provides a thin anchoring layer for cell wall assembly and in budding yeasts such as *Saccharomyces cerevisiae* it is the principal component of the separation device that allows newly formed daughter cells to detach from the parental cell. It is in this latter context that the biosynthesis of chitin presents an interesting question: how is it produced at a particular point on the cell surface?

Reference Fry, S.C. (1986) Cross-linking of matrix polymers in the growing cell walls of angiosperms, *Annual Review of Plant Physiology*, **37**, 165–86. A detailed survey of chemical and physical links contributing to the architecture of plant cell walls.

Reference Cabib, E. (1976) The yeast primary septum; a journey into three-dimensional biochemistry, *Trends in Biochemical Sciences*, **1**, 275–7. A brief and readable account of the biochemistry of the yeast cell wall during budding.

Box 4.3
Cross-linking of polymers:
the molecular bag
surrounding the plant cell

In addition to cellulose a wide range of polysaccharides are present in the plant cell wall. The principle polymers are xyloglucan, arabinoxylan, galacturonan and rhamnogalacturonan. The uronans carry short but often structurally complex side-chains of galactose, xylose, glucuronate and arabinose residues. In addition, there are the extensins, water-insoluble glycoproteins containing large amounts of hydroxyproline residues. Most of these polymers are cross-linked to one another through a variety of bonds to form the molecular bag that surrounds the plant cell. Extensin and galacturonan associate through salt links and cellulose binds to xyloglucan and arabinoxylan via hydrogen bonds. Covalent cross-links are also established. The phenolic groups of tyrosine side-chains in adjacent extensin molecules combine oxidatively to form a biphenyl bridge:

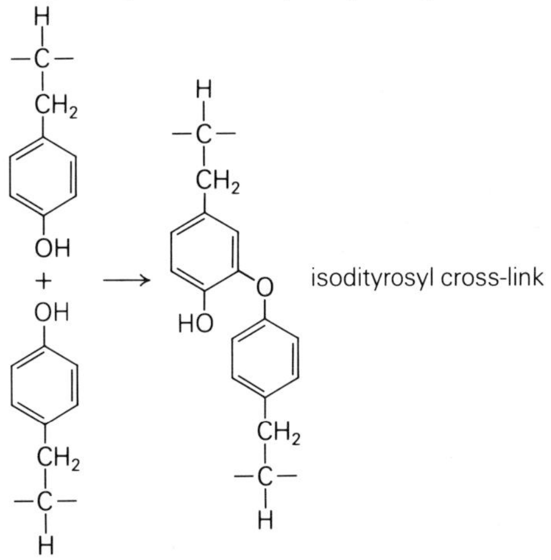

isodityrosyl cross-link

The C-6 hydroxyl of galactose and the C-3 hydroxyl of arabinose residues, both terminally located in side-chains of rhamnogalacturonan, can be esterified with ferulate. Oxidative coupling produces a C–C bond bridging the pectin molecules:

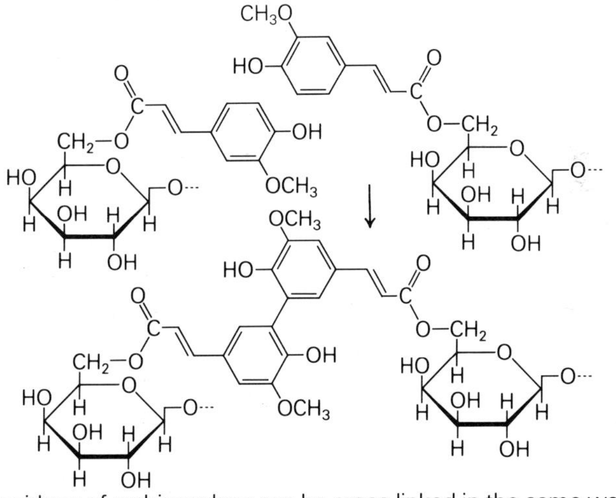

Arabinose residues of arabinoxylans can be cross-linked in the same way.

Fry, S.C. (1989) Dissecting the complexities of the plant cell wall, *Plants Today*, **2**, 126–32. Interesting account of how analysis of the products of partial digestions of the plant cell wall casts light on its intricate structure.

Box 4.4
DCB herbicide

Herbicides, like other chemicals that kill or inhibit the growth of cells, have many points of attack on plant metabolism. 2,6-Dichlorobenzonitrile (DCB) is effective in micromolar concentrations as an inhibitor of cellulose synthesis in higher plants and algae. An obvious mechanism would be action on cellulose synthetase but no such inhibition can be demonstrated *in vitro*. However, DCB has been found to bind to a protein of M_r 18 000 that associates with the synthetase. There is a possibility that this protein might control the activity or specificity of cellulose synthetase.

CHITIN SYNTHETASE, located in the cytoplasmic membrane, accepts UDP-GlcNAc at the internal face and extrudes chitin at the external face. The process is similar to the biosynthesis of cellulose. However, when the synthetase is synthesized and integrated in the membrane, it is initially in an inactive form as a zymogen. In this form it is distributed throughout the membrane and it is the next step that determines the particular site of chitin synthesis in the cell. The conversion of inactive to active synthetase is brought about by limited proteolysis in much the same way as mammalian serine proteases such as trypsinogen are activated. The cellular location of this very specific proteolysis thus determines where chitin synthesis will occur. Clearly it is the control and location of proteolysis that are the keys to understanding the spatial assembly of the yeast septum.

SITE OF SYNTHESIS. The protease is delivered to the internal face of the cytoplasmic membrane in a protective vesicle (Fig. 4.15). Combination of the vesicle and membrane releases the protease which duly cleaves the zymogen forming active synthetase. To prevent neighbouring zymogen from being activated, straying proteases are inactivated by a specific inhibitor. The protease-containing vesicle may well be directed to its site of action by being associated in some way with the internal cytoskeleton. The orientation of actin fibrils where they abut on to the internal face of the cytoplasmic membrane seems to coincide with assembly of the chitinous septum.

☐ The mechanism for the biosynthesis and deposition of chitin in filamentous fungi is broadly similar to that in yeasts. New cell wall is assembled at the advancing hyphal tip. To this site inactive chitin synthetase is delivered in bagged or vesicular form and activated by proteases present in the membrane. The vesicles containing the zymogen have been termed **chitosomes**, but it is not clear if they are present in all fungi. Upon fusion with the membrane and activation this package may be responsible for producing a chitin microfibril through the synchronous and closely orientated production of numerous single chitin molecules.

Box 4.5
Inhibitors of
polysaccharide synthesis

Inhibitors often mimic the structures of natural substrates of enzymes blocking binding or active sites through a non-productive association. Two inhibitors of chitin synthetase, polyoxin D and nikkomycin Z, work in this fashion.

UDPGlcNAc (substrate)

polyoxin D

nikkomycin Z

Other enzymes which use UDP–GlcNAc as substrate are not blocked by these inhibitors presumably because the accepting residue is different.

Tunicamycin is effective in blockading glycoprotein biosynthesis by inhibiting the transfer of N-acetylglucosamine-1-phosphate from UDP–GlcNAc to dolichyl phosphate. The blocking of viral glycoprotein synthesis in this fashion is the source of its antiviral action. Similarly, the inhibition of synthesis of the mannoprotein component of fungal cell walls explains its antifungal properties.

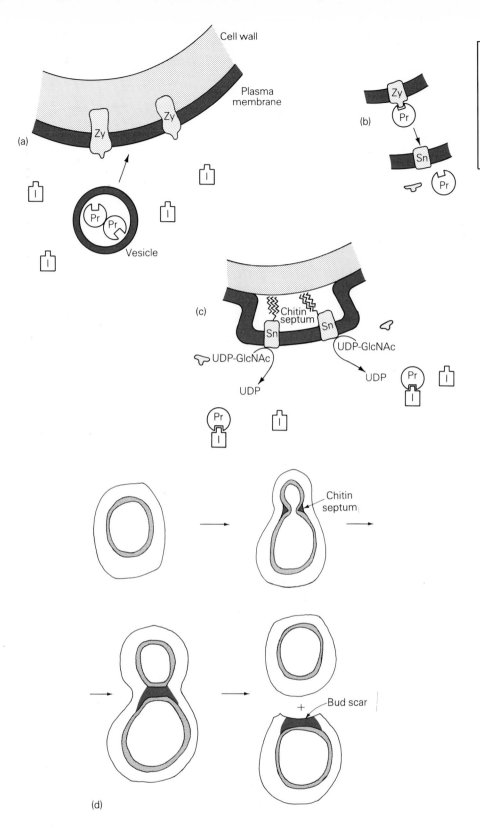

Exercise 5

A subcellular preparation catalyses the incorporation of *N*-acetylgalactosamine and glucuronate from UDP precursors into a high molecular weight product. What experimental evidence would show that this polymer has a glycosaminoglycan-like structure and is not a mixture of two homopolymers?

Fig. 4.15 Septum formation during yeast budding involves (a) vesicles containing protease (Pr) migrating to membrane bound, inactive chitin synthetase zymogen (Zy), and (b) released protease acting on the zymogen in a process of limited proteolysis to produce synthetase (Sn). (c) Free protease is inactivated by inhibitor (I) and commencement of chitin synthesis to produce a primary septum followed by (d) cell duplication in a budding yeast.

4.4 The biosynthesis of polysaccharides containing mixtures of sugars

Polysaccharides comprising more than one variety of sugar residue may be periodic, interrupted or irregular in their sequence. The assembly of mixed-sugar periodic polysaccharides is illustrated by the biosynthesis of glycosaminoglycans and that of the interrupted ones by the biosynthesis of agarose and the pectins. The biosynthesis of irregular oligosaccharides, such as those present in glycoproteins and glycolipids, are described in *Molecular Biology and Biotechnology*, Chapter 5 and in Section 8.7 respectively.

Glycosaminoglycans

In some ways the structures of glycosaminoglycans such as chondroitin 4-sulphate or chondroitin 6-sulphate and dermatan sulphate are similar to those of amylose, cellulose and chitin in that they have regular repeating units, but now of disaccharides rather than monosaccharides. Consequently their synthesis requires two enzymes to catalyse the transfer of sugar residues from sugar nucleotide precursors, each specific for the residue transferred and the accepting sugar. These enzymes are referred to as **transferases** rather than synthetases. Residues of glucuronate and N-acetylgalactosamine are alternately transferred from the uridine diphospho precursor to the growing chain. Sulphation of the C-4 or C-6 hydroxyl groups of N-acetylgalactosamine is catalysed by a sulphotransferase. Transfer of a *sulphate* group from 3'-phosphoadenosine-5'-phosphosulphate (Fig. 4.16) produces chondroitin 4- or 6-sulphate respectively. Epimerization of the carboxylate group at C-5 of the uronate residue (Fig. 4.17) soon after its incorporation into the chain changes D-glucuronate to L-iduronate and chondroitin 4-sulphate becomes dermatan sulphate.

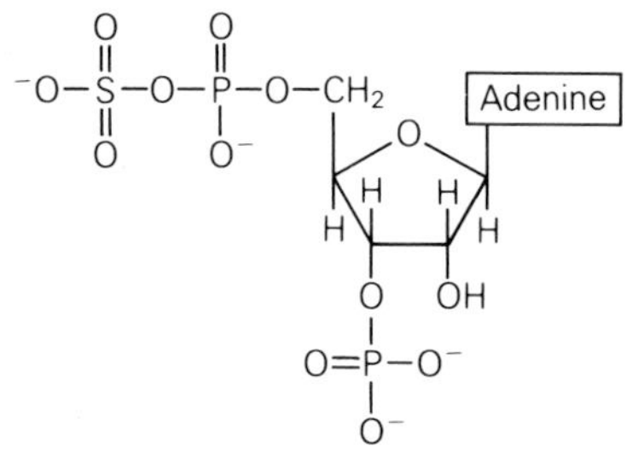

□ In describing polysaccharide biosynthesis the term **transferase** is used in two ways. During the assembly of glycogen or amylopectin, small sections of glucan chain are moved from one part of the molecule to another with no requirement for the supply of free energy from nucleoside triphosphate hydrolysis. The making of bonds is matched by the breaking of bonds and this process is catalysed by a transferase. However, the term is also used to describe any one of a series of enzymes requiring activated sugar precursors which assemble a polysaccharide containing a variety of sugar residues and glycosidic bonds. *Synthetase* is used when the polysaccharide has a simple structure with a single type of residue and glycosidic bond.

Fig. 4.16 3'-Phosphoadenosine-5'-phosphosulphate (PAPS), the sulphate donor for the transfer of sulphate to polysaccharides, is formed from ATP and inorganic sulphate.

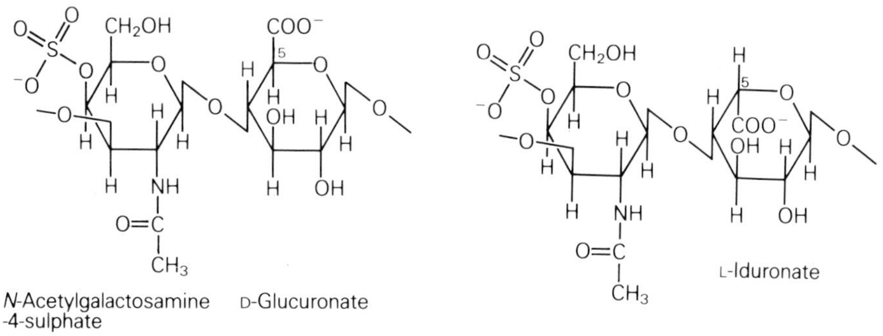

N-Acetylgalactosamine -4-sulphate   D-Glucuronate   L-Iduronate

Fig. 4.17 The epimerization of -COO⁻ in D-glucuronate produces L-iduronate, thus converting chondroitin 4-sulphate to dermatan-sulphate.

□ The biosynthesis of dermatan sulphate illustrates the common and the rare covalent modifications that can occur to already assembled polysaccharides. The derivitization of the functional groups of a residue already incorporated into a polymer such as sulphation of a hydroxyl or methylation of carboxyl is common. The more radical alteration of one sugar to another after polymerization, in this case by epimerization, is much rarer and is usually effected before incorporation into the polymer.

See *Cell Biology*, Chapter 7

ATTACHMENT TO THE CORE PROTEIN. Glycosaminoglycans are covalently bound to a protein backbone that is part of the vast proteoglycan complex. Synthesis starts not with the formation of the alternating sequence of residues described above, but with the assembly of a trisaccharide that links the glycosaminoglycan to the protein. The orderly sequential transfer of a xylose residue to the hydroxyl group of a serine residue followed by two galactose residues from their uridine diphosphodonors is catalysed by a xylosyltransferase and two galactosyltransferases, respectively. Two galactosyltransferases are necessary because there are two different accepting sugar residues, namely xylose and then galactose.

Sulphate: a term widely used in describing plant and mammalian polysaccharides such as algal fucan sulphate, human chondroitin 4- and 6-sulphates, although not strictly correct. Such groups are better thought of as sulphuryl radicals attached to the oxygen atom of the hydroxyl group.

Agarose

Agarose is one of a number of polysaccharides, found in the cell walls of red seaweeds. Its structure is based on the repeating disaccharide sequence:

$$\ldots \text{D-Gal-}\beta1\text{--}4\text{-L-Gal-}\alpha1\text{--}3 \ldots$$

This polymer is initially synthesized from UDP-D-Gal and GDP-L-Gal with separate transferases catalysing the transfer of each type of residue. The gelling properties of agarose are not shown by this galactan and only appear when it is structurally altered *in situ* in the cell wall. To produce a gel at the point of synthesis would have disastrous consequences for the cell. Sulphation of the C-6 hydroxy in the L-galactose residue followed by an internal bridging of the sugar ring to produce 3,6-anhydro-L-galactose produces a new disaccharide repeating sequence:

$$\ldots \text{D-Gal-}\beta1\text{--}4\text{-L-Gal(3,6-anhydro)-}\alpha1\text{--}3 \ldots$$

which can form double helices with neighbouring polymers. The removal of the sulphate drives the bridging. Stretches of polysaccharide where conversion to the new sequence has not occurred cannot form interhelical association with other polymers and so provide the interrupting domains important to gel formation. The degree to which the original galactan is modified depends on the pattern and extent of the bridging reaction. This in turn determines the physical nature of the gel in the cell wall.

Pectins

Pectins have a backbone of $\alpha1$–4-linked galacturonate residues. This regular periodic structure is interrupted in two ways. L-Rhamnosyl residues are added in place of the uronate during assembly of the polymer, and short but complex side-chains containing xylose, galactose, glucuronate and arabinose residues are attached to the assembled backbone. Assembly of the linear portion requires the activity of at least four transferases; two to transfer galacturonate residues from uridine diphosphogalacturonate (UDP-GalUA) to either galacturonate or rhamnose, and two to transfer rhamnose from uridine diphospho-L-rhamnose (UDP-L-Rha) to acidic or neutral residues. Esterification of the carboxyl group, another way of breaking up the regularity of the structure, occurs soon after incorporation of the galacturonate into the polymer. The methyl group is donated by *S*-adenosylmethionine (Fig. 4.18). It is possible that the side-chains are partially assembled before attachment to the backbone although it is more probable that residues are added sequentially.

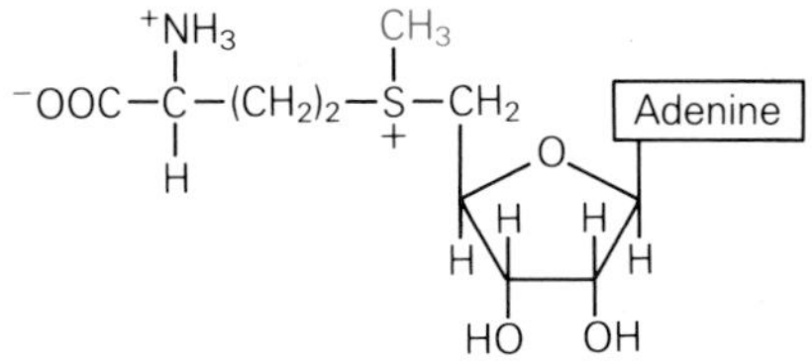

4.5 *The biosysthesis of polysaccharides using lipid pyrophosphoryl sugar donors*

Many bacterial polysaccharides have repeating structures, often containing two or more types of sugar residue. Examples of such polymers are the capsular and lipopolysaccharide coatings of cells and the teichoic acid and peptidoglycan structural polysaccharides of the wall.

The strategy for the biosynthesis of periodic extracellular bacterial poly-saccharides combines the transfer of single sugars to form oligosaccharide units followed by the transfer of oligosaccharide units to build the poly-saccharide.

The assembly of oligosaccharide units occurs within the cell. These are

☐ Rhamnose is 6-deoxymannose. L-Rhamnose is found in plant mucillages, plant cell walls, acidic pectin and rhamnogalacturonan. It is also a component of the lipopolysaccharide of the cell envelope of Gram-negative bacteria. The core structure of *Pseudomonas aeruginosa* lipopolysaccharide is:

$$\text{D-Glc} \rightarrow \text{L-Rha} \rightarrow$$

$$\text{D-Glc} \rightarrow \text{D-Glc}$$
$$\downarrow$$
$$\text{D-Glc} \rightarrow \text{GalN} \rightarrow \text{Hep} \rightarrow \text{Hep} \rightarrow \text{(KDO)}$$
$$\|$$
$$\text{L-Ala}$$

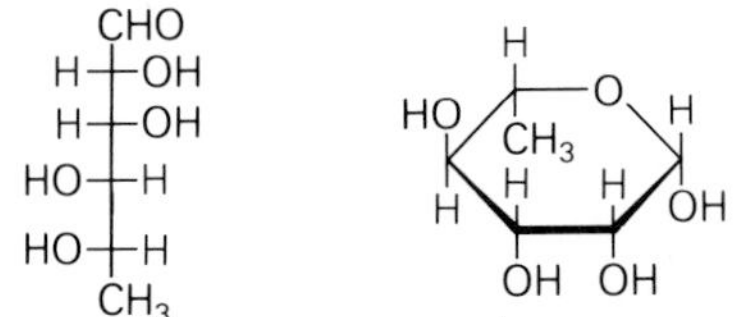

where L-Rha is L-rhamnose, Hep is heptose and (KDO) is 3-deoxy-D-manno-octulosonic acid.

Fig. 4.18 *S*-adenosylmethionine (SAM), the methyl donor for esterification of –COO⁻ in pectate. Other products are homocysteine and adenosine.

Box 4.6
Cyclodextrins

Cyclodextrins or Schardinger dextrins are water-soluble $\alpha1$–4-glucans containing six, seven or eight residues linked in a circle:

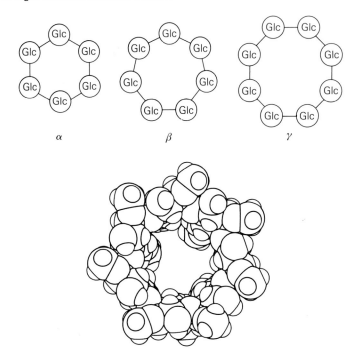

α   β   γ

Molecular model of a cyclodextrin. Courtesy of M.J. Dereham, Department of Chemistry, Manchester Polytechnic, Manchester, UK.

Lipophilic C–H bonds are presented at the interior surface of the doughnut and hydrophilic hydroxyl groups are on the outside. The dextrins are produced from the linear portions of starch molecules by the action of various microorganisms such as *Bacillus macerans* in a break-and-make bond mechanism. No free energy is required to be supplied. Originally a curosity, cyclodextrins are now receiving a great deal of attention because of their capacity to trap or absorb hydrophobic guest molecules into the central hydrophobic space. The soluble dextrins can thus be used to transport water-insoluble drugs around vascular systems or stabilize food flavour substances during food manufacture.

Box 4.7
Biosynthesis catalysed by hydrolases

Halling, P.J. (1984) Effects of water on equilibria catalysed by hydrolytic enzymes in biphasic reaction systems *Enzyme and Microbial Technology*, **6**, 513–6. A rather technical exposition of the theory behind reversing hydrolysis.

Common conditions, both *in vivo* and *in vitro*, where solute concentrations are typically between 1 and 20 mmol dm^{-3}, means that hydrolases catalyse the *hydrolysis* of chemical bonds. However, the release of free energy associated with the hydrolysis of a peptide or ester bond is not large. Thus, an increase in solute concentrations together with a constant removal of product, for example by the use of solvents to remove lipophilic products, will lead esterases to catalyse the creation of ester bonds rather than hydrolysing them. Modest success has been achieved using carbohydrases, such as glucosidase and mannosidase, to catalyse the production of disaccharides. The concentration of monosaccharides required for this reversal is high (e.g. 5 mol dm^{-3}). Nevertheless, the ability to produce oligosaccharides without the use of expensive sugar nucleotides is an attractive commercial proposition.

transported to the external face of the cytoplasmic membrane where they are added to the growing polysaccharide. This is by no means the only method used by bacteria for the synthesis of extracellular polysaccharide. For instance, the assembly of bacterial cellulose does not require a lipid intermediate. Some bacteria can use exogenous sucrose as a glycosyl donor for the production of extracellular glucans or fructans by enzymes in the periplasmic space. Nevertheless, lipid pyrophosphoryl sugars play a major role in the biosynthesis of most extracellular bacterial polysaccharides.

Lipopolysaccharides

The biosynthesis of the *O* side-chain (*O* antigen) of *Salmonella anatum* is typical of the sequence of reactions that produce an extracellular bacterial lipo-polysaccharide (Fig. 4.19). A series of sugars is donated from nucleotides, frequently differing in the purine or pyrimidine moiety, to undecaisoprenyl pyrophosphate bound to the cytoplasmic membrane. The first donation is different from the rest in that it forms the pyrophosphoryl bond between lipid and sugar. D-Galactose 1-phosphate is transferred in place of D-galactose. The remaining two sugars, L-rhamnose and D-mannose, are donated to the non-reducing end of the previously added sugar. The sequence of sugar residues and the configuration of the glycosidic bond are determined by the membrane-bound transferases. Depending on the bacterium, more complex oligosaccharides may be assembled at this stage. An example is the branched tetrasaccharides that form the capsular polysaccharides of some *Klebsiella* strains.

See *Cell Biology*, Chapter 2

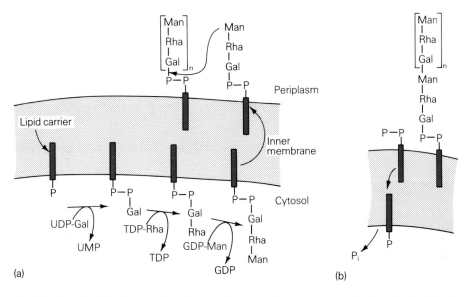

(a)

(b)

Fig. 4.19 Assembly of the trisaccharide unit, mannosyl–rhamnosyl–galactose, during the biosynthesis of the *O* antigen component of *S. anatum* lipopolysaccharide. (a) The sequential transfer of sugars to an undecaisoprenyl carrier and subsequent relocation from the cytoplasmic to the periplasmic face of the inner membrane is followed by insertion into partially assembled antigen. (b) Return of the carrier with loss of phosphate to the cytoplasmic face for another round of trisaccharide assembly. Transferases catalysing each step have been omitted for clarity.

The lipid pyrophosphoryl trisaccharide now moves to the external face of the membrane where the trisaccharide residues are polymerized to form the *O* antigen domain of the lipopolysaccharide. The polymerization occurs not by the usual transfer to the non-reducing distal end of the polymer, but by addition to the adjacent 'reducing' end. The reason for this is twofold. Firstly, the strong association of the lipid moiety for the membrane means that the

Reference Hammond, S.M., Lambert, P.A. and Rycroft, A.N. (1984) *The Bacterial Cell Surface*, Croom Helm, London and Sydney. A comprehensive survey of the biosynthesis and assembly of bacterial cell walls, coats and appendages in a little over 200 pages.

activated trisaccharide cannot move into the periplasmic space but remains bound to the external face of the membrane. Secondly, the partially formed acceptor polymer is still attached to a lipid pyrophosphoryl moiety and therefore is itself bound to the membrane. It is the bond between the pyrophosphoryl group and the partially formed polysaccharide that is broken and the exposed sugar residue that is the recipient of the nucleophilic attack by the newly arrived trisaccharide (Fig. 4.19). Thus, the lengthening of the polysaccharide is by an insertion mechanism and the lipid moiety of the incoming trisaccharide now acts as anchor to the extended polysaccharide. The displaced lipid pyrophosphate returns to the cytoplasmic side of the membrane during the course of which cleavage of the pyrophosphoryl group, catalysed by a pyrophosphatase, occurs. When the O side-chain is complete it is transferred to the core region and lipid A is embedded in the outer membrane.

Box 4.8
Bacitracin, vancomycin and phosphonomycin; inhibitors of bacterial wall assembly

These three inhibitors of bacterial cell wall synthesis each act in a different manner. Bacitracin blockades peptidoglycan assembly in bacteria but not by directly inhibiting any of the sugar or peptide transfer reactions. It prevents the loss of phosphate from the recycled polyisoprenyl pyrophosphate (Fig. 4.21) and so inhibits the formation of the acceptor of the pentapeptide N-acetylmuramate residue. In contrast, vancomycin inhibits the addition of the pentapeptide disaccharide unit to the partially assembled wall, while phosphonomycin prevents the synthesis of UDP-N-acetylmuramate from UDP-N-acetylglucosamine and phosphoenolpyruvate (Fig. 4.20).

Exercise 6

You have experimental evidence that galactose residues are being incorporated into polymeric material by a bacterial preparation. Without exhaustive structural analysis what evidence would strongly suggest that the residue was being used in the biosynthesis of lipopolysaccharide?

Peptidoglycan

The principal structural polymer of both Gram-negative and Gram-positive cell walls is peptidoglycan. The polysaccharide backbone is an alternating sequence of N-acetylmuramate and N-acetylglucosamine residues and is assembled in much the same way as the O side-chain of a lipopolysaccharide. The toughness of the peptidoglycan arises from covalent links with other

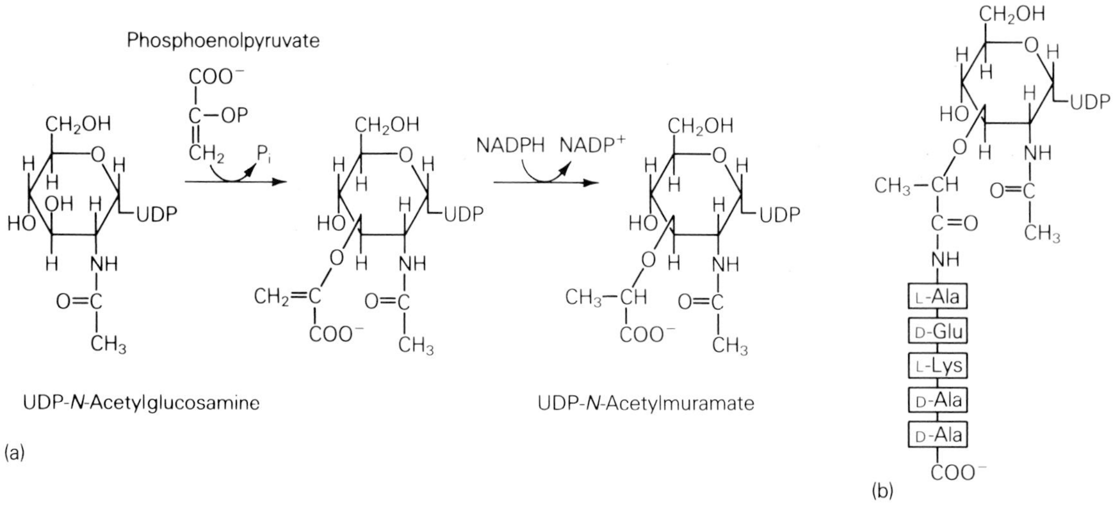

Fig. 4.20 The biosynthesis of the uridine diphosphomuramate pentapeptide. (a) addition of lactyl ether to uridine diphospho-N-acetylglucosamine, is followed by (b) pentapeptide formation by the sequential addition of amino acids catalysed by specific ATP-dependent ligases.

88 Biosynthesis

Muramic acids (muramates): and teichoic acids (teichoates) are compounds found in bacterial cell walls, from the Latin murus, a wall, Greek teichos, a wall.

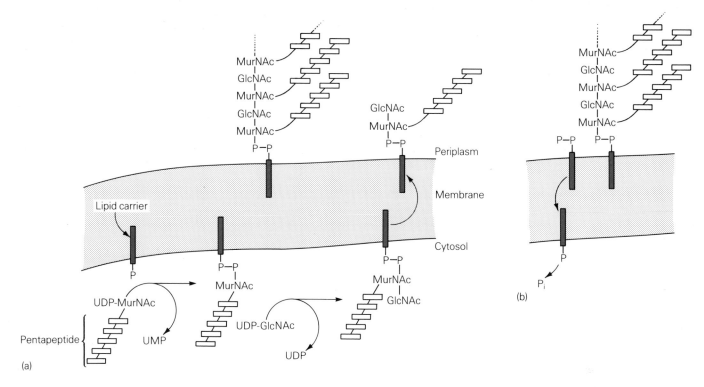

Fig. 4.21 Biosynthesis of bacterial cell wall peptidoglycan. (a) Formation of undecaisoprenyl diphospho-pentapeptide disaccharide and subsequent relocation from the cytoplasmic to the periplasmic face of the membrane. (b) Elongation by insertion and return of carrier with loss of phosphate.

polysaccharides through short peptide chains. Like the *O* side-chain polysaccharide, the repeating units, this time disaccharides covalently attached to a penta- or decapeptide side-chain, are built up on the inner surface of the cytoplasmic membrane (Fig. 4.20). They are transported to the external face still anchored to the lipid pyrophosphoryl group. There they are received by the activated terminus of the partially assembled peptidoglycan (Fig. 4.21). The lipid pyrophosphate is recycled after removal of the terminal phosphoryl group. The final step, forming the enveloping molecular bag round the bacterium, is the cross-linking of the glycan chains through the peptide bridges.

4.6 *Overview*

The sequence of sugars in a polysaccharide is dependent on the specificity of synthetases transferring sugars from activated precursors, nucleoside or lipid pyrophosphoryl sugars, to recipient residues in the growing chain. The greater the variety of sugars employed and glycosidic bonds formed, the more types of transferases are required. Each sugar transferred requires the hydrolysis of two phosphoanhydride bonds. Occasionally glycosidic bonds are produced without the participation of sugar pyrophosphoryl derivatives, but only at the expense of breaking existing bonds.

Polymerization may be followed by covalent modification. For instance, galacturonate residues already incorporated into growing pectate molecules may have their carboxylate group esterified thus removing the negative charge from the backbone. Production of some glycosaminoglycans requires

☐ The notion that oligo- and polysaccharide synthesis involves one transferase for each type of glycosidic bond produced is in general correct. However, there is one well-documented exception and there may be more. In mammalian cells the glycosylation of proteins occurs within the endoplasmic reticulum and the Golgi apparatus. In the latter stages of assembling the oligosaccharyl moiety, galactose residues are transferred from UDP-Gal to a terminal *N*-acetyl glucosamine residue forming a $\beta 1$–4 linkage. However, in response to hormonal changes associated with birth, the protein, **α-lactalbumin** (M_r 14 000) is produced. In its presence galactose residues are no longer transferred to glycoproteins but to glucose to form lactose which is secreted into milk.

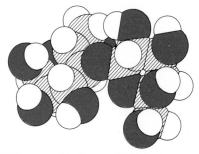

Molecular model of lactose. (Courtesy of C. Holloway, Biosyn Technologies Ltd.)

The α-lactalbumin-galactosyl transferase complex is called **lactose synthetase**.

incorporated residues to become negatively charged. Sulphation of *N*-acetyl-galactosamine increases the already highly anionic nature of these polymers. During the synthesis of glycogen and amylopectin oligosaccharyl portions of α1–4-glucan are transposed between neighbouring parts of the molecule. On occasions the sugar residue itself is changed. For example, incorporated D-glucuronate residues are epimerized to L-iduronate converting chondroitin-4-sulphate to dermatan sulphate. More usually, if one sugar is to be converted to another, this occurs before polymerization. This step may occur before or after the sugar has become part of the nucleotide or lipid intermediate. All of these changes alter the conformation of the polysaccharide formed and provide a way of controlling polymer function.

Although many polysaccharides such as chitin and cellulose are located extracellularly, their biosynthesis requires energy (ATP) which is only available in the cytoplasm. One solution to this problem is to orientate the polysaccharide synthetase in the cytoplasmic membrane such that it receives activated precursors at the cytoplasmic face and extrudes polymer from the external face. Controlling the location of the synthetase enables the cell to restrict production of polysaccharide to selected areas of the cell surface. Alternatively, in bacteria small pre-assembled and pre-activated oligosaccharides are translocated through the membrane on a lipophilic polyisoprenol carrier. Polymerization then occurs in the periplasmic space.

Answers to exercises

1. The distinction between radiolabelled low M_r precursor and high M_r polymer, both present in a reaction mixture, is made by exploiting differences in their chemical and physical characteristics. It is usually possible, and much easier, to use physical differences such as charge and size. Thus passage through an anion-exchange resin will resolve a neutral polymer from a nucleotide precursor. Passage down a gel-filtration column retards the low M_r precursor and separates it from the high M_r polymer which elutes earlier. Simple differential extractions may also be used. Thus, the application of samples of reaction mixture to 1 cm squares of thick filter paper followed by their extraction in an ethanol/water mixture will elute the precursor but leave the insoluble polymer on the paper, which is then assayed for radioactivity.

2. Disruption of the cell followed by centrifugation of the products at different forces and for different times will produce a series of fractions that can be incorporated into an assay for polysaccharide synthetase.

Incubation of whole cells or cellular fragments with radiolabelled sugars or precursors followed by electron microscopy and autoradiography can reveal the location of labelled polymers which are probably located at their site of synthesis.

3.

non-reducing terminal

reducing terminal

Recent findings suggest that the 'reducing' residue is bound to protein and cannot reduce.

4. Incubate the product with a β1–3-glucanase and look for the release of low M_r products (glucose, β1–3-oligoglucans). However, there may be structural features that resist enzymic hydrolysis and one would have to make sure that there was no cellulase activity in the β1–3-glucanase.

A better method would be to treat a sample with sodium metaperiodate followed by acidic hydrolysis of the oxidized material. Any glucose appearing in the hydrolysate indicates that it was originally linked 1–3 as this blocks periodate oxidation. The 1–4-linked glucosyl residues in cellulose would be oxidized.

5. Partial hydrolysis by enzyme or acid, followed by separation and characterization of the fragments with respect to composition should reveal that some, if not all, of the oligosaccharides contain mixtures of the two sugars.

6. Incubation of the reaction mixture with bacitracin would halt the incorporation of galactose from nucleotide precursor into lipid linked intermediates as no acceptor would be available.

FILL IN THE BLANKS

1. The control of glycogen biosynthesis is _____ by the concentrations of certain effector molecules including _____ _____ and _____ , as well as by hormones such as _____ , _____ and _____ . The enzymes involved, _____ _____ can exist in _____ interconvertible forms only one of which is active. The change from active to inactive state or _____ _____ involves a _____ modification to the protein and this control is co-ordinated with the process of glycogen _____ .

Choose from: adrenalin, AMP, breakdown, covalent, Glc 6-P, glucagon, glycogen synthetases, insulin, regulated, two, *vice versa*.

2. Sucrose is a _____ composed of _____ and _____ residues, and is produced by photosynthetic _____ of green plants and is subsequently _____ round the plant. Two enzymes are known to catalyse sucrose synthesis: _____ synthetase and _____ _____ synthetase. One of these produces sucrose 6-phosphate in photosynthetic tissues and sucrose formation from this compound is promoted by the action of _____ _____ . In _____ tissues, in contrast, the enzyme _____ _____ tends to catalyse the formation of UDP-Glc and _____ rather than sucrose _____ .

Choose from: cells, disaccharide, fructose (2 occurrences), glucose, non-photosynthetic, sucrose, sucrose 6-phosphate, sucrose-6-phosphate phosphatase, sucrose synthetase, synthesis, transported.

3. When the periodic extracellular polysaccharides are being synthesized, a typical process involves transfer of _____ sugars to form _____ units which are subsequently transferred themselves. The process occurs _____ the cell, and the oligosaccharides are then _____ to the _____ face of the cytoplasmic membrane to be added to the growing _____ . However, other pathways exist and some polysaccharides are produced by the action of _____ in the periplasmic space. _____ phosphoryl sugars play a major role in the biosynthesis of many bacterial polysaccharides although the assembly of bacterial _____ does not require a lipid intermediate.

Choose from: cellulose, enzymes, external, lipid, oligosaccharide, polysaccharide, single, transported, within.

MULTIPLE-CHOICE QUESTIONS

4. Which of the following may the biosynthesis of glycosidic bonds require?

A. hydrolysis of ATP to ADP and Pi
B. the participation of a polyisoprenol unit
C. the participation of UDP-Glc
D. the participation of CTP
E. the participation of NADPH

5. Which of the following modifications to sugar residues occurs *before* polymerization to form polysaccharides?

A. sulphation
B. epimerization
C. methylation
D. oxidation of $-CH_2OH-$ to $-CO_2^-$
E. introduction of $-NH_2$

6.   In the formation of starch, which of the following is the immediate donor of glucose units?

A. free glucose
B. glucose 6-phosphate
C. sucrose
D. ADP-Glc
E. mannose

7.   Which of the following is *true* of glycogen?

A. it contains β 1–4 glycosidic links
B. it contains α 1–6 glycosidic links
C. it has only one non-reducing end
D. it contains one reducing end for every branch point
E. it is found in bacteria as well as vertebrates

SHORT-ANSWER QUESTIONS

8.   The production of an insoluble polysaccharide within the cell would cause disruption of cytoplasmic functions, but why should some polysaccharides, composed of highly water-soluble sugars, be insoluble?

9.   Where are glucosyl residues transferred to during the biosynthesis of glycogen: (*a*) the reducing end, (*b*) the non-reducing end or (*c*) both ends depending on availability?

10.   Commencing with glucose, and naming the appropriate enzymes, write the metabolic pathway from glucose to glycogen. Which enzyme(s) control(s) the rate of glycogen biosynthesis in (*a*) mammals, (*b*) bacteria?

11.   Draw the structure of fructose 6-phosphate. Why is it a key intermediate in polysaccharide synthesis?

12.   Draw the structure of undecaprenyl diphosphogalactose. What are the structural differences between this lipid intermediate and those found in yeast?

13.   The bridging trisaccharide between the core protein and the polysaccharide chain of chondroitin 4-sulphate comprises a xylose and two galactose residues. How many transferases are required to build this link and why?

14.   In assembling an oligosaccharide unit for incorporation into a lipopolysaccharide, where are the sugar residue transfers made in the bacterial cell? Using symbols lipid, P, UDP-Gal, TDP-Rha and GDP-Man outline the biosynthesis of the lipopolysaccharide unit (Man–Rha–Gal)–.

15.   To which end of the partially assembled polymer are oligosaccharide units transferred during the biosynthesis of lipopolysaccharides: (*a*) the reducing end, (*b*) the non-reducing end or (*c*) both ends depending on availability?

16.   On which side of the bacterial cytoplasmic (inner) membrane are peptide–disaccharide units polymerized during the formation of peptidoglycan?

17.   Structural alterations that change one sugar to another usually occur before the sugars are incorporated into a polysaccharide. Describe two structural changes that occur to sugar residues that have already become part of a polysaccharide. What advantages might there be in these post-polymerization alterations?

Nitrogen fixation and incorporation of nitrogen into amino acids

Objectives

After reading this chapter, you should be able to:

☐ describe the range of different nitrogen sources that are used by microorganisms and plants;

☐ list the requirements for the bioconversion of N_2 to ammonia;

☐ explain the assimilation of ammonia into amino acids;

☐ discuss the regulatory mechanisms which govern the order of preference of nitrogen sources;

☐ appreciate the energy cost of nitrogen assimilation.

5.1 Introduction

Nitrogen occurs in many organic compounds in organisms, especially in amino acids and proteins and in nucleotides and nucleic acids. All organisms are capable of synthesizing proteins from the constituent amino acids, but only plants and certain microorganisms are able to synthesize the common amino acids from inorganic sources of nitrogen. In contrast, animals in general require dietary sources of some or all of the protein amino acids and this is also true of some microorganisms (Fig. 5.1).

The simplest form of nitrogen, **dinitrogen** (N_2), makes up 78% of the earth's atmosphere. It is a stable, unreactive gas, especially when compared to O_2, the other major component of the earth's atmosphere. Only the most reactive metals, for example, Na, K, and Mg combine directly with N_2 to form nitrides, whereas a wide range of metals combine with oxygen. The two atoms of nitrogen are combined in N_2 through a triple bond made of one σ-bond and two π-bonds, and this is a particularly stable arrangement. N_2 has a high ionization energy:

$$N_2 \rightarrow N_2{}^+ + e^- \quad \Delta G^0 = 1.5 \text{ MJ mol}^{-1}$$

Dinitrogen: used to distinguish N_2 from nitrogen in its more general sense in nitrogen metabolism, nitrogen repression, etc. In the same way, the terms dihydrogen and dioxygen should also be used, but this would be cumbersome and is usually unnecessary since there is unlikely to be any ambiguity.

Reference Gallon, J.R. and Chaplin, A.E. (1987) *An Introduction to Nitrogen Fixation*, Cassell, London, UK. A detailed and up-to-date account of all aspects of nitrogen fixation but at the same time giving a good background for the non-specialist.

□ Detailed discussions of σ and π bonds can be found in any textbook of structural chemistry. For example, Baikess, R.S. and Edelson, E. (1985) *Chemical Principles*, 3rd edn, Chapters 6 and 7, Harper and Row, London. The important point to note is the difference in electron density distribution between the two bonds. This is illustrated below in the formation of σ and π bonds using two C orbitals.

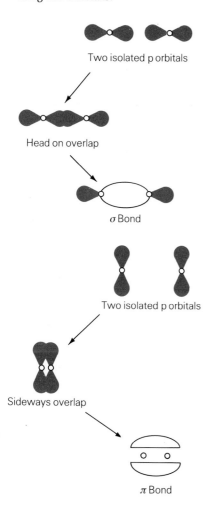

Two isolated p orbitals

Head on overlap

σ Bond

Two isolated p orbitals

Sideways overlap

π Bond

See *Cell Biology*, Chapters 1 and 5

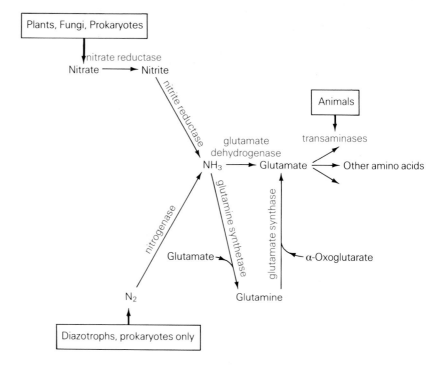

Fig. 5.1 Pathways leading to incorporation of nitrogen into amino acids: note that not all pathways occur in all organisms.

and a high bond enthalpy (the energy required to split N_2 into two nitrogen atoms):

$$N_2 \rightarrow 2N \quad \Delta H = 945 \text{ kJ mol}^{-1}.$$

Also, the electron density for the σ-bond is concentrated between the two atoms where it is less accessible for chemical reactions.

Whereas plants can use ammonia, nitrates, and nitrites to make organic nitrogen compounds, the ability to use this simplest form, dinitrogen, is restricted to a limited range of bacteria including the cyanobacteria. Some of these bacteria are free living, while others have symbiotic relationships with various plants. Plants are dependent on these bacteria to 'fix' atmospheric nitrogen, and all animal life is dependent in turn on plants (or micro-organisms) for its supply of inorganic nitrogen.

5.2 *Nitrogen fixation*

The term **nitrogen fixation** is generally used to encompass the steps in bacteria, involved in converting N_2 into ammonia. Indeed, the formation of ammonia from N_2 and H_2 has been studied in great detail by chemists. Nitrogen fixation in organisms occurs at moderate temperatures, at around neutral pH, and is driven by the hydrolysis of substantial amounts of ATP. ATP is required in part for the generation of reducing potential from water, but also, it is assumed, so that the reaction proceeds at appreciable rates under physiological conditions.

Nitrogen fixation by symbiotic microorganisms in plants has been known about for a long time, but it is only within the last 20 years that N_2 fixation by a wide range of independently living bacteria has been discovered. A breakthrough in the detection of nitrogen fixation was the discovery in 1966 that the

Reference Postgate, J. (1987) *Nitrogen Fixation* 2nd edn, Studies in Biology, No. 92, Edward Arnold, London. A good general account, less expensive and less detailed than Gallon and Chaplin.

Reference Leadbetter, E.R. and Poindexter, J.S. (eds) (1985) *Bacteria in Nature*, vol. I. Plenum, New York. Chapter 6 gives a detailed account of the history of the study of nitrogen fixation.

enzyme **nitrogenase**, which is responsible for catalysing the reduction of N_2 to ammonia, could also reduce acetylene (ethyne) to ethylene (ethene).

$$C_2H_2 + 2[H] \rightleftharpoons C_2H_4$$

The product of this reaction can be readily detected by gas chromatography. This is known as 'the acetylene reduction test', and it has made it easier to survey a range of organisms for their ability to fix N_2. To be absolutely sure that an organism can fix N_2 one has ultimately to measure nitrogen fixation (and this is technically rather difficult), but so far all organisms found to be able to reduce acetylene have also been found to be able to reduce N_2. All the evidence to date indicates that nitrogen fixation occurs exclusively in prokaryotes. These may be either free-living or symbionts living in association with plants, fungi, or even animals.

DIAZOTROPHS occur in most of the nutritional groups of prokaryotes, as shown in Table 5.1. **Diazotrophs** may be free-living or they may form associations with other organisms. In associating with other organisms there is a gradation from a very loose association to an intimate one, the partners in the latter having a high degree of interdependence. Some diazotrophs which are free-living tend to concentrate around the roots of plants, although they show virtually no morphological or biochemical modifications. At the other extreme is the intimate relationship between *Rhizobia*, the **root nodule bacteria**, and **leguminous plants** (Fig. 5.2). These plants are able to form nodules in response to invasion by the bacteria. The bacteria differentiate to form **bacteroids**, a process involving a change in the permeability of the bacteroid wall to allow exchange of metabolites (Fig. 5.3). An essential step in the integration is the production of the protein **leghaemoglobin** which ensures the low oxygen tension needed for nitrogen fixation to occur. Interestingly, the plant carries the gene for the globin but the haem prosthetic group is provided by the bacteroid.

The best-known of the symbionts are those associated with leguminous plants, but the actinomycete bacteria also form symbiotic associations with angiosperms.

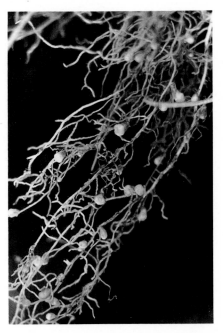

Fig. 5.2 Root nodules on clover (*Trifolium repens*) root.

Box 5.1 *The Haber process*	Nitrogen fixation by the **Haber process** is a major industrial concern. Out of an annual total of about 24×10^7 tonnes of N_2 fixed globally, around 16×10^7 tonnes are fixed by microorganisms, 6×10^7 tonnes by the Haber process, and 2.4×10^7 tonnes fixed by a variety of processes including the effects of ultraviolet light, lightning, fires, and the internal combustion engine. This last group generally involves the oxidation of N_2, the oxides formed eventually being reduced to ammonia. The Haber process is more significant agriculturally than these figures suggest, since any N_2 fixed in this manner is likely to be distributed where it is required, rather than randomly throughout the globe.

The reaction:

$$N_2(g) + 3H_2(g) = 2NH_3 \quad \Delta H^0 \; -92 \text{ kJ mol}^{-1}$$

is exothermic, and at 27°C and atmospheric pressure the equilibrium mixture contains 98.5% ammonia, but the rate of reaction is low. At higher temperatures the equilibrium does not lie so far to the right. For example, at 327°C, 8.7% ammonia is present. However, the optimum conditions for the process involve a balance between thermodynamic and kinetic considerations. A commercially acceptable rate of reaction is attained using a high temperature, usually about 500°C, and a pressure of approximately 1000 atmospheres, in the presence of an Fe catalyst.

Diazotroph: *The suffix -troph, as in autotroph, auxotroph, diazotroph, heterotroph, prototroph and phototroph all relate to the mode of nutrition, and is derived from the Greek word meaning nourish. A diazotroph is an organism which is able to use N_2 as its sole source of nitrogen. The N_2 becomes reduced to ammonia and is then incorporated into organic nitrogen.*

Bacteroid: *the name used for the irregularly-shaped or branched form of bacterium such as Rhizobium as it occurs in the root cells during nodule formation.*

Table 5.1 *Examples of diazotrophs*

Organisms	Mode of nutrition	Comment
Clostridium	Strict anaerobe	Several diazotrophic species
Klebsiella	Facultative anaerobe	One of the most studied species
Azospirillum	Microaerobe	Associates with grasses
Rhodospirillum	Phototroph	Fixes N_2 anaerobically, photosystem I only
Rhodopseudomonas	Phototroph	Fixes N_2 anaerobically, photosystem I only
Thiobacillus	Chemoautotroph	Oxidizes S compounds
Anabaena	Phototroph	Filamentous cyanobacteria, N_2 fixation in heterocyst, photosystems I and II
Nostoc	Phototroph	Filamentous cyanobacteria, N_2 fixation in heterocyst, photosystems I and II
Gloeothece	Phototroph	Unicellular cyanobacteria, photosystems I and II
Rhizobium	Symbiont	Occurs in root nodules of legumes
Bradyrhizobium	Symbiont	Occurs in root nodules, slower growth than *Rhizobium*
Frankia	Symbiont	Occurs in non-legumes, e.g. alder

Understanding the mechanism of nitrogen fixation was hampered for several years because attempts to demonstrate nitrogen fixation *in vitro* were unsuccessful. A successful *in vitro* experiment is generally an important prerequisite for analysing any biochemical pathway at the molecular level. The development of *in vitro* experimental techniques was easy in the cases of the major metabolic pathways such as glycolysis, the tricarboxylic acid cycle and β-oxidation of fatty acids. However, when extracts were prepared from most nitrogen-fixing bacteria, nitrogen fixation could not be detected. It was only when the obligate anaerobe, *Clostridium pasteurianum*, was studied that success was achieved. This organism will only grow and survive in the absence of oxygen, or at least only at very low oxygen tensions, and so extracts were prepared under conditions in which oxygen was excluded. That did the trick, and active extracts were obtained. It has since been shown in

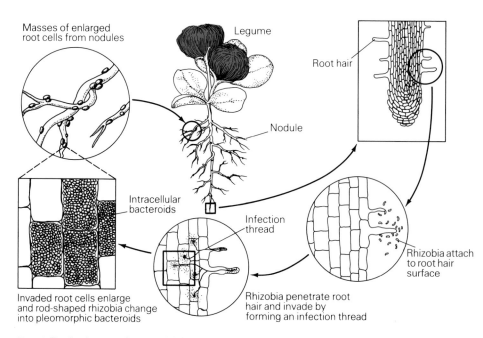

Fig. 5.3 The development of a root nodule in a legume. Redrawn from P.J. Van Denmark and B.L. Batzing (1987), *The Microbes: An introduction to their nature and importance*, Benjamin-Cummings Publishing Co., p. 860.

most nitrogen fixers that the enzymes required for nitrogen fixation are irreversibly inhibited and inactivated by oxygen. Since this realization in 1960, rapid progress has been made in understanding nitrogen fixation. A second stumbling block was the observation that the addition of ATP to extracts resulted in an initial surge in activity which then rapidly decreased. It transpired that although ATP is essential for fixation, the ADP produced inhibits nitrogen-fixing activity. Thus to sustain activity *in vitro* it is necessary to use an ATP generating system, so that any ADP formed is rapidly converted back to ATP, for example:

$$\text{ADP} + \text{Phosphoenolpyruvate} \rightarrow \text{ATP} + \text{Pyruvate}$$

Another problem encountered in studying nitrogen fixation is that the nitrogenase from many organisms is a **cold labile** enzyme. The 'standard technique' of routinely preparing extracts and of keeping them at near 0°C in order to reduce autolysis and hinder bacterial growth, has an adverse effect in this case. Nitrogenase has been isolated from several diazotrophs and its properties and role in nitrogen fixation are described in the next section.

☐ Cold-labile enzymes are enzymes which are more readily denatured at near 0°C than at room temperature. These are enzymes in which hydrophobic interactions are important for maintaining their tertiary structure and they are often associated with membranes. The stability of hydrophobic interactions increases with temperature over a limited range.

5.3 Nitrogenase

Nitrogenase comprises two distinct components, called Components I and II, which are directly involved in nitrogen fixation, although many more proteins may be involved less directly. For example, in *Klebsiella pneumoniae* 17 genes are known to be involved either directly or indirectly in nitrogen fixation. The functions of the gene products are shown in Table 5.2. In *Klebsiella* these genes are arranged in eight transcriptional units or operons. It is clear that these constitute all the genes essential for nitrogen fixation: they can be transferred in a single piece of DNA (total length 24 000 base pairs), to a non-diazotroph such as *E. coli*, which then becomes capable of nitrogen fixation. The genetics of diazotrophs belonging to other genera such as *Azotobacter*, *Rhizobium*, *Bradyrhizobium*, *Azospirillum*, *Rhodopseudomonas* and *Anabaena* have also been studied, although less extensively. The genes for nitrogen fixation in all of these organisms show a high degree of homology with those of *Klebsiella pneumoniae*, although they differ generally in that their nitrogen-fixing genes are dispersed in a few clusters throughout the genome rather than being together in one contiguous unit as in *K. pneumoniae*.

Nitrogenase is a major protein in diazotrophs and may account for up to 10% of the cellular protein. The enzyme is a relatively inefficient catalyst when compared to other enzymes (see Table 5.3). Because of its low **turnover**

Table 5.2 *Genes for nitrogen fixation in* Klebsiella

Gene (symbol)	Function of gene products
K, D	α and β polypeptide of Component I
H	γ polypeptide of Component II
B, V, N, E	Mo/Fe cofactor
Q	Mo uptake
S, U	Processing of Component I
M	Processing of Component II
A, L	Regulation (positive and negative)
J	Oxidoreductase
F	Flavodoxin
X, Y	?

Turnover number: *a measure of the catalytic activity of an enzyme. It is defined as the number of moles substrate transformed to product per mole of enzyme per second* (see *Biological Molecules, Chapters 4 and 5*).

Table 5.3 *Comparison of enzyme turnover numbers*

Enzyme	Turnover number (k_{cat}, s^{-1})
Carbonic anhydrase	1×10^6
Catalase	4×10^7
Fumarase	8×10^2
Triose phosphate isomerase	4.3×10^3
Nitrogenase	8×10^{-1}

☐ Acid-labile sulphur is sulphur that is released as hydrogen sulphide upon treatment with HCl. It includes sulphur that is bonded to Fe as shown in Fig. 5.4, but not sulphur present in cysteine or methionine residues in proteins, peptides or the free amino acids.

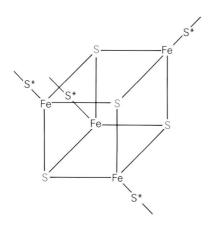

Fig. 5.4 Structure of the Fe–S clusters which occur in nitrogenase (acid-labile sulphurs shown in red).

number a high concentration of nitrogenase in the cell is necessary to fix N_2 at a sufficiently high rate.

The structures and properties of both components of nitrogenase are similar in all the bacteria from which they have been isolated. Hybrids may be made by mixing Component I and Component II from different species; most of these hybrids are fully functional in nitrogen fixation, although a few function less well than the homologous systems.

Component I has an M_r of 220 000–250 000 and is a tetramer of structure $\alpha_2\beta_2$. The protein contains iron, molybdenum (Mo) and **acid labile sulphur(s)**. The enzyme contains between 24 and 32 Fe atoms and two atoms of Mo. Each enzyme molecule contains four **iron–sulphur clusters** (Fig. 5.4) each of composition Fe_4S_4, and two Mo/Fe cofactors each containing six S, eight Fe and one Mo. The Mo/Fe cofactor can be dissociated from the enzyme and when it is added to an inactive mutant, nitrogenase activity is restored.

In *Azotobacter* there is an alternative pathway of nitrogen fixation, which is most readily detected in molybdenum deficiency. The nitrogenase necessary for this pathway is a vanadium-containing enzyme.

Component II is smaller, having a structure γ_2 in which each polypeptide has an M_r of 30 000. The protein contains Fe but not Mo, the former being in the form of an Fe_4S_4 complex. Nitrogenase has two binding sites for ATP.

The mechanism of nitrogen fixation

The mechanism of nitrogen fixation in general terms is now reasonably well understood, but many of the mechanistic details are still controversial. The

---

Box 5.2
Transfer of nitrogen-fixing genes

There is currently much interest in the organization of the nitrogen fixation genes, especially because of the possibility of introducing these genes into plants to enable them to fix nitrogen. This is of great potential importance, since nitrogen could thus become available for agriculturally important plants such as rice. It would also save on the fossil fuels which are presently used in the production of nitrogenous fertilizers. There has been some success in transferring the genes for nitrogen fixation (*nif* genes) between species: *nif* genes have been transferred from *K. pneumoniae* to related Gram-negative bacteria such as *E. coli*, where they are expressed, and to less closely related bacteria and to yeast where they have not been found to be expressed. There is still much to be done before it might be possible to transfer *nif* genes to plants and to have successful expression. Problems generally arise when trying to express prokaryote genes in eukaryote hosts. For nitrogenase to function it requires the exclusion of oxygen, the generation of ATP and sufficient reducing potential even after any problems of expression have been overcome.

See *Molecular Biology and Biotechnology*, Chapter 9

precise stoichiometry for the overall reaction is still uncertain, but is close to:

$$N_2 + 8H^+ + 8e^- + 16\,MgATP + 16H_2O = 2NH_3 + H_2 + 16\,MgADP + 16P_i$$

This overall reaction may be regarded as a coupling of reaction (i) with that of (ii) and (iii).

☐ Many reactions involving phosphoryl transfer requiring Mg^{2+} use MgATP as the 'true' substrate, not ATP.

(i) $MgATP + H_2O \rightarrow MgADP + P_i$
(ii) $2H^+ + 2e^- \rightarrow H_2$
(iii) $N_2 + 6H^+ + 6e^- \rightarrow 2NH_3$

In vivo, electrons are donated to Component II by the proteins ferredoxin or flavodoxin. However, it is possible and often more convenient *in vitro* to use a non-physiological electron donor such as sodium dithionite ($Na_2S_2O_4$).

$$Na_2S_2O_4 + 2H_2O \rightarrow 2NaHSO_3 + 2H^+ + 2e^-$$

The reducing potential (that is, the ability to donate electrons) may be expressed as a redox potential. Electrons tend to be transferred to increasingly positive redox couples. Both ferredoxin and flavodoxin have highly negative redox potentials and are therefore good electron donors.

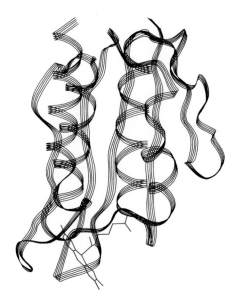

Fig. 5.5 Representation of flavodoxin. The bound FAD is shown in red. Courtesy Dr E.E. Eliopoulos, Department of Biochemistry and Molecular Biology, University of Leeds, UK.

FLAVODOXIN is a protein of M_r 16 000 containing an FAD cofactor (Fig. 5.5). It has two markedly different redox potentials for the two stages of its reduction. This means that it can act as both an electron donor and acceptor according to the circumstances. The flavin moiety of FAD which undergoes the reduction may exist as the quinone, semiquinone or hydroquinone. The structures of the three forms are shown below.

Quinone → semiquinone   E −150 mV;   Semiquinone → hydroquinone   E −440 mV

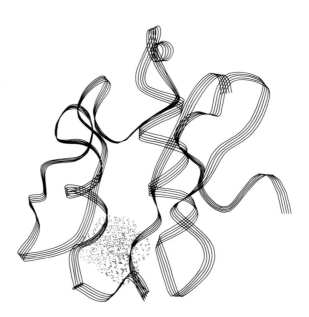

Quinone   Semiquinone   Dihydroquinone

In nitrogen fixation it is the highly negative redox couple that is used.

FERREDOXINS are also small redox proteins. However, in these an Fe–S cluster is the redox component (Fig. 5.6). They also have highly negative redox potentials. The reductions involve the Fe–S clusters; the Fe interconverts between Fe^{2+} and Fe^{3+}.

Electrons are transferred from ferredoxin or flavodoxin to Component II, which has a redox potential of −290 mV. The reduced Component II then binds MgATP. This produces a conformational change in Component II, so that it now has the more negative redox potential of −400 mV. This enables an electron transfer to occur from Component II to Component I causing, in turn, the reduction of N_2 to NH_3 (Fig. 5.7). Eight successive electron transfers are necessary for the complete reduction to NH_3. The details of this process are still unclear. Although intermediates such as HN : NH and $H_2N : NH_2$ have been postulated to participate in the reaction, none has yet been proven to be involved. However, an enzyme-bound intermediate that release hydrazine ($H_2N . NH_2$) has been detected.

The conversion of N_2 to NH_3 requires a source of ATP to interact with

Fig. 5.6 Representation of ferredoxin. The bound iron–sulphur centre is shown in red. Courtesy of Dr E.E. Eliopoulos, Department of Biochemistry and Molecular Biology, University of Leeds, UK.

Component II (Fig. 5.7). It also requires reducing potential for the initial reduction of ferredoxin or flavodoxin. The mechanism for these reductions varies between organisms, but it is believed that a process of reversed electron transport driven by the hydrolysis of ATP may be important in most diazotrophs. This accounts for the high ATP requirement accompanying nitrogen fixation. Some diazotrophs are photosynthetic and in these ferredoxin becomes reduced by electrons released by the photosystem as a result of light capture.

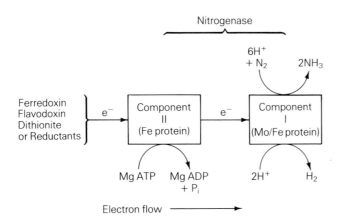

Fig. 5.7 Diagrammatic representation of the likely flux of electrons through nitrogenase.

CHEMICAL MODELS have been used to try to understand the chemistry of the reduction of N_2 to ammonia, and in particular the role of the **transition metals** iron and molybdenum in catalysis by nitrogenase. Perhaps the two most important features of transition metals in this context are their variable states of oxidation, for example, Fe^{2+}, Fe^{3+}; Mo^{3+}, Mo^{4+}, Mo^{5+}, Mo^{6+}, and their ability to form complexes with a variety of ligands. A familiar example of the latter is the way in which Fe^{2+} complexes with the porphyrin ring in haemoglobin.

The complexes formed between transition metal ions (M) and N_2 are probably of a linear type (M–N≡N). This type of bonding would be expected to weaken the triple bonds between the nitrogen atoms and also cause the nitrogen atom most distant from the metal ion to have a partial negative charge, which in turn would cause it to attract a proton. In the nitrogenase enzyme it is the Mo/Fe cofactor which is involved in binding N_2. Spectroscopic evidence suggests that it is the molybdenum rather than the iron to which the N_2 is bound, but the details are still somewhat speculative.

EFFICIENCY OF NITROGEN FIXATION is usually defined by the number of ATP molecules used per N_2 conversion to two molecules of NH_3. This ratio is difficult to estimate accurately. When the ratio is measured *in vitro* using sodium dithionite as an artificial electron donor, 12 to 15 molecules of ATP are consumed per N_2 converted to ammonia. The variation between 12 and 15 molecules of ATP is probably due to endogenous phosphatases in the extracts which degrade a portion of the ATP. The results obtained *in vivo* are more

Exercise 1

What are the functions of ATP in nitrogen fixation?

variable: *Rhizobium*, 25–35 ATP/N$_2$; *Klebsiella*, 30 ATP/N$_2$; *Clostridium*, 20 ATP/N$_2$; *Azotobacter*, 6 ATP/N$_2$. Most authorities now favour a stoichiometry of 16 ATP as indicated earlier. However, a factor which affects the efficiency is the amount of H$_2$ generated by the nitrogenase activity, and whether that H$_2$ is used by the organism. Some diazotrophs such as *Azotobacter* possess a hydrogenase which catalyses the oxidation of H$_2$ to water, and this is coupled to ATP production. This lessens the net amount of ATP required per N$_2$ fixed.

Nitrogen fixation requires much energy and often occurs under anaerobic conditions, so it is not surprising that the ***molar growth yield*** is lower when organisms are using N$_2$ rather than most other sources of nitrogen. The growth rate is often much slower. For example, when *Bacillus macerans* grows aerobically using ammonia as a nitrogen source its generation time is about 3 hours, whereas when growing anaerobically using N$_2$ its generation time is increased to about 23 hours.

Importance of oxygen exclusion during nitrogen fixation

Nitrogenase is both inhibited (possibly by uncompetitive inhibition) and rapidly inactivated by oxygen, and it can only function *in vivo* at low oxygen tensions. Some diazotrophs are obligate anaerobes, for example, *Clostridium* and *Desulphovibrio* which grow only at low oxygen tensions, but others are facultative anaerobes, for example, *Klebsiella*, *Bacillus* and *Propionobacteria*. These organisms fix nitrogen only when growing anaerobically. The green and purple photosynthetic bacteria, for example, *Chromatium* and *Chlorobium* are also anaerobes. In all these examples, nitrogen fixation occurs under conditions of low oxygen tension. There are, however, several aerobes that fix nitrogen. These have evolved methods of insuring low oxygen tensions in the intracellular region of nitrogenase.

For example, *Azotobacter* is a free-living aerobe and maintains a high rate of respiration which consumes the oxygen in the vicinity of the nitrogenase. The photosynthetic Cyanobacteria possess photosystems I and II and evolve oxygen during the course of photosynthesis. In some of these, such as the filamentous Cyanobacteria, *Anabaena* and *Nostoc*, nitrogen fixation occurs in specialized cells called **heterocysts**, while the evolution of oxygen during photosynthesis occurs in the vegetative cells. Heterocysts are surrounded by a thick envelope, the innermost layer of which is composed of glycolipid (Fig. 5.8). It is thought that the envelope restricts the diffusion of oxygen into the heterocyst. The separation of nitrogen and carbon metabolism between the vegetative cells and the heterocysts represents an interesting division of labour (Fig. 5.9). In some non-heterocystous Cyanobacteria, of which *Gloeothece* has been extensively studied, there is a *temporal* separation between nitrogen fixation and the evolution of oxygen by photosynthesis. This is most clearly shown when the organism is grown on a cycle of 12 hours light followed by 12 hours dark: about 95% of nitrogen fixation occurs in the dark phase.

The most important examples of nitrogen fixation from an agricultural standpoint are the root-nodule bacteria occurring in legumes. There is a symbiotic association between the bacteroid, *Rhizobium* and the host legume. *Rhizobium* is an obligate aerobe and heterotroph, able to fix nitrogen. It acquires organic carbon from the legume derived from photosynthesis, and provides the legume with nitrogen which, in the case of lupins, is transported from the nodule in the form of asparagine (Fig. 5.10). Soya beans and tropical legumes transport nitrogen from the nodules in the form of allantoin and allantoate (Fig. 5.10).

Oxygen is maintained in the cells of the root nodule at a low tension by the presence of leghaemoglobin. The free oxygen concentration in the nodules is

Exercise 2

Which important metabolic processes require segregation from nitrogenase if both are to operate concurrently?

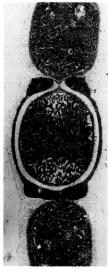

Fig. 5.8 Electronmicrograph of *Anabaena cylindrica*. ×6720. Note thich wall of the heterocyst. Courtesy of Professor G.A. Cod and Ms G. Alexandre, Department of Biological Sciences, University of Dundee, UK.

Molar growth yield: *an index of the efficiency of a substrate to support growth. It is defined as the increase in dry weight of biomass per mole of substrate consumed.*

about 11 nmol dm^{-3}, with over 99.9% of the oxygen bound to leghaemo-globin. Leghaemoglobin is a haem protein, rather like myoglobin (Fig. 5.11), having a high affinity for oxygen. It constitutes 20–30% of the protein of the nodule cells. It appears to maintain a steady but low oxygen tension for the bacteroid, when the supply of oxygen in the cells fluctuates.

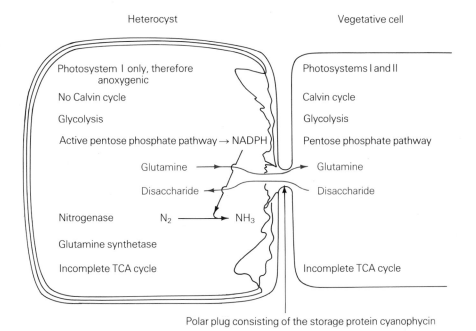

Fig. 5.9 Division of labour between heterocysts and vegetative cells in the Cyanobacterium *Anabaena*.

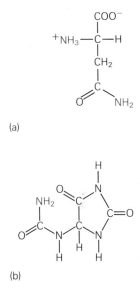

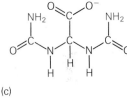

Fig. 5.10 (a) Asparagine, (b) allantoin and (c) allantoate.

(a)   (b)

Fig. 5.11 Representations of molecules of (a) sperm whale myoglobin and (b) lupin leghaemoglobin. Note the similarity in overall shape. Redrawn from Lesk, A.M. (1984) Themes and contrasts in protein structures. *Trends in Biochemical Sciences*. **9**, 290–1.

Reference Brewin, N.J. and Vandenbosch, K.A. (1988) Inside the legume root nodule, *Plants Today*, **1**, 114–21. Superb short account of nitrogen fixation in leguminous plants. Nicely illustrated.

5.4 Nitrate as a source of nitrogen

Nitrate is the most widely used source of nitrogen in plants and is also used by many bacteria and fungi. In all these organisms the nitrate is reduced to ammonia before incorporation into organic compounds. In plants, other than those associating with diazotrophs, the sources of inorganic nitrogen are nitrate or ammonia. In most soils nitrate is much more abundant than ammonia. Any ammonia formed in soil is generally oxidized by **nitrifying bacteria** to nitrate.

$$NH_3 + 1.5\,O_2 = NO_2^- + H_2O + H^+$$
$$NO_2^- + 0.5\,O_2 = NO_3^-$$

□ Nitrifying bacteria are chemoautotrophic bacteria (*Energy in Biological Systems*, Chapter 2) which obtain their energy from the oxidation of ammonia to nitrate. They fall into two groups, those which oxidize ammonia to nitrite and include the genera, *Nitrosoamonas*, *Nitrosococcus*, *Nitrosolobus* and *Nitrosovibrio*, and those which oxidize nitrite to nitrate and include the genera *Nitrobacter* and *Nitrococcus*.

Most plants take up nitrate readily. However, in climax conifer forests, and in grasslands where the low pH inhibits nitrification, nitrogen is taken up mainly as ammonia. For most fungi that have been investigated, although ammonia is preferred, nitrate is also a good source of nitrogen.

Whereas the enzymes for ammonia metabolism are constitutive, those for nitrate reduction are inducible. Nitrate reduction to ammonia requires two enzymes, nitrate reductase and nitrite reductase which catalyse the following overall reactions:

$$NO_3^- + 2H^+ + 2e^- = NO_2^- + H_2O$$
$$NO_2^- + 8H^+ + 6e^- = NH_4^+ + 2H_2O$$

NITRATE REDUCTION may occur either in the roots or in the stem depending on the plant. Nitrate reductase, which is generally NADH-requiring, occurs in the cytosol. Nitrate reductases from fungi such as *Aspergillus* and *Neurospora* have been most thoroughly studied. In *Neurospora* at least eight genes are involved, either directly or indirectly, in nitrate reduction. Two are required for the nitrate reductase, four for the synthesis and incorporation of a molybdenum cofactor into the enzyme and two genes are involved in regulation. The nitrate reductase in *Neurospora* is a dimeric protein made up of identical subunits each of M_r 145 000 and requiring a molybdenum-containing cofactor. In *Aspergillus nidulans* the subunits are slightly smaller with an M_r of 91 000. The same molybdenum cofactor occurs in a number of other molybdenum-containing enzymes such as xanthine dehydrogenase.

Exercise 3

What are the likely sources of NADH in the cytosol?

Box 5.3 Non-molybdenum nitrogenases

Bishop, P.E. and Joerger, R.D. (1990) Genetics and molecular biology of alternative nitrogen fixation systems. *Annual Review of Plant Physiology and Plant Molecular Biology*, **41**, 109–25. Covers the genetics, including the gene organization, of the Mo-containing nitrogenases as well as brief comments on the regulation of their expression.

When nitrogenase was isolated from a variety of microorganisms it was found to be an iron–sulphur protein containing molybdenum (Mo). Therefore when such microorganisms were grown on Mo-free media they should not be able to fix N_2. This was almost true, but not quite, and the ability to fix N_2 under these conditions was attributed to their scavenging traces of Mo from media and culture vessels. However, in 1986 it was found that certain *Azotobacter* spp. could produce a vanadium (V)-containing nitrogenase using a genetically distinct system. This V-containing nitrogenase consists of a V–Fe protein and a Fe protein. The V–Fe protein is an $\alpha_2\beta_2$ tetramer of M_r 210 000 containing two V, 23 Fe and 20 acid-labile sulphides per tetramer. The Fe protein has M_r 60 000 and contains four Fe and four acid-labile sulphurs. Interestingly, V-nitrogenase reduces acetylene to ethylene and then ethane, whereas Mo-nitrogenase only reduces acetylene to ethylene.

Even more recently a third nitrogenase that appears to use simply Fe alone has been reported. It is not very active in converting acetylene to ethylene which may be why its presence has not previously been suspected.

Reference Lewis, O.A.M. (1986) *Plants and Nitrogen*, Studies in Biology, No. 166, Edward Arnold, London. A useful account of nitrogen assimilation in plants is given in Chapter 2.

In most soils nitrate is more abundant than ammonia. Nitrate is most readily used by growing plants. However, because of its high solubility and its positive charge (the same as most soil particles) it is readily leached out of the soil. Nitrogenous fertilizers added to the soil to increase fertility increase the nitrate content. The nitrate leached out of the soils often increases the nitrate levels in rivers and lakes. This promotes the increased growth of algal blooms. The level of nitrate in the environment has also risen as a result of fuel combustion and the volatile oxides of nitrogen contribution to acid rain. Concern has grown about the increased levels of nitrate in the environment and its toxicity, particularly to humans. The main sources of nitrate in the human diet is in drinking water and vegetables.

Nitrate and nitrites may also be added to meat products as a preservative to prevent botulism, and as a flavour enhancer. Nitrates in the diet may be reduced to nitrites by bacteria present in the body. Nitrites and related nitroso compounds are known carcinogens. For this reason there is concern to keep nitrate levels in the diet to a minimum.

Plant nitrate reductases are associated with FAD and cytochromes as well as molybdenum. It appears that nitrate reductase is the only essential molybdenum-containing enzyme in plants that do not fix N_2, since if such plants are grown with ammonia as source of nitrogen, they can be grown on molybdenum-free medium. Nitrite reductase occurs in the chloroplasts of leaves and the proplastids of roots. It contains an Fe_2S_2 centre and one molecule of **sirohaem**, a partially reduced iron porphyrin (see Chapter 9). Nitrite formed in the cytosol by nitrate reductase activity has to be transported into these organelles for reduction. In the chloroplasts, photosynthesis generates electrons which are captured by ferredoxin. The reduced ferrodoxin then transfers electrons to nitrite causing its reduction.

5.5 *Incorporation of ammonia into amino acids*

Animals are dependent on organic nitrogen in the form of amino acids. They are able to interconvert certain amino acids and synthesize non-essential amino acids from essential amino acids (see *Energy in Biological Systems*, Chapter 7), but they are unable to synthesize amino acids from inorganic nitrogen. In contrast, plants, certain fungi and bacteria can directly incorporate ammonia into glutamate. However, N_2, nitrate or organic nitrogenous compounds such as urea, acetamide or purines (which can be utilized by many fungi) must first be converted to ammonia before the nitrogen can be incorporated into amino acids. Two routes are used by bacteria, fungi, and plants for this conversion. (The three enzymes involved are also present in animals tissues, but in animal the net flux is not generally in favour of glutamate formation from ammonia.) Both routes involve glutamate as an intermediate. The first involves glutamate dehydrogenase (Fig. 5.12), and the second involves two enzymes, glutamine synthetase and glutamate synthase (often combined and the second of which is known by the abbreviation to GOGAT, Glutamine OxoGlutarate Amino-Transferase).

The equilibrium for the glutamate dehydrogenase catalysed reaction is not favourable for glutamate formation under physiological conditions, and the K_m for ammonia is high. The equilibrium for the coupled reactions involving glutamine synthetase and glutamate synthase is much more favourable for glutamate formation. The glutamine synthetase-catalysed reaction has a high negative $\Delta G'$ since it involves the hydrolysis of ATP; the enzyme also has a low K_m for ammonia of between 10^{-4} and 10^{-5} mol dm^{-3}. In bacteria such as

---

Exercise 4

What are purines and how do they arise as metabolites? Hint: it may be helpful to consult Chapter 7.

Reference Solomonson, L.P. and Barber, M.J. (1990) Assimilatory nitrate reductase: functional properties and regulation. *Annual Review of Plant Physiology and Plant Molecular Biology*, **41**, 225–53. A comprehensive review of nitrate reductase.

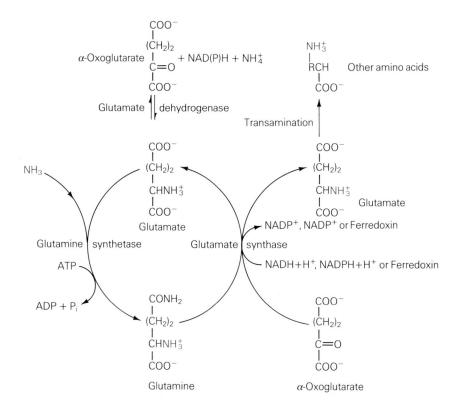

Fig. 5.12 The incorporation of ammonia (NH_3) into amino acids showing the glutamine–glutamate cycle.

Klebsiella aerogenes, glutamate dehydrogenase is the principal enzyme involved in amination when the ammonia concentration is high, whereas the reactions catalysed by glutamine synthetase and glutamate synthase are used at the low ammonia concentrations more usually encountered in natural environments. These coupled enzymes form a 'glutamate cycle' in which one of the two glutamates formed in the second reaction can be used as a source of other amino acids (Fig. 5.12), while the other glutamate is recycled.

Glutamate synthase has been purified from a variety of leaves and from roots. There are two distinct types of glutamate synthase, one using NADH or NADPH, and the other ferredoxin. Many plant tissues have both enzymes, but generally the ferredoxin-requiring enzyme accounts for the bulk of synthase activity.

Glutamate dehydrogenases from fungi such as *Neurospora*, *Aspergillus*, *Candida* and *Saccharomyces* have been studied for several years. However, the presence of glutamate synthase in fungi has only become apparent recently. In *Neurospora*, where the most detailed study has been made, there are two glutamine synthetase ***isozymes*** called the α and β forms. These are in addition to two glutamate dehydrogenases (NAD^+- and $NADP^+$-requiring) and a glutamate synthase. In a series of experiments using a **chemostat**, where *Neurospora* was grown at different steady state levels of ammonia, the results showed that at high ammonia concentrations, $NADP^+$-glutamate dehydrogenase and the β isozyme of glutamine synthetase were principally responsible for ammonia assimilation. However, under ammonia limitation, the α isozyme of glutamine synthetase and glutamate synthase partially replaced the β isozyme and $NADP^+$-glutamate dehydrogenase, but the two pathways were not mutually exclusive. The NAD^+-glutamate dehydrogenase is responsible for the oxidative deamination of glutamate, particularly when the organism uses glutamate as a source of nitrogen and carbon.

Exercise 5

The equilibrium of the glutamate dehydrogenase-catalysed reaction will depend upon pH. Why is this?

☐ A **chemostat** is a vessel used for growing microorganisms under conditions in which one chemical composition of the grown medium is held constant. This generally entails a continuous input of nutrient matching the rate at which it is consumed by the microorganisms and also the continuous removal of the waste materials produced.

Isozyme: *a contraction of isoenzyme. Two enzymes may be regarded as isoenzymes if they catalyse the same reaction but differ in their amino acid sequence (Chapters 4 and 5).*

5.6 Transamination

From the previous sections it is apparent that the central pathways leading to amino acid biosynthesis involve ammonia and glutamate. Many other amino acids, for example, proline, arginine, asparagine, tryptophan and histidine, are formed from glutamate using both its amino group and carbon skeleton (after modification). An important group of amino acids are those formed from α-oxoacids by transamination with glutamate. There are a number of different transaminases. Most catalyse the transfer of the amino group of glutamate to an α-oxoacid acceptor:

$$^-OOC\ CH_2CH_2CH-COO^- + RCH_2COCOO^- \rightleftharpoons {}^-OOCCH_2CH_2\overset{\underset{||}{O}}{C}-COO^- + RCH_2CHCOO^-$$

Glutamate α-Oxoglutarate

A number of different transaminases have been identified which catalyse key steps in the synthesis of aspartate, alanine, histidine, serine and valine. All these enzymes have a **pyridoxal phosphate cofactor** which undergoes a cycle in the catalytic procedure:

Transamination is also involved in the catabolism of amino acids (see also *Energy in Biological Systems*, Chapter 7).

5.7 Regulation of incorporation of nitrogen into amino acids

The main routes of incorporation of nitrogen into amino acids are shown in Fig. 5.1. There is evidence for control mechanisms operating at most of the steps in this scheme. Regulation of enzyme activity can be brought about in many different ways. Most regulatory mechanisms fit into one of two categories. Long-term regulation in which the *amount* of enzyme is controlled and short-term regulation in which the *activity* of a given amount of enzyme is controlled. The former is slow to respond but is capable of producing larger changes. Examples of both mechanisms have been found in the regulation of amino acid biosynthesis from N_2 and nitrate.

Most plants and microorganisms have a preference for a particular source of nitrogen although the preferred source is almost always ammonia. If ammonia is plentiful, it will act to repress all routes leading to ammonia formation. Ammonia prevents both the induction of nitrate reductase and the utilization of a whole range of nitrogen metabolites such as purines, proline, asparagine and acetamide which are potential sources of ammonia.

Nitrogenase regulation

Ammonia has no direct effect on extracts containing nitrogenase, but it is able to repress the biosynthesis of nitrogenase. The regulation of nitrogenase has not been studied in detail although it is clearly regulated by ammonia, oxygen and adenine nucleotides. Oxygen has the effect of initially inhibiting the enzyme and then irreversibly inactivating it. Ammonia causes the repression of nitrogenase, but whether ammonia is the active corepressor is not completely clear. It is possible that glutamine synthetase or the product of its catalytic action, glutamine, is responsible for the repression of nitrogenase. The evidence for the involvement of glutamine comes from both studying the effects of a glutamine synthetase inhibitor, methionine sulphoxime, and from mutations in the structural gene for glutamine synthetase. Both these methods of inhibiting glutamine synthesis prevent the repression of nitrogenase by ammonia. It is possible that methionine sulphoxime may inhibit other enzymes as well.

The genetic basis for regulating the expression of the nitrogenase gene has been studied most extensively in *Klebsiella pneumoniae*. Regulation is complex. Two groups of genes are able to regulate the expression of nitrogenase: the *nif* (nitrogen fixation) genes and the *ntr* (nitrogen regulation) genes. Of the 17 *nif* genes in *Klebsiella* (see Table 5.2) *nifA* and *nifL* are thought to code for an activator and a repressor respectively. The latter is thought to combine with ammonia, when present at concentrations of about 4 mmol dm^{-3} to produce the inactive inducer (Fig. 5.13).

The second group of genes, *ntr*, is also thought to produce an activator, which is able to combine with higher concentrations of ammonia (about 20 mmol dm^{-3}) and then act as a repressor. Adjacent to the *ntr* genes is the gene coding for glutamine synthetase, *glnA*. The latter is believed to be cotranscribed with the *ntr* genes. It is not yet clear whether mutations in the *glnA* gene in *Klebsiella* prevent repression of nitrogen fixation because of the production of inactive glutamine synthetase, or because mutations in *glnA* prevent expression of the *ntr* genes.

In addition to regulation by ammonia, nitrogenase may also be regulated by nitrate, carbamoyl phosphate (an intermediate in pyrimidine synthesis), urea and amino acids. Adenine nucleotides also regulate nitrogenase activity. ATP is necessary for nitrogen fixation, but any build up of ADP causes inhibition of nitrogenase activity. There is evidence from photosynthetic bacteria that Component II of nitrogenase is inactivated by covalent modification by ADP-ribosylation.

Nitrogen repression in fungi

In fungi, the enzymes for several catabolic pathways which result in ammonia production, together with nitrate reductase, are regulated by 'nitrogen repression' (Fig. 5.14). The concentration of ammonia regulates the production of these enzymes. In *Aspergillus* and *Neurospora* nitrate reductase is not synthesized constitutively but requires the presence of nitrate in the growth medium as an inducer. If, however, both nitrate and ammonia are present, ammonia overrides nitrate induction and represses the synthesis of nitrate reductase. Similar observations have been made on the catabolic systems shown in Fig. 5.14. In each case the presence of ammonia leads to repression of the catabolic pathway. This phenomenon is known as nitrogen (or ammonia) repression and is the equivalent of carbon catabolite repression.

Nitrogen repression requires both an active NADP$^+$-glutamate dehydrogenase and glutamine synthetase. Mutants deficient in either of these enzymes are derepressed. This has led to the suggestion that the active corepressor is glutamine:

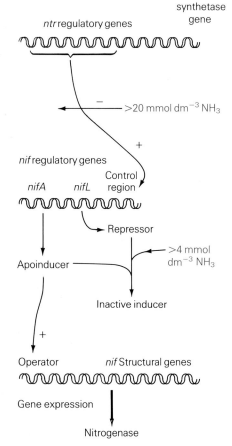

Fig. 5.13 Regulation of nitrogenase expression.

Reference Orme-Johnston, W.H. (1985) Molecular basis of biological nitrogen fixation. *Annual Review of Biophysics and Biophysical Chemistry*, **14**, 419–59. Excellent review covering the structure and mechanisms of nitrogenase, together with an introduction to the genetic basis of nitrogenase expression.

$$\text{Ammonia} \xrightarrow[\text{dehydrogenase}]{\text{glutamate}} \text{Glutamate} \xrightarrow[\text{synthetase}]{\text{glutamine}} \text{glutamine}$$

This is positive repression, where the product of the regulatory gene is an active **apoinducer** (compare with the lactose operon which is repressed in the *absence* of inducer, *Molecular Biology and Biotechnology*, Section 6.2). The apoinducer binds to the operator to switch on the contiguous structural genes. The corepressor, which in this case is believed to be glutamine, combines with the apoinducer to prevent it combining with the operator, thus preventing expression of the structural genes. A key feature of the scheme

☐ Apoenzyme, coenzyme, holoenzyme; apoinducer, coinducer, and aporepressor and corepressor are all terms concerned with the reversible binding of a small molecule or ligand to a protein.

$$P + A \rightleftharpoons PA$$

In each case the protein lacking the ligand is given the prefix apo- meaning 'away from or detached' and the corresponding ligand is given the prefix co- to indicate that it operates 'with' the apoprotein. The prefix 'holo' is then used to indicate 'whole or complete'. It is, however, rarely used with inducer or repressor. The suffix in each case indicates the function.

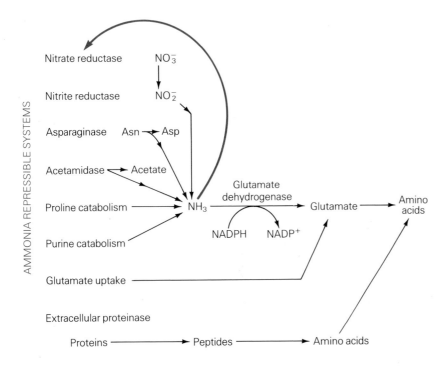

Fig. 5.14 Scheme to show how NH_3 exerts nitrogen repression in fungi.

Box 5.5
Nitrogen utilization by wood-rotting fungi

A group of fungi which are particularly efficient at extracting nitrogeneous compounds from their natural substrate are the wood-rotting fungi. Wood has a particularly low ratio of nitrogen to carbon (N : C is 1 : 200 to 1 : 1600) which means organisms colonizing it have a shortage of nitrogen and an abundance of carbon. The N : C ratio in the mycelia of the fungi varies with the available nitrogen. For example when the N : C ratio in an artificial growth medium is changed from 1 : 16 to 1 : 1600, the percentage nitrogen in the mycelia falls from about 6% to about 0.2%. At low N : C ratios the extraction of nitrogen from the medium is virtually complete. At these low ratios, the proportion of nitrogen in the fungal proteins is reduced preferentially to that of nucleic acids. These organisms survive and grow in this low nitrogen environment by a combination of efficient nitrogen uptake, together with reutilization of the intracellular nitrogen. Nitrogen-containing compounds such as proteins and nucleic acids are concentrated at the hyphal tip where the bulk of protein synthesis occurs. Behind the hyphal tip there is extensive vacuolation and the nitrogen content is extremely low. The mechanism of uptake has not been extensively studied but wood-rotting fungi have active, $NADP^+$-requiring, glutamate dehydrogenases and glutamine synthetases. The amounts of the enzymes are increased during nitrogen deficiency.

Reference Magsanik, B. (1982) Genetic control of nitrogen assimilation in bacteria, *Annual Review of Genetics*, **16**, 135–68. Rather old, but a useful and detailed account of nitrogen assimilation in bacteria.

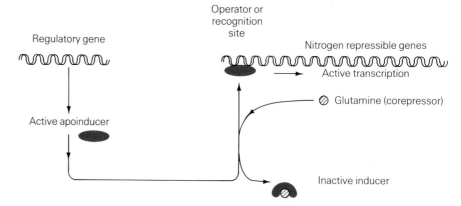

Fig. 5.15 Proposed mechanism of nitrogen repression.

would be that the structural genes for the separate nitrogen repressible systems shown in Fig. 5.15 must all be switched on by the apoinducer. Since these genes for the nitrogen repressible systems are dispersed throughout the genome, often on different chromosomes, it would be expected that each separate operon having a common operator sequence or recognition site would be identified by the apoinducer. This part of the scheme has still to be firmly established.

Glutamine synthetase regulation

Another step in the pathway which has been studied in detail is the regulation of glutamine synthetase in *E. coli*. Glutamine synthetase (GS) is an important enzyme in the uptake of ammonia as mentioned in Section 5.5. It is also a donor of amino groups in the biosynthesis of histidine and tryptophan together with several other metabolic intermediates such as glucosamine, CTP, AMP and carbamoyl phosphate. The enzyme is oligomeric, made up of 12 identical subunits (Fig. 5.16) arranged in two hexagonal layers. Glutamine synthetase activity is regulated both by covalent modification and allosterically. Its activity is susceptible to a wide range of effectors: inhibition by several end-products of glutamine metabolism including glycine, alanine, tryptophan, histidine, CTP and CMP. Each product exerts a cumulative inhibition of activity by binding to regulatory sites. In addition, GS is subject to covalent modification by adenylylation of a tyrosine residue (Fig. 5.17). The adenylylation can occur on all 12 subunits leading to a complete loss of activity

$$12\,\text{ATP} + \text{GS}_0\,(active) \rightarrow \text{GS--(AMP)}_{12}\,(inactive) + 12\,\text{PP}_i$$

Intermediate states exist *in vivo* which have partial activity. The intermediate states show different sensitivities towards effectors. For example, GS_0 is sensitive to alanine, glycine and AMP, whereas GS--(AMP)_{12} is most sensitive to histidine, tryptophan and CTP. Adenylylation is catalysed by adenylyl transferase in the presence of a second protein, P_{II}, (see Fig. 5.17). In contrast to glycogen phosphorylase, where phosphorylation and dephosphorylation are catalysed by separate enzymes (Chapter 4), both adenylylation and deadenylylation of GS are catalysed by a single enzyme but the separate reactions are catalysed by modified and unmodified forms of P_{II} protein respectively. The latter can be covalently modified by UTP at a tyrosine residue. Both the enzyme-catalysed adenylylation and deadenylylation and the uridylylation and deuridylylation processes are modified by effectors as shown in Fig. 5.17. Both processes are particularly sensitive to

Reference Marzluf, G.A. (1981) Regulation of nitrogen metabolism and gene expression in fungi, *Microbiological Reviews*, **45**, 437–61. A detailed review giving biochemical and genetic approaches used in fungi.

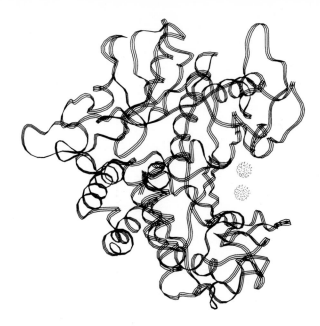

Fig. 5.16 Representation of the backbone of a single subunit of glutamine synthetase of *E. coli*. The dotted spheres are Mn^{2+}. Courtesy of Dr E.E. Eliopoulos, Department of Biochemistry and Molecular Biology, University of Leeds, UK.

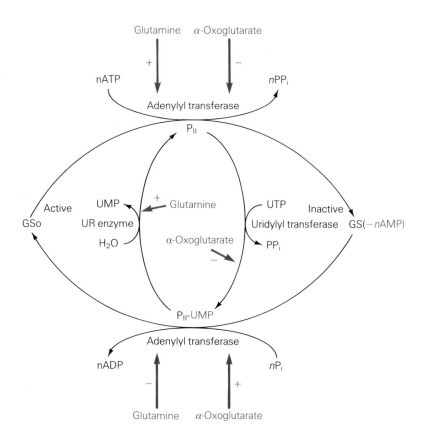

Fig. 5.17 Regulation of glutamine synthetase. A positive sign indicates stimulation and a negative sign indicates inhibition.

changes in the concentrations of α-oxoglutarate and glutamine which respectively increase and decrease the active forms of the enzyme. Thus glutamine synthetase is responsive to a wide range of effectors and hence to the needs of the cell.

5.8 Overview

The incorporation of two of the most important sources of inorganic nitrogen, namely nitrate and dinitrogen (N_2) into amino acids and hence proteins requires their prior reduction to ammonia. In both cases metabolic energy is required to generate the necessary reducing potential. The ability to fix N_2 is restricted to certain prokaryotes which contain nitrogenase, the key enzyme required to fix N_2. Nitrogenase is particularly sensitive to oxygen and is only active at low oxygen tensions and this hindered its study experimentally. Diazotrophs have evolved a range of strategies to exclude oxygen from the intracellular site of nitrogenase activity.

Glutamate and glutamine are central to the biosynthesis of amino acids, since most amino acids are formed via transamination from glutamate. Many organisms show a preference for particular sources of nitrogen. Generally ammonia is the preferred source of nitrogen. This preference is manifest in the control mechanisms which, for example, repress nitrate reduction and nitrogen fixation when the supply of ammonia is abundant.

--- Answers to exercises ---

1. Binds to component II, generate reducing potential.
2. TCA cycle, β-oxidation, photosynthesis involving photosystems I and II aerobic processes.
3. Glycolysis (glyceraldehyde-3-phosphate dehydrogenase), possibly pentose phosphate pathway together with transhydrogenase.

4. Purines have a general structure containing purine rings:

They arise from catabolism of purine nucleotides such as ATP, GTP etc.

5. H^+ released during glutamate oxidation.

QUESTIONS

FILL IN THE BLANKS

1. Glutamate synthase catalyses the reaction:

L-glutamine + α-oxoglutarate + _____ + _____ $\rightleftharpoons$ α-L-glutamate + _____ .

Nitrogenase is generally detected using the reaction:

_____ + _____ = C_2H_4.

Pyridoxal phosphate is a cofactor concerned with the transfer of _____ groups in

_____ reactions.

A root nodule bacterium such as *Rhizobium* requires a _____ source in the form of

_____ or its hydrolysis products from its leguminous host. A heterocyst supplies

_____ in the form of _____ to the adjacent vegetative cells.

Choose from: amino, C_2H_2, carbon, glutamine, H^+, H_2, NAD^+, NADH, nitrogen, sucrose, transamination.

MULTIPLE-CHOICE QUESTIONS

Indicate the most appropriate alternative.

2.   The Haber process operates at a high temperature because:

A. the reaction is exothermic
B. the equilibrium position is more favourable
C. NH_3 is more stable at the higher temperature
D. the rate of product formation is faster

3.   Ferredoxins are important in nitrogen fixation as a source of:
A. Fe^{2+}
B. Fe^{3+}
C. electrons
D. auxins

4.   Transition metals are important components of nitrogenase to:
A. transport electrons
B. stabilize the enzyme against denaturation by oxygen
C. form nitrides with N_2
D. absorb O_2

5.   The activity of glutamine synthetase is regulated by:
A. glutamate dehydrogenase
B. adenylyl transferase
C. glutamyl regulase
D. uridylyl transferase

SHORT-ANSWER QUESTIONS

6.   Give three methods used by diazotrophs to circumvent the oxygen-sensitivity of nitrogenase.

7.   A mutant of *Neurospora* lacking glutamine synthetase, synthesizes nitrate reductase when grown on nitrate and ammonia, but the corresponding prototroph when grown on nitrate and ammonia does not. Explain.

8.   Place the following pairs of redox couples in order of redox potentials starting from the most positive: FAD/FADH$_2$, O_2/H_2O, ferredoxin Fe^{3+}/Fe^{2+}, NO_3^-/NO_2^-, and NAD^+/NADH, $2H^+$/H_2.

9.   Which of the following metal ions are required for the biochemical conversion of $N_2 \rightarrow NH_3$: Fe^{2+}, Mo^{2+}, Ru^{2+}, Mg^{2+}, Mn^{2+}, Na^+?

10.   Glutamate dehydrogenase has a K_m for NH_4^+ of 11 mmol dm^{-3}, glutamine synthetase has a K_m for NH_4 of 0.3 mmol dm^{-3}. The intracellular concentration of NH_4^+ in a microorganism is 6.2 mmol dm^{-3}. Are either of these enzymes operating near their maximum velocity? Use the Michaelis–Menten equation to solve this problem.

6
Amino acid interconversions

Objectives

After reading this chapter, you should be able to:

☐ describe how non-essential amino acids are synthesized and explain their roles as precursors of other molecules;

☐ describe how serine arises from glycolytic precursors and explain its role as a precursor of glycine and one-carbon units;

☐ explain how one-carbon units are transferred and modified by oxidation or reduction as necessary;

☐ outline how polyamines are formed, partly from *S*-adenosylmethionine and partly from ornithine;

☐ explain how proline, ornithine, glutamine, arginine and γ-aminobutyrate arise from L-glutamate and discuss their roles as precursors of other cell constituents;

☐ discuss the importance of individual enzymes in amino acid biosyntheses by considering inherited metabolic conditions in which key enzymes are lacking.

6.1 Introduction

Organisms vary widely in their abilities to synthesize amino acids. In general, microorganisms and plants can synthesize all the amino acids they require using chemically simple forms of nitrogen such as ammonia or nitrate. However, the biosynthetic routes used are often tortuous and involve the activities of many enzymes as well as a substantial input of metabolic energy. Consequently, many microorganisms will use preformed amino acids from the environment, whenever available, rather than synthesize them *de novo*. Mammals lack the ability to synthesize about half of the 20 amino acids they need for making proteins. These amino acids are therefore called **essential amino acids** and those which can be synthesized are called **non-essential amino acids**. As a rule essential amino acids are chemically the most complex types with aromatic rings or branched hydrocarbon side-chains.

Amino acids are not used solely to form proteins although this is a major role: they are also precursors of other cell constituents. For example, purine nucleotides are synthesized in part from glycine (Chapter 7), while methionine and ornithine are the precursors of polyamines (Section 6.8). The carbon skeletons of all the amino acids may be used as metabolic fuels.

Table 6.1 Essential amino acids

Histidine
Isoleucine
Leucine
Lysine
Methionine
Phenylalanine
Threonine
Tryptophan
Valine

Reference Herrmann, K.M. and Somerville, R.L. (eds) (1983) *Amino acids: Biosynthesis and Genetic Regulation*, Addison–Wesley, New York. A comprehensive survey of biosynthetic pathways of amino acids and their control.

Box 6.1
Some herbicides are inhibitors of amino acid biosynthesis

Kishore, G.M. and Shah, D.M. (1988) Amino acid biosynthesis inhibitors as herbicides. *Annual Review of Biochemistry*, **57**, 627–63. A comprehensive and detailed review. For reference only.

Several herbicides, for example *glyphosate* and *sulfometuron methyl*, act by inhibiting the biosynthesis of amino acids necessary for plant health and growth. The metabolic reactions necessary for the biosyntheses of essential amino acids are unique to plants and microorganisms. Therefore the use of inhibitors which block these pathways are likely to be non-toxic to animals.

Herbicides which act by inhibiting amino acid biosynthesis usually function by binding to and inhibiting enzymes involved in these biosyntheses or by competing with the usual substrate of the enzyme. However, some may also function by preventing the translocation of the enzymes from their site of synthesis, the cytosol, to their site of activity, the chloroplast. Treatment with herbicide initially blocks amino acid synthesis in the rapidly growing regions or meristems, and this is followed by a systemic effect with inhibition occurring throughout the plant. Eventually, virtually all the cells of the plant are killed, although death may take several days or weeks to occur.

Glyphosate

Sulfometuron methyl

6.2 Essential and non-essential amino acids

Humans and rats are unable to synthesize nine out of the 20 standard amino acids used in protein synthesis. These essential amino acids (Table 6.1) can be made by plants and microorganisms using complex metabolic routes described in some of the background reading.

Mammals obtain essential amino acids mostly from dietary proteins which are digested in the intestine to release amino acids and subsequently absorbed. These amino acids can serve as precursors of proteins or other biological materials. The amino groups of amino acids surplus to requirements are removed by transamination and the carbon skeletons remaining are catabolized to intermediates that may be oxidized to release energy or may be converted into metabolic fuels such as glucose or **ketone bodies**. These processes occur mainly in the liver.

The non-essential amino acids that mammalian cells synthesize are glutamate, glutamine, aspartate, asparagine, alanine, proline, tyrosine, cysteine, glycine, and serine. Many of them are precursors of other, non-protein, cell constituents. Histidine is regarded as an essential amino acid for both rats and humans. However, it is required in the diet only during the growth of young individuals. Some amino acids can only be described as non-essential because they can be formed from essential amino acids. An example is tyrosine which can be made from phenylalanine. Thus, if there is only sufficient phenylalanine in the diet to satisfy the requirement for phenylalanine, then a deficiency of tyrosine may occur. The sulphur atom of cysteine (a non-essential amino acid) originates from methionine (an essential

See *Energy in Biological Systems*, Chapters 7 and 6

Ketone bodies: these are acetoacetate, β-hydroxybutyrate and acetone. They are chemical compounds rather than bodies and unfortunately β-hydroxybutyrate is not even a ketone. However, the name has stuck and they certainly form a closely related metabolic group.

amino acid) via cystathionine, an intermediate formed during the breakdown of methionine (Fig. 6.1).

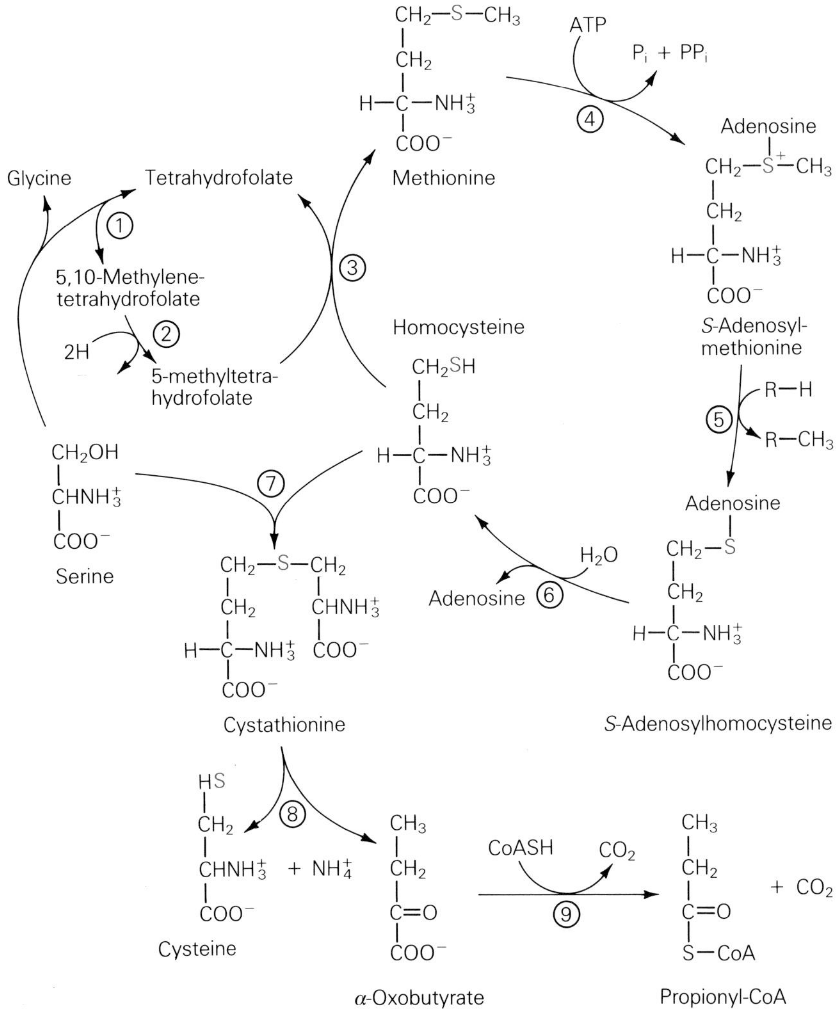

Fig. 6.1 Formation of cysteine from methionine. This composite diagram contains material that will reappear in Figs 6.4, 6.7, 6.12, 6.13. The enzymes involved are: 1, serine hydroxymethyltransferase; 2, methylene-tetrahydrofolate reductase; 3, homocysteine methyltransferase; 4, methionine adenosyltransferase; 5, various methyltransferases; 6, adenosylhomocysteinase; 7, cystathionine β-synthase; 8, cystathionine γ-lyase; 9, pyruvate dehydrogenase system.

Box 6.2
Lignin

Lignin is an aromatic, high M_r, insoluble polymer of the woody tissues of plants. It is derived by enzymic dehydrogenation and subsequent polymerization of coumaryl, coniferyl and sinapyl alcohols:

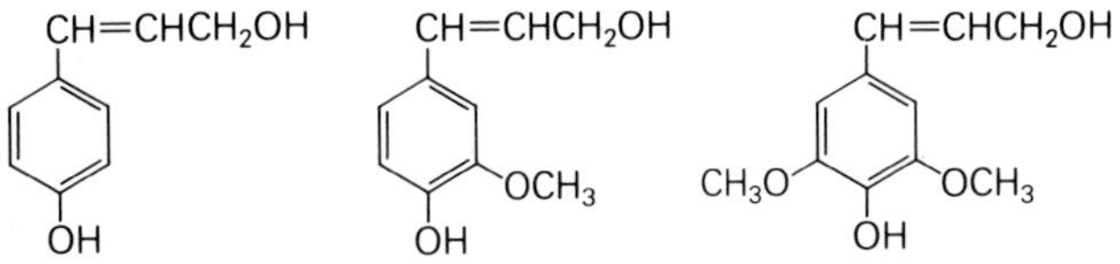

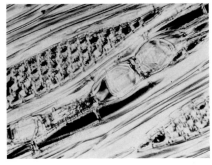

Coumaryl alcohol

Coniferyl alcohol

Sinapyl alcohol

A sample of nutwood viewed by polarized light. Photograph kindly supplied by Wild–Leitz.

The relative proportions of these alcohols taking part in the synthesis of lignin varies in different plants. Consequently, the most characteristic feature of a particular lignin is its methoxyl content. The enormous polymers are extremely difficult to analyse and are, in any case, to some extent random. Part of a structure might be as follows:

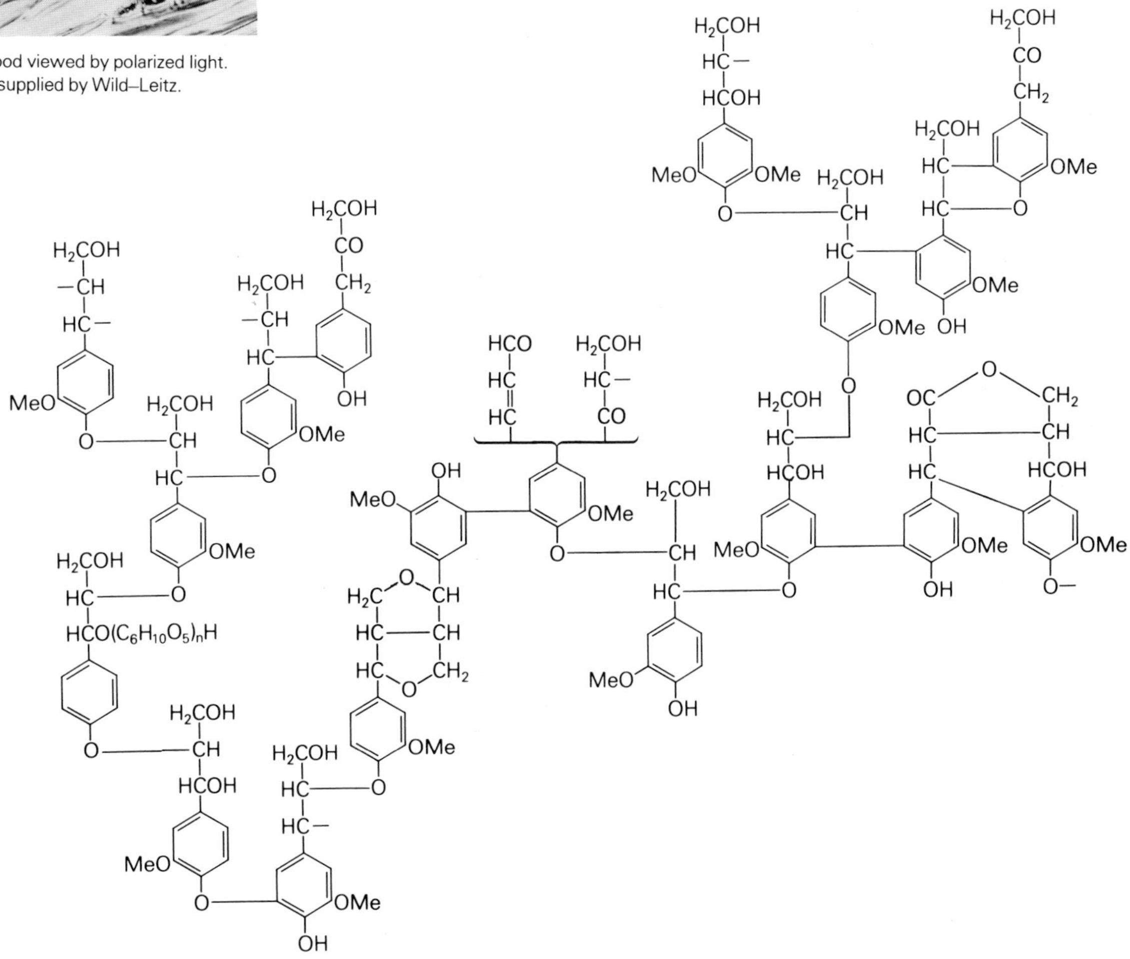

If lignin is oxidized with nitrobenzene it will give a mixture of *p*-hydroxy-benzaldehyde, vanillin and syringaldehyde. In histochemistry, lignin is detected by the use of phloroglucinol or safranin.

The alcohols involved in lignin production are related to the amino acids phenylalanine and tyrosine, which themselves are synthesized from shikimic acid in plants:

Phosphoenolpyruvate
+
D-Erythrose
4-phosphate

⋯⋯→

Shikimate

⋯⋯→

Phenylalanine
Tyrosine
Tryptophan

^{14}C-Shikimate is incorporated into lignin in wheat, maples and sugar-cane plants.

Lignin appears to be produced from phenylalanine and tyrosine. A first step removes ammonia and subsequent steps introduce further ring hydroxyls some of which become methylated. The polymerization process is referred to as a *reductive polymerization* and is mediated by a peroxidase. Artificial lignin may be produced by incubating the various alcohols, under strongly aerobic conditions, with a phenol oxidase such as fungal laccase. Polymerization probably takes place between free radical species to produce a highly heterogeneous three-dimensional polymer.

The economics of lignin and cellulose are very topical subjects. The river Rhine contains nearly 2 mg lignin per dm^3 from industrial effluent. Only about 15% of the total mass of a tree eventually appears as the cellulose product of chemical pulping and there are millions of tonnes of waste in the world as sawdust. Although sugar-cane waste (bagasse) is used to fuel the distillation of 'Gasohol' in Brazil, there remains a large excess of bagasse.

Lignin is highly resistant to both chemical and biological breakdown. For example, it is insoluble in hot 70% sulphuric acid. However, the white rot fungus, *Phanerochaete chrysosporium*, can degrade it, to many a houseowners dismay. Probably such fungi produce monomers from the lignin which are then degraded by the typical metabolic routes for dealing with aromatic amino acids. The family of enzymes that break down lignin are being studied intensively: some are haem-containing peroxidases and some are manganese-dependent peroxidases. Free radicals are probably involved, but the source of the hydrogen peroxide required for the reaction is unknown.

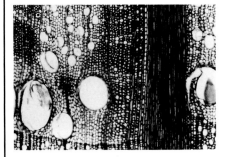

A sample of oakwood stained for lignin.

Broda, P., Sims, P.F.G. and Mason, J.C. (1989) Lignin biodegradation: a molecular biological approach. *Essays in Biochemistry*, **24**, 82–114. Explains how the techniques of molecular biology are being exploited to try to find ways of breaking down lignin.

Harborne, J.B. (ed.) (1989) *Plant Phenolics* (Methods in Plant Biochemistry, Vol. 1, eds P.M. Dey and J.B. Harborne), Academic Press. Phillipson, J.D., Ayres, D.C. and Baxter, H. (1989) *Plant Polyphenols: vegetable tannins revisited*, Cambridge University Press. These two books give an up-to-date coverage of the isolation, characterization and biological roles of plant phenols. For reference only.

6.3 The biosynthesis of alanine, aspartate, asparagine, glutamate and glutamine

These five non-essential amino acids are formed from intermediates of glycolysis or the TCA cycle simply by a transamination reaction (Fig. 6.2).

Pyruvate   Glutamate   Alanine   α-Oxoglutarate

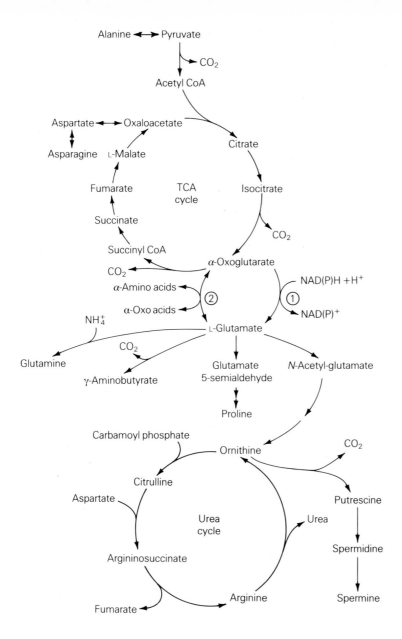

Fig. 6.2 The key role of glutamate in nitrogen metabolism.

In this example, pyruvate is a product of glycolysis and the α-oxoglutarate formed could be fed into the TCA cycle.

Amides are made by the addition of an amide group:

$$\begin{array}{c} COO^- \\ | \\ (CH_2)_2 \\ | \\ CHNH_3^+ \\ | \\ COO^- \end{array} + ATP + NH_3 \longrightarrow \begin{array}{c} CONH_2 \\ | \\ (CH_2)_2 \\ | \\ CHNH_3^+ \\ | \\ COO^- \end{array} + ADP + P_i$$

Glutamate   Glutamine

as summarized in Fig. 6.2. Aspartate can contribute part of its molecule in the biosynthesis of pyrimidine nucleotides (Chapter 7).

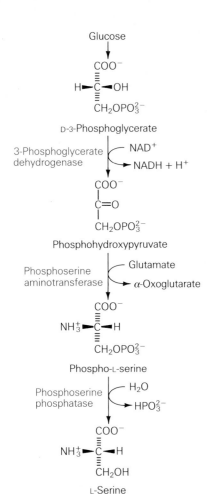

Glucose

D-3-Phosphoglycerate

3-Phosphoglycerate dehydrogenase — NAD⁺ → NADH + H⁺

Phosphohydroxypyruvate

Phosphoserine aminotransferase — Glutamate → α-Oxoglutarate

Phospho-L-serine

Phosphoserine phosphatase — H₂O → HPO₃²⁻

L-Serine

Fig. 6.3 The biosynthesis of serine from glycolytic intermediates.

Table 6.3 Cell constituents formed from glycine

Bile acids (*Energy in Biological Systems*, Chapter 6)
Creatine (*Cell Biology*, Chapter 10)
Glutathione (*Biological Molecules*, Chapter 2)
Purines (Chapter 7)
Porphyrins (Chapter 9)

☐ The last two reactions leading to glycine formation shown in Fig. 6.4 are catalysed by dimethylglycine dehydrogenase and sarcosine dehydrogenase respectively. These enzymes are mitochondrial flavoproteins that bind tetrahydrofolate very tightly. They use the mitochondrial electron-transferring flavoprotein as electron acceptor and transfer the former methyl groups of dimethyl-glycine and sarcosine to tetrahydrofolate.

6.4 The biosynthesis of serine

Serine can be synthesized completely from glycolytic precursors and is, in turn, the source of a number of important compounds (Table 6.2). In the majority of organisms, the biosynthesis of serine (Fig. 6.3) from 3-phosphoglycerate involves: (1) oxidation of the 2-hydroxyl group to an oxo group, forming phosphohydroxypyruvate, a reaction catalysed by **phosphoglycerate dehydrogenase**; (2) transamination of the phosphohydroxypyruvate using glutamate to yield an amino group (catalysed by **phosphoserine aminotransferase**), and (3) the hydrolytic removal of the *O*-phosphate, in a reaction catalysed by **phosphoserine phosphatase**.

Table 6.2 Cell constituents formed from serine

Membrane lipids	Others
Sphingosine	Cysteine (carbon skeleton)
Sphingolipids	Glycine and its derivatives (see Table 6.3)
Phosphatidylserine	One-carbon units for other biosynthetic routes
Phosphatidylethanolamine	(see section 6.6)

6.5 Interconversion of serine and glycine

Serine is converted into glycine by transfer of its C-3 atom to a tetrahydrofolate coenzyme (Fig. 6.4). This reaction is catalysed by serine hydroxymethyltransferase.

Serine + Tetrahydrofolate ⇌ Glycine + 5,10-Methylenetetrahydrofolate

The glycine formed from serine is required for the synthesis of a number of cell constituents (Table 6.3). It is also a **conjugating agent**, that is, it is a vehicle for the excretion of certain foreign compounds in less toxic and more soluble forms by the body. For example, benzoate is excreted in conjugation with glycine as *N*-benzoylglycine (hippurate). Glycine can also give rise to 5,10-

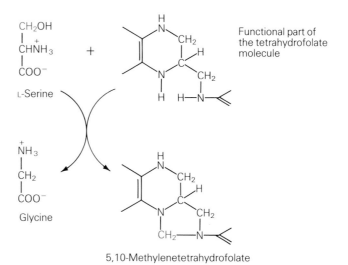

Fig. 6.4 Formation of glycine from serine using a tetrahydrofolate coenzyme.

Serine: first isolated from sericin, the protein associated with silk. It is usually removed from the silk fibres (fibroin) by boiling with alkali.

methylenetetrahydrofolate in an NAD^+-dependent cleavage which also yields ammonia and carbon dioxide:

$$NH_3^+CH_2COO^- + NAD^+ + Tetrahydrofolate$$
$$\rightleftharpoons NH_4^+ + CO_2 + 5,10\text{-Methylenetetrahydrofolate} + NADH$$

This reaction (which is reversible) is catalysed by a group of enzymes called the **glycine cleavage system**. This system is probably the major breakdown route for glycine and serine in the body. This supposition is supported by the existence of a hereditary defect of glycine metabolism called **non-ketotic hyperglycinaemia** in which one of the components of this enzyme system is absent.

The glycine cleavage system consists of four proteins (Fig. 6.5), designated P, H, T and L, and four cofactors: (i) pyridoxal 5'-phosphate bound to the P-protein (which is also known as glycine dehydrogenase [decarboxylating]), (ii) tetrahydrofolate, which interacts with the T-protein (which is also known as aminomethyltransferase), (iii) FAD which is bound to the L-protein, which has lipoamide dehydrogenase activity, and (iv) NAD^+. The aminomethyl group is carried on the H-protein as shown in Fig. 6.5.

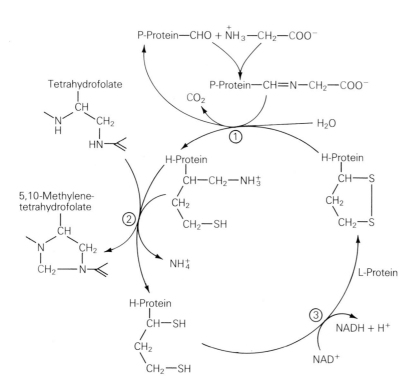

Fig. 6.5 The intermediate stages in the glycine cleavage enzyme system. The enzymes involved are: 1, glycine dehydrogenase (decarboxylating) (P-protein); 2, aminomethyltransferase (T-protein); 3, dihydrolipoamide dehydrogenase (L-protein).

□ Children born with non-ketotic hyper-glycinaemia lack the H-protein of the glycine-cleavage enzyme system (see Fig. 6.5). They have high levels of glycine in the blood and body fluids and die shortly after birth or survive with mental handicap.

6.6 Carriers of one-carbon units in the cell

One-carbon compounds, apart from CO_2, are, in general, chemically active or even toxic. For example, formaldehyde, formic acid and methanol are all compounds that are deleterious to life. In order to handle such compounds, or units derived from them safely, and also in order to achieve conversion between different oxidation states, they are dealt with on carriers such as folate.

Compounds containing isolated carbon atoms (usually attached to a sulphur, a nitrogen or an oxygen atom), are synthesized via various kinds of **active one-carbon units** or **groups**. These units are usually derivatives of tetrahydrofolate coenzymes (Table 6.4 and Fig. 6.6) and are extremely important as one-carbon group carriers in the cell. They are derived from the vitamin ***folic acid*** by reduction and usually with the addition of further glutamyl residues to the single glutamate tail of the tetrahydrofolate molecule (Fig. 6.6). The structure and functions of tetrahydrofolate coenzymes are described in *Energy in Biological Systems*, Chapter 7.

Table 6.4 *One-carbon units carried by tetrahydrofolate*

Group	Structure
Methyl	—CH$_3$
Methylene	—CH$_2$—
Formyl	—CHO
Formimo	—CHNH
Methenyl	—CH=

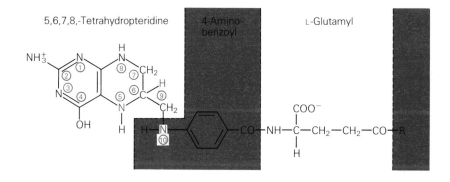

Fig. 6.6 Structure of tetrahydrofolate coenzymes. The nitrogen atoms in colour represent the functional parts of the molecule involved in one-carbon unit transfers. The hydrogen atom in colour is that donated during the action of thymidylate synthase. In tetrahydrofolate –R is –OH. (In most coenzyme forms –R is one or more glutamate residues up to a maximum of six.)

The functional parts of the tetrahydrofolate molecule are the nitrogen atoms at positions 5 and 10 (Fig. 6.6). The one-carbon units attached to tetra-hydrofolate can be oxidized or reduced to give formyl or methyl groups, respectively or further oxidized to CO_2.

FORMYL GROUPS (either 5,10-methenyltetrahydrofolate or 10-formyltetra-hydrofolate) are required as precursors of parts of the purine molecule and of N-formylmethionine. Formyltetrahydrofolate derivatives can also be formed directly from formate in the presence of tetrahydrofolate, ATP and the enzyme **formyltetrahydrofolate synthetase** (Fig. 6.7).

HYDROXMETHYL GROUPS (5,10-methylenetetrahydrofolate) are required to make serine from glycine (Fig. 6.4) and as the precursor of thymidine (see later) and methionine. The formation of 5,10-methylenetetrahydrofolate by cleavage of glycine has been discussed earlier.

Folic acid: was first isolated from 4 tons of spinach leaves in 1941, as an acid that stimulated the growth of Streptococcus faecium. *From the Latin,* folium, *leaf.*

Reference Blakely, R.L. and Benkovic, S.V. (1984) *Folates and Pterins*, vol. 1: *Chemistry and Biochemistry of Folates*, John Wiley, New York. A comprehensive survey of folates and the reactions which they undergo.

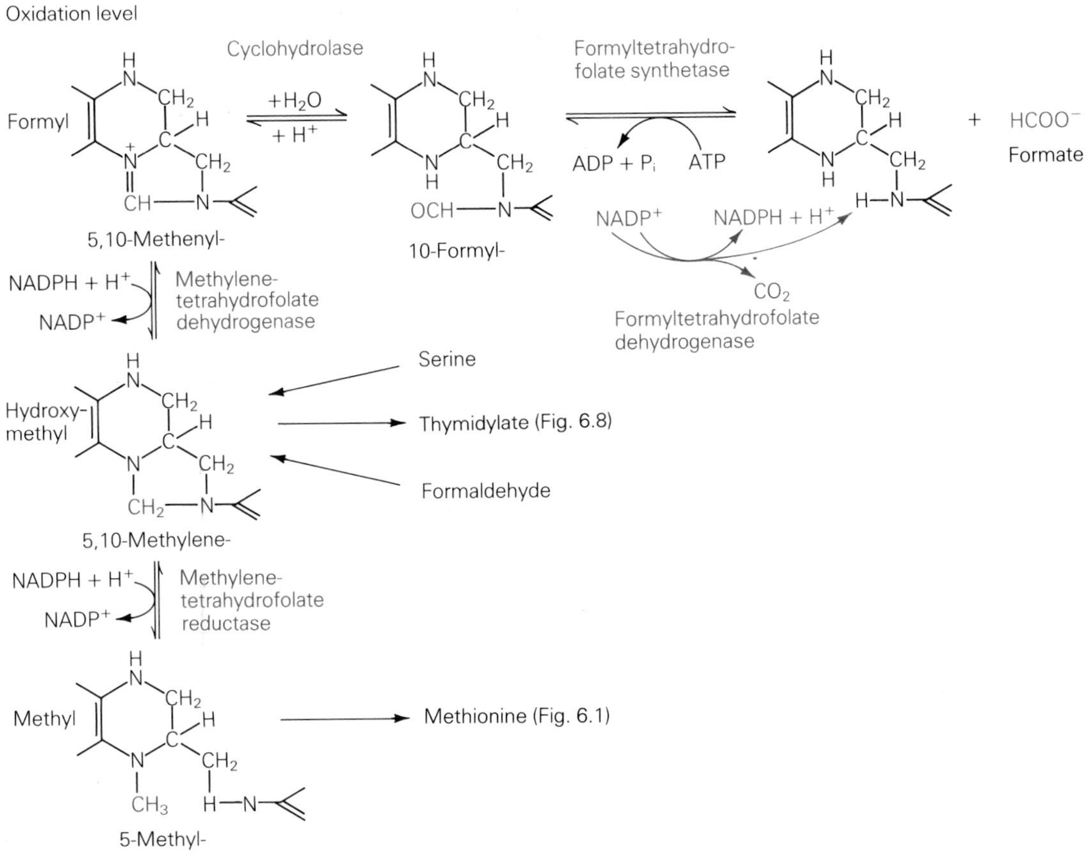

Fig. 6.7 Formation and interconversion of active one-carbon units.

METHYL GROUPS are formed from 5,10-methylenetetrahydrofolate by two distinct routes. One is by direct reduction to 5-methyltetrahydrofolate (Fig. 6.7). This reduction is catalysed by **methylenetetrahydrofolate reductase** and requires the electron donor NADPH:

5,10-Methylenetetrahydrofolate + NADPH + H$^+$ →

5-Methyltetrahydrofolate + NADP$^+$

This reaction is followed by transfer of the methyl group to homocysteine to give methionine (see Section 6.7).

The other route for the formation of methyl groups involves tetrahydrofolate acting as an electron donor as well as a one-carbon unit donor. This occurs in the formation of the methyl group of thymidylate (see Fig. 6.8). The dihydrofolate produced during the synthesis of thymidylate needs to be reconverted to tetrahydrofolate before further methylation can occur. This is achieved by the action of NADPH-dependent enzyme, **dihydrofolate reductase** (Fig. 6.9). Catabolic process also contribute to the one-carbon pool. Oxidation of two of the methyl groups of choline gives methylenetetrahydrofolate (Fig. 6.14 and marginal note on p. 121), while the breakdown of histidine produces formiminoglutamate which can be converted into a tetrahydrofolate derivative. Tetrahydrofolate derivatives that are not required for biosynthetic purposes can be oxidized to CO_2 by the NADP$^+$-linked enzyme **formyltetrahydrofolate dehydrogenase** (Fig. 6.7).

Exercise 1

The compound aminopterin shown below is a non-competitive inhibitor of dihydrofolate reductase. Malignant cells are characterized by rapid growth rates which are dependent on high rates of synthesis of proteins and nucleic acids. Suggest reasons why aminopterin is an effective drug for the treatment of leukaemia.

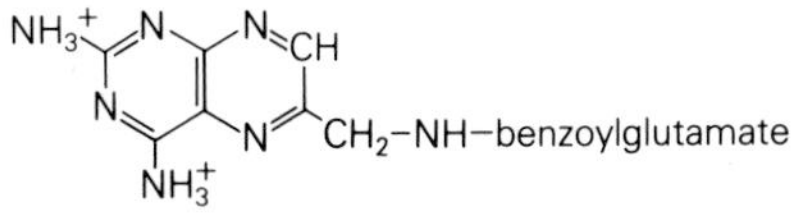

Homocystinuria and homocystinaemia are the names given to the conditions where abnormally high levels of homocysteine occur in urine and blood, respectively. These symptoms can arise as a consequence of one of at least three hereditary defects:

1. A deficiency of the enzyme cystathionine β-synthase, which is involved in the conversion of homocysteine into cysteine and homoserine (Fig. 6.1). This particular enzyme deficiency is also accompanied by hypermethioninaemia (see Box 6.5).

2. A less common cause is a deficiency of the enzyme 5,10-methylenetetra-hydrofolate reductase (equation on p. 123), involved in the remethylation of homocysteine to methionine.

3. The third cause is a deficiency of methylcobalamin (see Box 6.4), which may be nutritional in origin or may result from a failure of the metabolic conversion of the hydroxocobalamin form of vitamin B_{12} to methyl-B_{12}.

Irrespective of cause, the conditions associated with these deficiencies are severe physical impairments as well as mental deficiency.

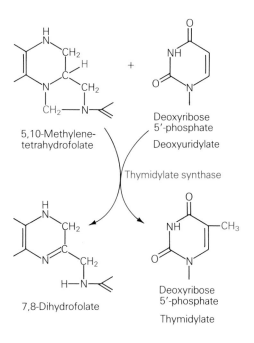

5,10-Methylene-tetrahydrofolate

Deoxyribose 5'-phosphate

Deoxyuridylate

Thymidylate synthase

7,8-Dihydrofolate

Deoxyribose 5'-phosphate

Thymidylate

Fig. 6.8 Formation of the methyl group of thymidylate.

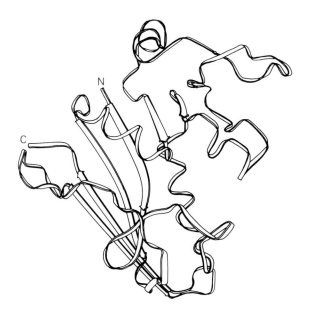

Fig. 6.9 The structure of dihydrofolate reductase of *E. coli*. Redrawn from Stryer, L. (1988) *Biochemistry*, 3rd edn, p. 616, W.H. Freeman, New York.

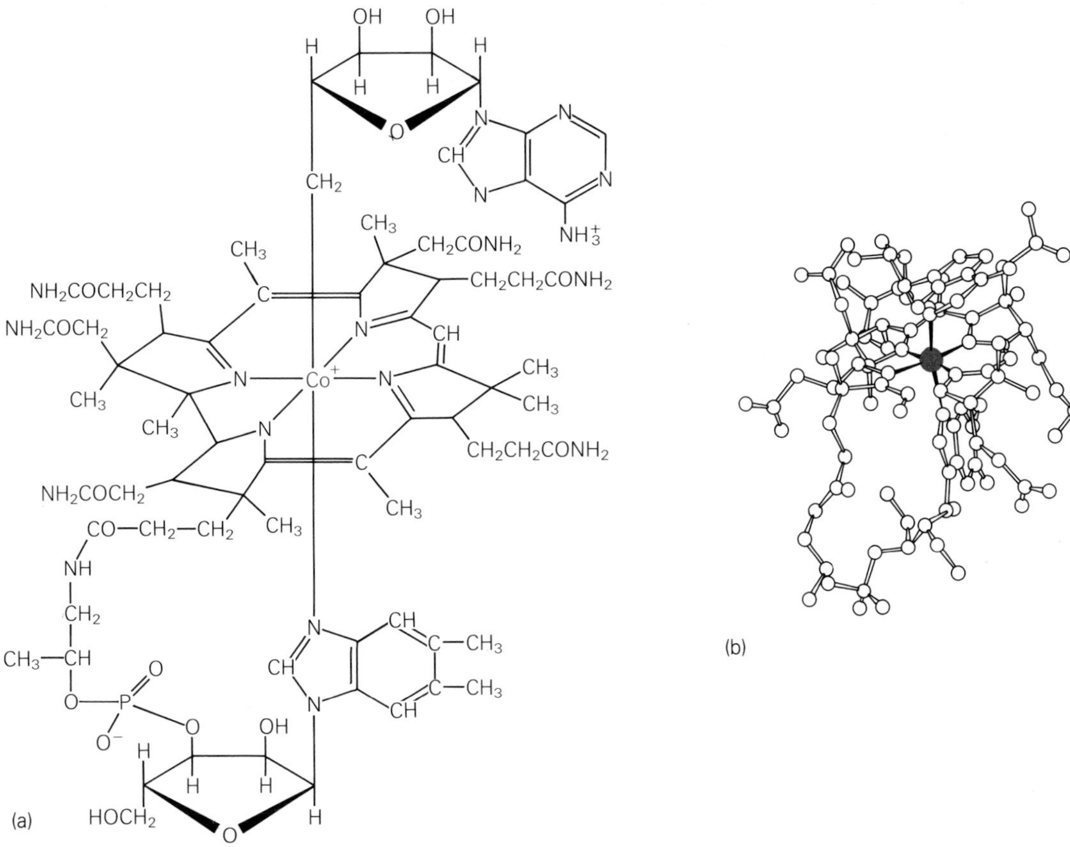

Fig. 6.10 (a) Structure of coenzyme B_{12}. (b) Molecular model of coenzyme B_{12}. The cobalt ion is shown in red.

6.7 S-Adenosylmethionine as a methyl donor

The methyl group of 5-methyltetrahydrofolate is the only *de novo* source of methyl groups in mammalian tissues. Structural methyl groups (Table 6.5) arise by transfer from **S-adenosylmethionine**; the sole function of 5-methyltetrahydrofolate is to supply the methyl group of methionine. The methylation of **homocysteine** involves methylcobalamin, a derivative of vitamin B_{12}, and this is one of only two proved metabolic roles of B_{12} coenzymes in mammalian tissues (see Fig. 6.10 and Box 6.4). Some bacteria possess an alternative B_{12}-independent pathway, but this is not found in mammals.

Methionine is not an essential amino acid *per se* since it can be replaced in the diet by homocysteine $HS–(CH_2)_2–CH(NH_3^+)–COO^-$ (Fig. 6.1). However, mammalian tissues appear to be incapable of producing this compound.

Methylated compounds other than methionine (Table 6.5), arise by donation of methyl groups from *S*-adenosylmethionine (Fig. 6.11). *S*-Adenosylmethionine is formed from methionine and ATP in a reaction catalysed by methionine adenosyltransferase (Fig. 6.12). The immediate products of this reaction are both orthophosphate and pyrophosphate: two of the phosphate bonds of the ATP have been cleaved. The pyrophosphate undergoes hydrolysis to phosphate, as is also observed in other reactions in which pyrophosphate is a product (Section 1.4). Thus in total, all three phosphate bonds of ATP have been hydrolysed. The presence of two positive

Fig. 6.11 Molecular model of *S*-adenosylmethionine. The methyl group transferred during methylation reactions is shown in red.

Reference Usdin, E., Borchardt, R.T. and Creveling, C.R. (eds) (1982) *Biochemistry of S-adenosylmethionine and related compounds*, Macmillan, London. Patchy, but very useful collection of short reviews and research work on the many roles of *S*-adenosylmethionine.

Homocysteine: *'homo-' means that this is the higher homologue of cysteine, that is, the compound of similar structure but having an additional methylene group.*

Cobalamin is based on a tetrapyrrole with a central cobalt atom (see Fig. 6.10). This is bonded to the four nitrogen atoms of the pyrrole units. The two additional co-ordination positions of the cobalt atom are filled by (i) a derivative of dimethylbenzimidazole, and (ii) either a methyl or a deoxyadenosyl group. In the isolated vitamin form of cobalamin, this last substituent is replaced by a cyanide ion, during the isolation procedure. The active coenzyme has the cobalt ion in a +1 valence state.

Cobalamin-dependent enzymes catalyse two types of reactions. **Methylations**, as for example, in the transfer of the methyl group of 5-methyltetrahydrofolate to homocysteine to form methionine:

$$5\text{-Methyltetrahydrofolate} + HS\text{-}(CH_2)_2\text{-}CH(NH_3^+)\text{-}COO^- \rightarrow$$
$$CH_3S\text{-}(CH_2)\text{-}CH(NH_3^+)\text{-}COO^- + \text{Tetrahydrofolate}$$

This is the most important biological function involving methylation of cobalamin in mammalian tissues. In bacteria it has many other functions such as the synthesis of the methyl group of acetate in homoacetogenic bacteria. The other type of reaction in which cobalamin derivatives are involved is in chemical **rearrangements**. For example, the conversion of L-methylmalonylCoA to succinyl CoA which occurs in mammalian tissues:

L-Methylmalonyl CoA   Succinyl CoA

Cobalamin cannot be synthesized by mammalian tissues and must be supplied in the diet or be synthesized by intestinal bacteria. A deficiency of vitamin B$_{12}$ is associated with the condition known as pernicious anaemia, where a glycoprotein called **intrinsic factor** is not produced by the body. Intrinsic factor is secreted by the stomach and binds vitamin B$_{12}$ in the intestine and ensures its absorption. The symptoms of cobalamin deficiency usually include homocystinuria (see Box 6.2) and methylmalonate acidaemia.

charges on the methionine portion of *S*-adenosylmethionine results in a great enhancement in the reactivity of the groups attached to the sulphur atom.

S-Adenosylmethionine can transfer its methyl group to a variety of acceptors to yield *S*-adenosylhomocysteine. One example of this is shown in Fig. 6.13. Other examples such as methylated proteins are described below. The *S*-adenosylhomocysteine formed can be subsequently hydrolysed to give

Hypermethionaemia is an inborn error of metabolism which can be caused by two possible enzymic defects: a deficiency in methionine adenosyltransferase or a deficiency in cystathionine β-synthase (Fig. 6.1), in which case it is also accompanied by homocystinaemia (see Box 6.2). The former condition does not seem to produce severe clinical abnormalities. However, a deficiency in cystathionine β-synthase produces a number of physical abnormalities, including damaged eye lenses, thrombosis, skeletal deformities and, in some cases, slight mental deficiency.

Table 6.5 *Compounds containing methyl or methylene groups arising from S-adenosylmethionine*

Monomers	Polymers
Adrenalin	Lignin
Betaine	tRNA
Chlorophyll	rRNA
Creatine	mRNA
Cyclopropane fatty acids	DNA
Ergosterol (and other methylated steroids)	Pectin, Proteins
S-Methylmethionine	
Phosphatidylcholine	
Sarcosine	

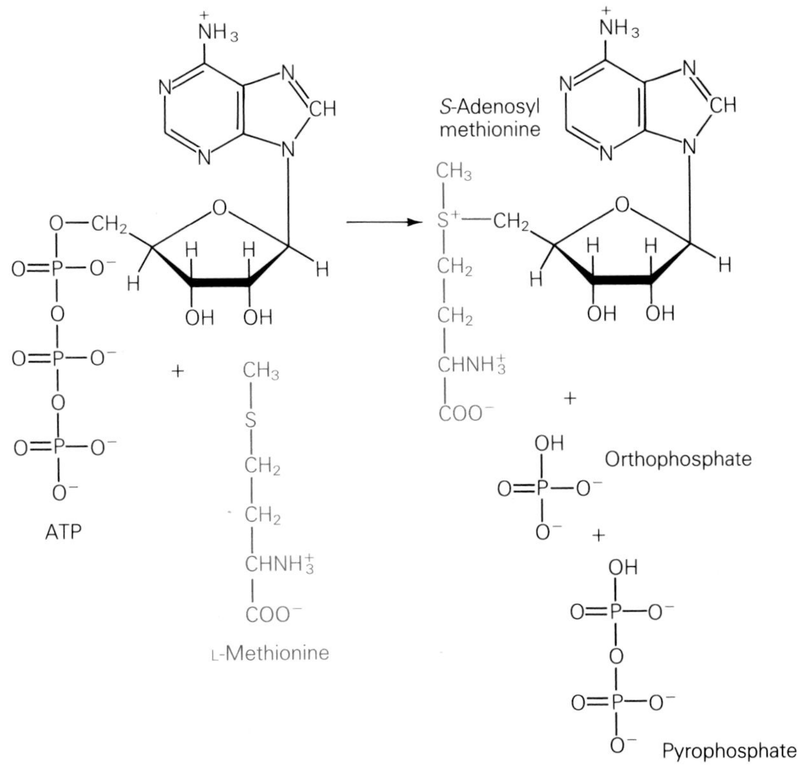

Fig. 6.12 Formation of *S*-adenosylmethionine.

adenosine and homocysteine, a reaction catalysed by **adenosyl homocysteinase** (Fig. 6.1).

Homocysteine can be remethylated by 5-methyltetrahydrofolate to yield methionine, or can give rise to cysteine via cystathionine as shown in Fig. 6.1. Homocysteine can also be remethylated to methionine by a methyl transfer from **betaine**, an intermediate in choline degradation (Fig. 6.14) in a reaction catalysed by betaine–homocysteine methyltransferase.

See *Energy in Biological Systems*, Chapter 9

Betaine: *beet leaves may contain up to 3% of this compound.*

Reference Benson, P.F. and Fensom, A.H. (1985), *Genetic Biochemical Disorders*, Oxford University Press, Oxford. A comprehensive survey of genetic disorders, not merely those due to defects in amino acid metabolism.

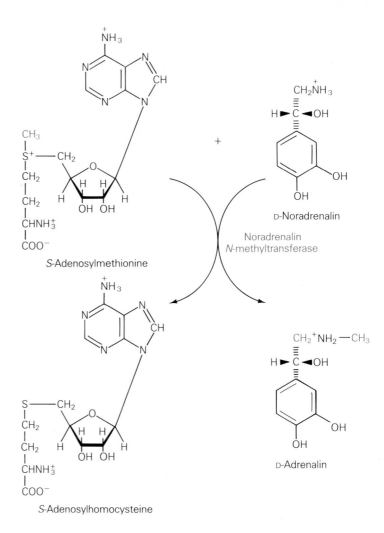

Fig. 6.13 S-Adenosylmethionine as a methyl donor in the formation of adrenalin.

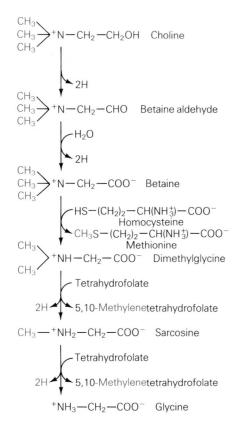

Fig. 6.14 The breakdown of choline, formation of betaine and the role of betaine in the regeneration of the methyl group of methionine.

6.8 Products derived from S-adenosylmethionine

Methylated proteins

Methylation is a form of post-translational modification in a variety of proteins in both prokaryotes and eukaryotes. The side-chain nitrogen atoms of a number of amino acid residues in proteins may be modified by methylation. Residues modified include lysine, arginine, histidine, proline and glutamine and the side-chain acid groups of glutamate and aspartate residues. N-Methylated amino acids tend to occur in specialized proteins, such as the histones of chromatin, flagella proteins, the skeletal muscle proteins myosin and actin, ribosomal proteins, opsin in the retina of the eye, calmodulin, elongation factor EF-Tu, the AI basic protein of myelin, and fungal and plant cytochromes c. Protein-bound ε-N-trimethyllysine is the precursor of **carnitine**. Enzymic methylesterification of protein carboxyl groups plays a crucial role in bacterial chemotactic behaviour. Protein methylation is also thought to be involved in DNA repair.

Carnitine: previously called 'vitamin B_T', identified as an essential component of the meal worm, Tenebrio molitor. Vertebrates can synthesize carnitine and thus it is not a vitamin for them.

Polyamines

Polyamines are a small group of compounds that contain several amino groups and which are normal constituents of both prokaryotic and eukaryotic cells. They were first observed in human semen in a crystalline form by Leeuwenhoek as long ago as 1678 (Fig. 6.15). However, it is only in recent years that their biological roles have started to be elucidated and the understanding of their function is still far from complete. Polyamines are polycations and appear to have functions in stabilizing biological structures which possess multiple negative charges such as lipid bilayers and nucleic acids. For instance, in bacteriophage T4 about 40% of the negative charges associated with the phosphate groups of the nucleic acid are neutralized by association with polyamines. This is possible because a molecule of a polyamine such as spermine has the appropriate size and charge distribution to bind to two phosphate groups in each strand of the double helix. Thus, spermine may stabilize the double helix by spanning the minor and major grooves of DNA, binding the two strands together.

Polyamines have critical roles in cell growth and cell division. If the biosynthesis of polyamines is inhibited, cells cease dividing. If, however, the polyamine-depleted cells are given a supplement of polyamines then cell proliferation resumes.

An outline of the biosynthesis of the most frequently encountered polyamines, putrescine, spermidine and spermine is shown in Fig. 6.16.

See *Molecular Biology and Biotechnology*, Chapter 5

Exercise 2

O-Methyl groups (–OCH$_3$) arise in nature by methyl transfer from *S*-adenosylmethionine to a hydroxyl group. Suggest two mechanisms by which the *O*-methyl group might be removed, to give rise to products at two different oxidation levels.

$^+$NH$_3$—CH$_2$—CH$_2$—CH$_2$—CH$_2$—NH$_3^+$ Putrescine

$^+$NH$_3$—CH$_2$—CH$_2$—CH$_2$—CH$_2$—CH$_2$—NH$_3^+$ Cadaverine

$^+$NH$_3$—CH$_2$—CH$_2$—CH$_2$—NH$_2^+$—CH$_2$—CH$_2$—CH$_2$—CH$_2$—NH$_3^+$ Spermidine

$^+$NH$_3$—CH$_2$—CH$_2$—CH$_2$—NH$_2^+$—CH$_2$—CH$_2$—CH$_2$—CH$_2$—NH$_2^+$—CH$_2$—CH$_2$—CH$_2$—NH$_3^+$ Spermine

(a)

(b)

Fig. 6.15 (a) The major polyamines. (b) Electron micrograph of a crystal of the polyamine spermine phosphate from human seminal fluid. Courtesy Dr D. Walters, Department of Plant Sciences, The West of Scotland College, UK.

Putrescine is formed by decarboxylation of ornithine, catalysed by **ornithine decarboxylase** (see Section 6.9) and may be converted to spermidine by transfer of a 3-aminopropyl group derived from 'decarboxylated *S*-adenosylmethionine'. Substitution of the other nitrogen atom of spermidine by an additional 3-aminopropyl group yields spermine. The enzymes catalysing these two reactions are called spermidine and spermine synthases respectively (Fig. 6.16).

Polyamines: *so named because they carry several amino groups. In fact, putrescine and cadaverine are diamines. Spermidine and spermine were given their names because they were first 'discovered' in semen.*

Box 6.6
Inhibition of polyamine biosynthesis

The enzyme ornithine decarboxylase (ODC) is necessary for polyamine biosynthesis. It is strongly inhibited by α-difluoromethylornithine (DFMO) which therefore inhibits polyamine biosynthesis.

$$
\begin{array}{l}
NH_3^+ \\
| \\
CH_2 \\
| \\
CH_2 \\
| \\
CH_2 \\
| \\
CF_2H-C-NH_3^+ \\
| \\
COO^-
\end{array}
$$

2-Difluoromethylornithine (DFMO)

Because of the association between rapidly proliferating cells and polyamines, many ornithine decarboxylase inhibitors have been synthesized as potential antitumour agents. Practically all of them have been unsuccessful. DFMO works reasonably well on some tumours because it is an irreversible inhibitor of ornithine decarboxylase. The action of the enzyme produces an active species that combines covalently with the active site of the enzyme. In fact, DFMO seems likely to be more useful in the treatment of trypanosomal infections than tumours.

For example, treatment with DFMO cures 95% of late-stage sufferers of African sleeping sickness caused by the protozoan *Trypanosoma brucei gambiense*. Other protozoa susceptible to DFMO are *Pneumocytis carinii*, an opportunistic infective agent in AIDS patients, and *Plasmodium falciparum*, the malarial parasite.

Plants possess a second pathway for synthesizing polyamines which uses the enzyme arginine decarboxylase (ADC) while fungi appear to have only the ODC-dependent pathway. Spraying plants with DFMO has been shown to give protection against some plant fungal diseases presumably because while the ODC-dependent pathway is blocked in both organisms, the plant survives using of the ADC-dependent pathway. Fungi, indeed, appear susceptible to a variety of inhibitors of polyamine synthesis.

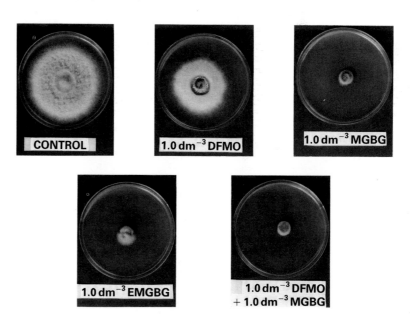

The effects of inhibitors of polyamine biosynthesis on fungal growth is clearly shown by comparing the growth of the untreated (control) fungus *Pyrenophora avenae*, a pathogen of oats, with inhibitor-treated fungus. DFMO (α-difluoromethylornithine) is an inhibitor of ornithine decarboxylase while MGBG (methylglyoxal*bis*(guanylhydrazone)) and EMGBG (ethylmethylglyoxal*bis*(guanylhydrazone)) are inhibitors of *S*-adenosylmethionine decarboxylase. Courtesy Dr D. Walters, Department of Plant Sciences, The West of Scotland College, UK.

Walters, D. (1987) Polyamines: the Cinderellas of cell biology. *Biologist*, 34, 73–6. Heby, O. and Persson, L. (1990) Molecular genetics of polyamine synthesis in eukaryotic cells. *Trends in Biochemical Sciences*, **15**, 153–8. These two short, yet comprehensive, reviews together form the perfect introduction to polyamines.

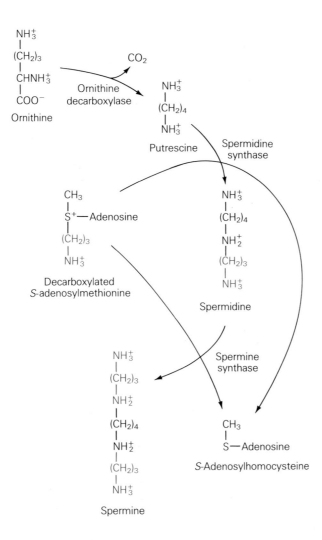

□ Decarboxylation of *S*-adenosylmethionine in a reaction catalysed by *S*-adenosyl-methionine decarboxylase results in the formation of *S*-methyl-*S*-3-amino-propylthioadenosine (usually referred to as 'decarboxylated *S*-adenosyl-methionine'). This is a feeble methyl donor but acts as an effective aminopropyl donor in polyamine biosynthesis (and also participates in the formation of ethylene, see *Cell Biology*, Chapter 8).

Fig. 6.16 The formation of the polyamines, putrescine, spermidine and spermine.

6.9 *Amino acids derived from glutamate*

Glutamate has a dual role in amino acid metabolism (Fig. 6.2). In addition to being the precursor of the amino acids glutamine, proline, ornithine and arginine, and the neurotransmitter γ-aminobutyrate, GABA, it is also the precursor of the amino groups of amino acids formed from the corresponding α-oxoacids by transamination (see equation on p. 132). Glutamate is formed from α-oxoglutarate, which is an intermediate of the TCA cycle in a transformation catalysed by glutamate dehydrogenase:

$$\begin{array}{cccc}
\text{COO}^- & & \text{COO}^- \\
| & & | \\
(\text{CH}_2)_2 + \text{NH}_4^+ & & (\text{CH}_2)_2 \\
| & \longrightarrow & | \quad + \text{NAD(P)}^+ + \text{H}_2\text{O} \\
\text{C}{=}\text{O} + \text{NAD(P)H} + \text{H}^+ & & \text{CHNH}_3^+ \\
| & & | \\
\text{COO}^- & & \text{COO}^- \\
\alpha\text{-Oxoglutarate} & & \text{Glutamate}
\end{array}$$

This reaction is the main route for the conversion of ammonia-nitrogen atoms into amino acid-nitrogen atoms and *vice versa*.

Glutamine

Glutamine is the source of the nitrogen atom(s) of a large number of compounds (Table 6.6). It is formed from glutamate by an ATP-dependent amination reaction catalysed by glutamine synthetase (see equation on p. 118). In bacteria this enzyme is controlled by a double mechanism of cumulative feedback inhibition and reversible covalent modification (see Fig. 6.17 and

Table 6.6 *Examples of compounds deriving one or more nitrogen atoms from the amide nitrogen atom of glutamine*

Tryptophan (bacteria, plants)
Histidine (bacteria, plants)
Carbamoyl phosphate
Glucosamine 6-phosphate
UTP
AMP
GMP
Glutamate (microorganisms and plants)

Exercise 3

Why are glutamate and glutamine regarded as key compounds in nitrogen metabolism? Hint: it may be helpful to consult Chapter 5.

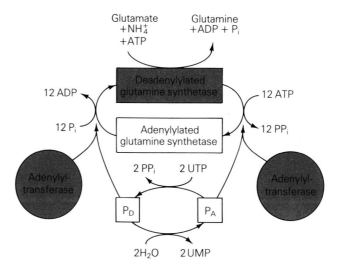

Fig. 6.17 Activation and deactivation of glutamine synthetase. See also Fig. 5.17 and Box 6.7.

Box 6.7
Glutamine synthetase

Glutamine synthetase of *E. coli* is controlled in its activity by **reversible covalent modification**. The enzyme molecule consists of 12 subunits of M_r 50 000 arranged in two facing hexagonal rings (see Section 5.7). Its activity is partially regulated by the covalent attachment of an adenylyl (AMP) unit to the hydroxyl group of a particular tyrosine residue in each subunit. The adenylylated enzyme is more susceptible to cumulative feedback inhibition by the metabolites that are formed from glutamine (Table 6.6) than is the unmodified form of the enzyme. The AMP unit can be removed from the adenylylated enzyme by phosphorolysis.

Adenylylation and phosphorolysis are catalysed by the same adenylyltransferase. The direction in which this enzyme works is controlled by a regulatory protein P. This can exist in two forms, P_A and P_D. The association of P_A with the adenylyltransferase promotes the adenylylation reaction, whereas P_D promotes deadenylylation. P_A is converted to P_D by attachment of a uridylyl (UMP) group in a reaction catalysed by a uridylyltransferase. This reaction is stimulated by ATP and α-oxoglutarate and inhibited by glutamine. Under conditions of high ammonia availability adenylylation is promoted and the form of the synthetase that predominates is that which is more susceptible to cumulative feedback inhibition. This regulatory cascade is summarized in Fig. 6.17 (see also Fig. 5.17).

Box 6.7). In mammalian cells, glutamate has the role of trapping and thereby allowing the transport of ammonia in the plasma. This role is vital because free ammonia is toxic. In the kidney, hydrolysis of glutamine releases NH_4^+, which are secreted into the urine as a means of neutralizing excreted anions. Given the ATP requirement for glutamine synthetase activity, the transport and excretion of NH_4^+ consumes metabolic energy.

Proline

Proline is formed from glutamate, retaining the nitrogen atom and its five carbon atoms as shown in Fig. 6.18. A feature of this metabolic route is that the reduction product of 5-glutamyl phosphate, glutamate 5-semialdehyde, 'bites its own tail'. Δ^1-**Pyrroline carboxylate** is therefore formed in a non-enzyme-catalysed reaction. This product is then reduced to proline, a reaction which requires NADPH as an electron donor.

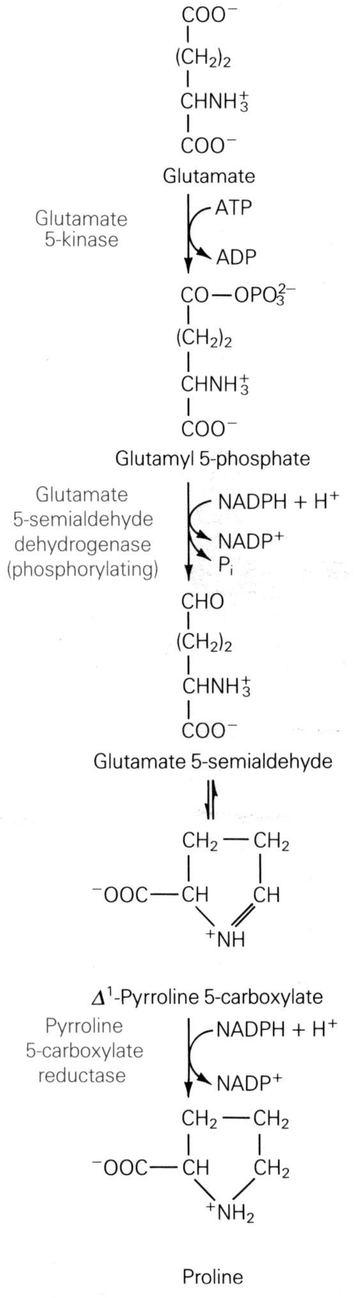

Ornithine and arginine

The key intermediate in the biosynthesis of arginine is ***N*-acetylglutamate**. This is required for two purposes, as an activator of carbamoyl phosphate synthetase 1, the key enzyme of the urea cycle, and as a precursor of ornithine. The latter is necessary when the urea cycle is being used to synthesize arginine from ornithine for use in protein synthesis (Fig. 6.2), and also in the biosynthesis of polyamines (Fig. 6.16), two situations where ornithine may become limiting.

N-Acetylglutamate is formed by the action of *N*-acetylglutamate synthase, which catalyses the reaction:

$$\text{L-Glutamate} + \text{Acetyl CoA} \rightarrow N\text{-Acetylglutamate} + \text{CoASH}$$

N-Acetylglutamate synthase is found in the mitochondria of liver and yeast cells. One case of hyperammonaemia has been reported that was shown to be due to lack of *N*-acetylglutamate synthase. In mammalian tissues the enzyme is *activated* by arginine and so stimulates urea formation under conditions of nitrogen excess. In bacteria, the enzyme is *inhibited* by arginine via feedback inhibition, because the ornithine cycle occurs in bacteria only as a route for arginine metabolism.

The synthesis of *N*-acetylglutamate is required for ornithine biosynthesis because free glutamate 5-semialdehyde gives rise exclusively to proline (Fig. 6.18). The cell prevents spontaneous cyclization of the semialdehyde by blocking the amino group of glutamate by acetylation, so allowing the aldehyde group to undergo transamination. In bacteria, *N*-acetylglutamate kinase, *N*-acetylglutamate 5-semialdehyde dehydrogenase, acetylornithine aminotransferase, and acetylornithinase, the enzymes which catalyse these reactions (Fig. 6.19), have been partially characterized (Table 6.7).

In eukaryotes and some bacteria acetylornithinase is replaced by ornithine acetyltransferase which allows the acetyl group to be conserved by being transferred to glutamate:

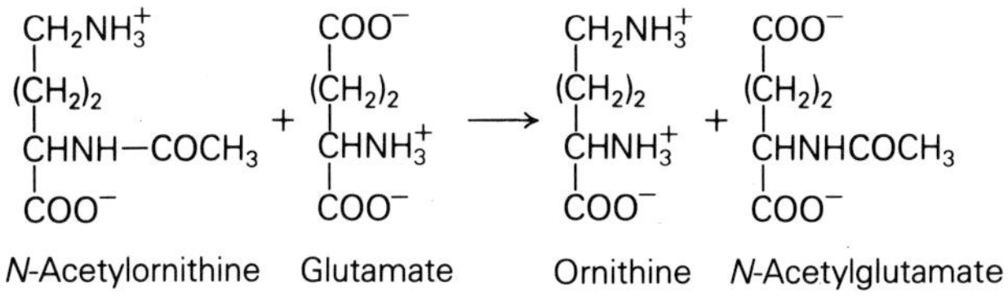

N-Acetylornithine   Glutamate   Ornithine   N-Acetylglutamate

Fig. 6.18 The biosynthesis of proline from glutamate.

References Davis, R.H. (1986) Compartmental and regulatory mechanisms in the arginine pathways of *Neurospora crassa* and *Saccharomyces cerevisiae*. *Microbiological Reviews* **50**, 280–313. Cunin, R., Glansdorff, N., Pierard, A. and Stalon, V. (1986) Biosynthesis and metabolism of arginine in bacteria. *Microbiological Reviews*, **50**, 314–52. These two reviews cover the biosynthesis and degradation of arginine in fungi and bacteria respectively.

Table 6.7 *Enzymes of ornithine biosynthesis in bacteria*

Enzyme	EC Number	Source	$M_r \times 10^{-3}$	Subunit $M_r \times 10^{-3}$
N-Acetylglutamate synthase	2.3.1.1	E. coli	300	50
N-Acetylglutamate kinase	2.7.2.8	P. aeruginosa	230	29
N-Acetylglutamate semialdehyde dehydrogenase	1.2.1.38	E. coli		47
N-Acetylornithine aminotransferase	2.6.1.11	P. aeruginosa	110	55
N-Acetylornithinase	3.5.1.16	E. coli		62
Ornithine acetyltransferase	2.3.1.35	Not yet purified		

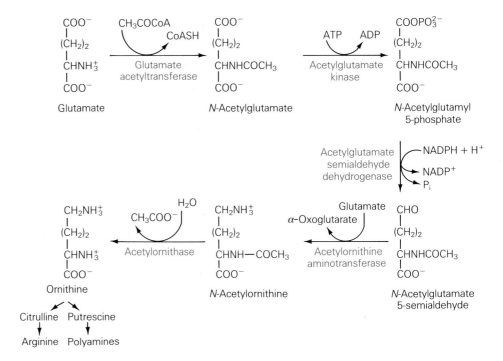

Fig. 6.19 Biosynthesis of ornithine and arginine from glutamate. In eukaryotes, the acetylornithinase-catalysed reaction is replaced by an acetyltransferase-catalysed reaction, resulting in the transfer of the acetyl group back to a molecule of glutamate (equation on p. 133).

The enzymes involved in the biosynthesis of ornithine in mammalian tissues have not been characterized. For many mammals, arginine is not an essential amino acid. Arginine biosynthesis in mammals probably does not involve the acetylated pathway, but detailed evidence is lacking. The reactions between ornithine and arginine have been described elsewhere as part of the urea cycle.

γ-Aminobutyrate

γ-Aminobutyrate (GABA) is an inhibitory transmitter in the central nervous system. It is formed by loss of the C-5 carboxyl group of L-glutamic acid.

$$\begin{array}{l} COO^- \\ | \\ (CH_2)_2 \\ | \\ CHNH_3^+ \\ | \\ COO^- \end{array} \longrightarrow \begin{array}{l} COO^- \\ | \\ (CH_2)_2 \\ | \\ CH_2NH_3^+ \end{array} + CO_2$$

Glutamate   γ-Aminobutyrate

Exercise 5

Write equations for the transamination steps involved in the biosynthesis of serine and of ornithine. Why is a transamination step unnecessary in biosynthesis of the proline?

The enzyme which catalyses this reaction, **glutamate decarboxylase**, a pyridoxal phosphate-dependent enzyme, is found in the mitochondria of brain tissue.

6.10 Overview

Microorganisms and plants can synthesize all the amino acids they require starting from CO_2 and NH_4. In contrast, animals can only synthesize some of the amino acids and must obtain the essential amino acids in their diets. About half of the 20 amino acids found in proteins fall into the category of 'essential'. Mammals do, however, possess the enzymes necessary to synthesize non-essential amino acids, probably because these amino acids are used in the synthesis of other cell constituents in addition to being components of proteins. For example, both glycine and serine can give rise to a wide range of cell components, including some lipids found in biomembranes and the one-carbon units are used in a variety of biosyntheses.

    S-Adenosylmethionine is apparently the only methyl donor in metabolism. Decarboxylated S-adenosylmethionine functions as an aminopropyl donor in the biosynthesis of polyamines. Glutamine, proline and arginine are all derived from glutamate, which is also the precursor of γ-aminobutyrate and, via ornithine, the C_4 portions of polyamines.

Answers to exercises

1. Aminopterin as an inhibitor of dihydrofolate reductase prevents the regeneration of tetrahydrofolate, so decreasing the amount of 5,10-methylene-tetrahydrofolate available for thymidylate synthase and reducing the supply of dTMP required (in the form of dTTP) for DNA synthesis (see also Chapter 7).

2. Hydrolysis to give methanol or by oxidative demethylation (O_2 + NAD(P)H + mono-oxygenase activity) to give formaldehyde.

3. Because of the wide variety of compounds which arise from them. For glutamate see Fig. 6.2, for glutamine see Table 6.6.

4. Reactions catalysed by the following enzymes. Proline: glutamate semialdehyde dehydrogenase, pyrroline 5-carboxylate reductase. Arginine: acetylglutamate semialdehyde dehydrogenase. Methionine methyl: (from serine) methylenetetrahydrofolate reductase; (from formate) also methylene-tetrahydrofolate dehydrogenase.

5. 3-Phosphohydroxypyruvate +
   L-Glutamate → O-Phospho-
   L-serine + α-Oxoglutarate

N-Acetylglutamate 5-semialdehyde
   + L-Glutamate → Acetylornithine
   + α-Oxoglutarate

Because proline does not have a primary amino group.

QUESTIONS

FILL IN THE BLANKS

1. One-carbon units are used in various biosynthetic processes. Purine synthesis involves incorporation of two _____ groups from tetrahydrofolate derivatives, while thymidine and _____ arise from _____ groups. 5-_____ tetrahydrofolate donates the _____ group of methionine, which is important as it is the origin (via _____) of _____ groups in a wide range of cell constituents. Most of the one-carbon units in the cell arise from the _____ group of serine, which is formed by transfer from serine to _____ with the formation of glycine. Serine, in turn, is formed from _____ , an intermediate of the glycolytic pathway, by successive oxidation, transamination and _____ reactions.

Choose from: S-adenosylmethionine, dephosphorylation, formyl, hydroxymethyl (2 occurrences), methyl (3 occurrences), 3-phosphoglycerate, serine, tetrahydrofolate.

2. Glycine is a precursor of a large number of non-protein cell constituents including _____ , essential in coenzymes and nucleic acids, and of various excretion products such as _____ , formed from benzoate. Any glycine ingested in the diet that is not required for such purposes is broken down via an enzyme system that involves _____ different proteins, one of which, the _____ protein is absent in the congenital condition called non-ketotic _____ . This condition is characterized by high concentrations of glycine in the blood. Apart from CO_2, the other product of this system is _____ , which can be used for biosynthetic purposes, such as the synthesis of purines, or can be oxidized to CO_2 via the enzymes _____ _____ , cyclohydrolase and formyltetrahydrofolate dehydrogenase. The electron acceptor for the last-mentioned enzyme is _____ .

Choose from: four, H-, hippurate, hyperglycinaemia, methylenetetrahydrofolate, methylenetetrahydrofolate dehydrogenase, $NADP^+$, purines.

3. Glutamate is the parent compound from which the following amino acids are derived: glutamine, γ-aminobutyrate, _____ and _____ . Ornithine not only gives rise to _____ , but also to putrescine and the C_4 portions of _____ and _____ . _____ of the amino group of glutamate is an essential preliminary to ornithine formation to prevent the spontaneous cyclization of _____ _____ to give Δ^1-pyrroline carboxylate, the precursor of _____ .

Choose from: acetylation, arginine (2 occurrences), glutamate semialdehyde, proline (2 occurrences), spermidine, spermine.

4. Different polyamines have two, three or four _____ charges. They consist of C_3 and _____ portions joined by secondary _____ groups. The C_3 portion arises from _____ _____ , while the _____ portion arises by a similar process from _____ . Inhibitors of _____ _____ have clinical applications.

Choose from: amino, C_4 (2 occurrences), decarboxylated S-adenosylmethionine, ornithine, positive, polyamine synthesis

MULTIPLE-CHOICE QUESTIONS

5.   Which of the following compounds contain carbon atoms *not* derived from glutamate or glutamine?

A. arginine   E. urea
B. cysteine
C. spermine
D. serine

6.   Which of the following compounds contain carbon atoms that do *not* arise from the C-3 of serine (including routes via *S*-adenosylmethionine)?

A. carbamoyl phosphate   D. thymidylate
B. methionine   E. noradrenalin
C. adrenalin

7.   Which of the following compounds is formed by a biosynthetic pathway which does *not* involve a decarboxylation?

A. serine   D. putrescine
B. proline   E. arginine
C. γ-aminobutyrate

SHORT-ANSWER QUESTIONS

8.   Using textbooks and reference books if necessary, classify the compounds in Table 6.5 into those containing *O*-methyl, *N*-methyl and *C*-methyl groups.

9.   Write equations for enzymic reactions by which all three carbon atoms of serine can be converted into carbon dioxide without involving the TCA cycle.

7

Purines and pyrimidines

Objectives

After reading this chapter, you should be able to:

□ explain the general strategy of purine and pyrimidine biosynthesis;

□ describe how pyrimidine bases are synthesized *de novo*;

□ describe how purine bases are synthesized *de novo*;

□ explain the importance of salvage pathways;

□ discuss the usefulness of some nucleoside derivatives in clinical practice.

7.1 Introduction

The biosynthesis of purine and pyrimidine nucleotides (Fig. 7.1, Table 7.1) is an essential adjunct to nucleic acid production in the cell. If the supply of these compounds is halted, replication of DNA and cell multiplication cannot take place. Consequently, it is not surprising to find that inhibitors of purine and pyrimidine biosynthesis may be effective inhibitors of cell, and hence of tumour, growth.

In addition to their roles as nucleic acid precursors, nucleoside phosphates have a number of other important roles in the cell. Examples are their role in the formation of ATP as a universal energy currency, and nicotinamide adenine dinucleotide (NAD^+) as a hydrogen and electron acceptor. Many plant products that have pharmacological properties are purines, as are certain plant hormones.

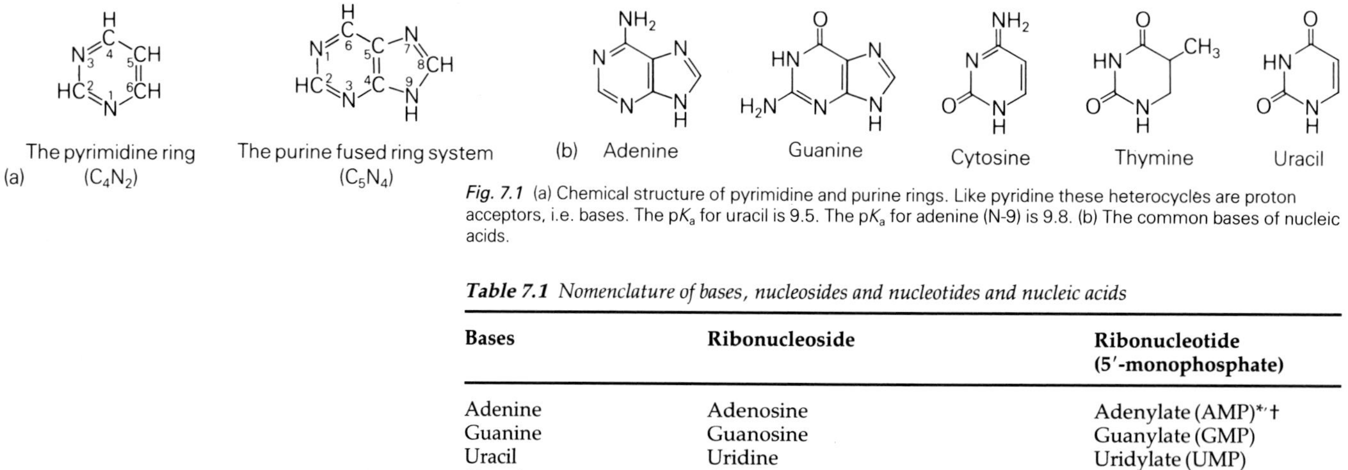

□ NAD^+ and $NADP^+$ are called dinucleotides, and FAD is called flavin adenine dinucleotide, although, strictly speaking, they are not really dinucleotides. In FAD, one of the sugars is ribitol (a sugar alcohol) rather than ribose. In NAD^+ and $NADP^+$, as well as in FAD, the nucleotides are joined tail-to-tail: for example, nicotinamide–ribose–P–P–ribose–adenine. This is in contrast with a two-base section of RNA which will be (base) sugar–P–(base) sugar–P.

The pyrimidine ring
(a)   (C_4N_2)

The purine fused ring system
(C_5N_4)

(b)   Adenine   Guanine   Cytosine   Thymine   Uracil

Fig. 7.1 (a) Chemical structure of pyrimidine and purine rings. Like pyridine these heterocycles are proton acceptors, i.e. bases. The pK_a for uracil is 9.5. The pK_a for adenine (N-9) is 9.8. (b) The common bases of nucleic acids.

Table 7.1 Nomenclature of bases, nucleosides and nucleotides and nucleic acids

Bases	Ribonucleoside	Ribonucleotide (5′-monophosphate)
Adenine	Adenosine	Adenylate (AMP)[*][†]
Guanine	Guanosine	Guanylate (GMP)
Uracil	Uridine	Uridylate (UMP)
Cytosine	Cytidine	Cytidylate (CMP)

Bases	Deoxyribonucleoside	Deoxyribonucleotide (5′-monophosphate)
Adenine	Deoxyadenosine	Deoxyadenylate (dAMP)
Guanine	Deoxyguanosine	Deoxyguanylate (dGMP)
Thymine	Deoxythymidine	Deoxythymidylate (dTMP)
Cytosine	Deoxycytidine	Deoxycytidylate (dCMP)

[*] Any other ester must be specified, e.g. 2′ UMP for Uridine 2′-phosphate.
[†] The di- and triphosphates are specified in same way, e.g. GTP for guanosine 5′-triphosphate.

See *Molecular Biology and Biotechnology*, Chapters 1 and 3

Nucleosides and nucleotides: *a nucleoside has the structure base–sugar. For example, adenine–ribose is adenosine. A nucleotide has the structure base–sugar–phosphate. For example, adenine–ribose–phosphate is adenosine monophosphate. There may be more than one phosphate group. For* example ATP (adenosine triphosphate) is a nucleoside triphosphate.

The stimulatory effect of coffee is due mainly to the **caffeine** (1,3,7-trimethyl-xanthine) which, like other methylxanthines acts as an inhibitor of phosphodiesterases and thereby raises the cyclic AMP concentration. In this way it mimics the physiological action of the neurotransmitter adrenalin (*Cell Biology*, Chapter 9) which increases cyclic AMP concentration by activating the enzyme adenylate cyclase. Similarly, theophylline (1,3-dimethylxanthine), also found in trace amounts in tea and coffee, is used therapeutically in inhalers to promote bronchodilation in asthmatic patients on the basis of its ability to imitate the action of β-receptor stimulants to raise cyclic AMP.

Caffeine

The methylxanthines are antagonists to the purinergic transmitter adenosine (but not to ATP); this is probably another factor in their anti-asthmatic action.

Substances regulating cell division and development in plants (phytohormones) include a class of compounds called the **cytokinins**. These are substituted purines and the first to be isolated (an artefact produced in autoclaved DNA) was 6-furfuryladenine (kinetin). This promoted cell division in tobacco pith cells at a concentration of 1 μg dm^{-3}. Letham isolated 1 mg of hydroxyisopentenyladenine (zeatin) from 60 kg of immature maize kernels (*Cell Biology*, Chapter 8). This compound has also been isolated from cocoanut milk, sycamore sap and crown gall tumour amongst other plant tissues. Zeatin also occurs as the nucleoside (9-riboside) and the nucleotide.

7.2 Strategy of biosynthesis

The cellular strategies for the **de novo** biosynthesis of purine and pyrmidine nucleotides are different. For purine nucleotides, the purine ring is assembled as the ribose sugar unit from the outset, but for pyrimidine nucleotides the nitrogenous pyrimidine ring is constructed first and then coupled to the ribose sugar unit. Both pathways occur largely in the cytosol and both are linked with glycolysis and the pentose phosphate pathway.

SALVAGE PATHWAYS. As well as *de novo* pathways, so-called **salvage pathways** exist by which the purine and pyrimidine bases (or their nucleoside derivates) can be rescued from further breakdown and excretion and re-incorporated into nucleotides (Section 7.5).

DIETARY NUCLEIC ACIDS. One possible source of nucleotides is the nucleic acids and nucleotides in food. In the intestine there are enzymes that catalyse the hydrolysis of nucleic acids to nucleosides and even to the free purine and pyrimidine bases. However, the mucosae in the mammalian gastrointestinal tract further degrade these products. For example, the purine ring may be oxidized to uric acid in a reaction involving xanthine oxidase (Fig. 7.2).

This means that orally ingested purines or pyrimidines are mainly metabolized and excreted without entering endogenous nucleotide pools and are not really effective precursors of endogenous nucleic acids. If the gut is bypassed, incorporation may take place. Thus, injected ^{3}H-deoxythymidine

Exercise 1

Test yourself now to see whether you can draw the purine and pyrimidine rings.

□ Xanthine oxidase is one of a group of molybdenum-containing enzymes (others include aldehyde oxidase, sulphite oxidase and the microbial nitrogenase). The enzyme is found in lactating mammary gland (and in milk) and liver. It is a dimer (M_r about 300 000) with each subunit containing a molybdopterin unit as well as one FAD molecule and two Fe_2S_2 clusters. The enzyme is alternately reduced by its substrate (hypoxanthine or xanthine) and then reoxidized by molecular O_2.

De novo: means 'from new'. In other words, the synthesis of compounds from small, simple precursors, rather than by metabolic conversion from related compounds, or, in this case, by the breakdown of dietary nucleic acids.

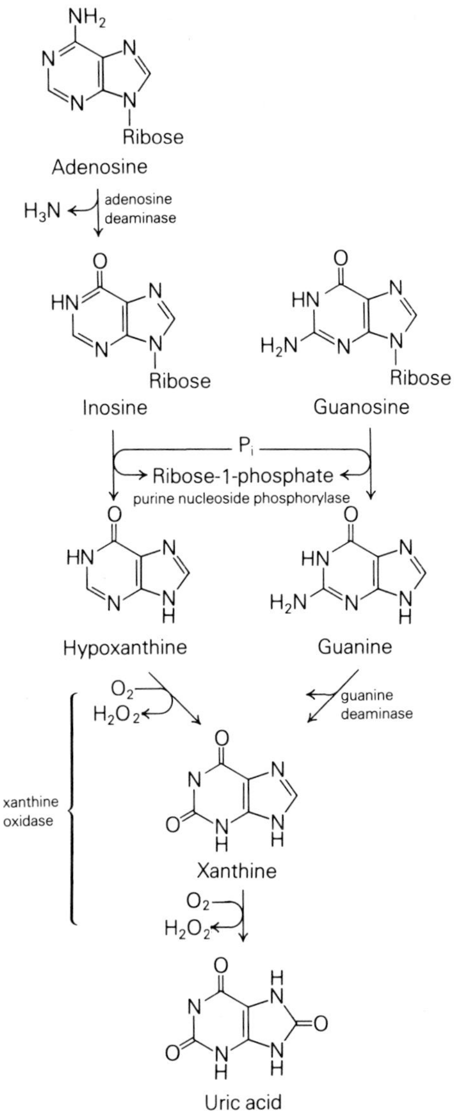

is incorporated into newly synthesized DNA and this technique is used as a way of determining the rate at which new DNA is produced.

In normal circumstances, nucleotides are synthesized *de novo* to meet metabolic needs with help from the salvage pathways.

7.3 Biosynthesis of nucleotides de novo

Although the point at which the ribose is added differs in purine, compared with pyrimidine, biosynthesis, in each case the ribose is derived from the pentose phosphate pathway. Ribose is subsequently converted to 2-deoxyribose when deoxyribonucleotides are being synthesized.

Activated ribose

The ribose sugar ring which is associated with the purine or pyrimidine base is produced from glucose 6-phosphate using the enzymes of the pentose phosphate pathway to generate ribose 5-phosphate (Chapter 3). The sugar ring is introduced in an 'activated' state as the pyrophosphate derivative, with the pyrophosphate moiety deriving from the terminal phosphate of ATP (Fig. 7.3). The pathways of nucleotide biosynthesis need to be regulated. The intracellular concentration of phosphoribosylpyrophosphate (PRPP) plays a major regulatory role for both the purine and pyrimidine *de novo* synthetic pathways. Its concentration is determined largely by the flux of glucose 6-phosphate through the pentose phosphate pathway. In addition, the enzyme PRPP synthase is activated by inorganic phosphate, and inhibited by GDP or ADP competitively with respect to ATP.

Fig. 7.2 Purine nucleoside degradative pathways. The metabolic pathways involved in the degradation of adenosine and guanosine to uric acid are shown here.

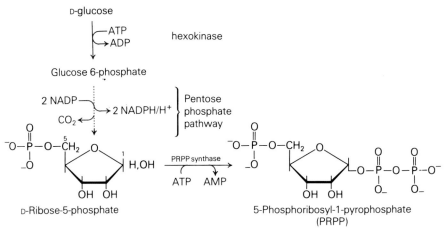

Fig. 7.3 Pathway for the formation of 5-phosphoribosyl-1-pyrophosphate (PRPP) from glucose. The pentose phosphate pathway is shown by a broken arrow to indicate that the individual steps are not depicted.

Pyrimidine nucleotide biosynthesis *de novo*

As was previously mentioned, in pyrimidine nucleotide biosynthesis, the nitrogen-containing heterocyclic ring is formed first and the ribose ring added only after its completion. Studies using radioisotopically labelled compounds have enabled the origin of all the atoms to be identified. The culmination of this process is the formation of **uridine 5-monophosphate (UMP)**, which is the precursor of all the other pyrimidine nucleotides.

Exercise 2

Describe the 'inputs' and 'outputs' of the pentose phosphate pathway.

Reference Adam, R.L.P., Knowler, J.T. and Leader, D.P. (1992) *The Biochemistry of the Nucleic Acids*, 11th edn, Chapman & Hall, London. This is a standard reference text in this area and is a good source of data and information on nucleotides, nucleosides and bases, as well as the polynucleotides.

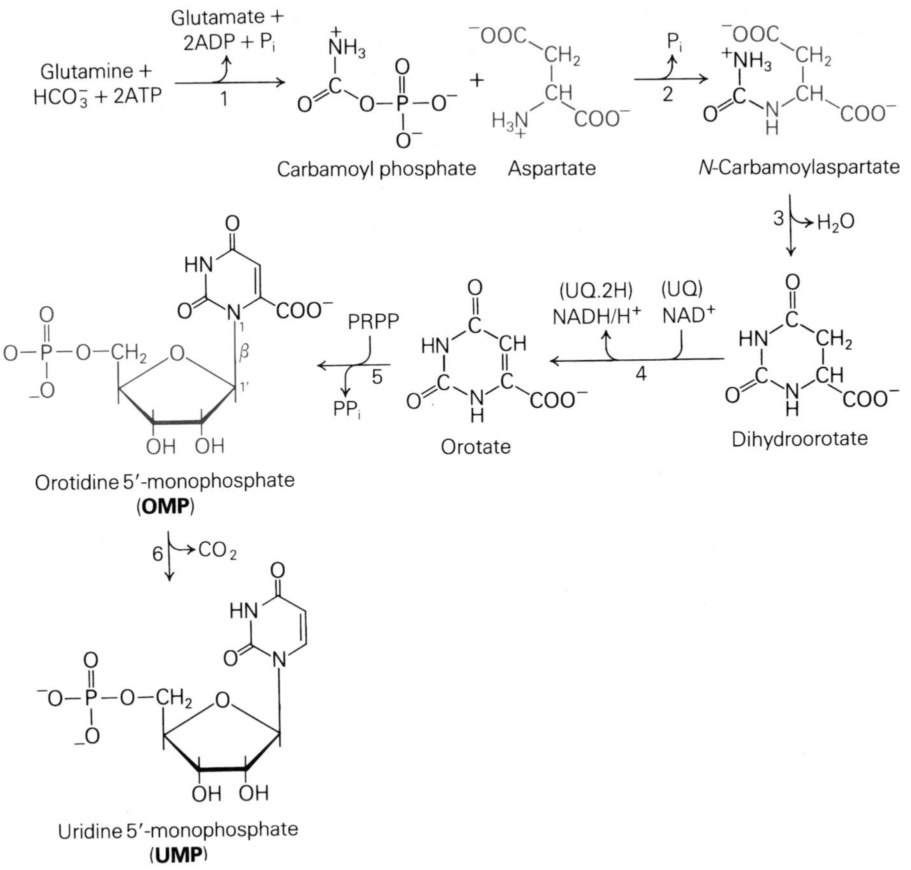

Fig. 7.4 *De novo* pathway for pyrimidine (UMP) biosynthesis. The enzymes involved in each of the six steps are: (1) carbamoyl phosphate synthase II (although utilizing HCO_3^-, this enzyme does not contain biotin); (2) aspartate carbamoyltransferase (also called aspartate transcarbamylase); (3) dihydroorotase; (4) dihydro-orotate dehydrogenase (a flavoprotein which can donate electrons directly into the electron transport chain, thus accounting for its localization on the inner mitochondrial membrane); (5) orotate phosphoribosyl-transferase; (6) orotidine monophosphate decarboxylase (or carboxylase). The new atoms introduced at each step are shown in red.

UMP FORMATION. Figure 7.4 shows the outline of UMP biosynthesis. The pyrimidine ring is constructed using the amide nitrogen of glutamine, the complete aspartate molecule and HCO_3^-. The first step of the *de novo* pathway utilizes glutamine as a nitrogen donor (for C–N bond formation), the reaction being catalysed by the cytosolic enzyme carbamoyl phosphate synthase II. In eukaryotic cells this enzyme is distinct from the mitochondrial carbamoyl phosphate synthase I, which in mammals utilizes ammonia directly for arginine synthesis and the urea cycle.

In prokaryotes, such as *E. coli*, there is only a single carbamoyl phosphate synthase enzyme and a single pool of carbamoyl phosphate. This common pool can be used for arginine biosynthesis or for the pyrimidine pathway. In the latter case regulation is centred upon reaction (2) in Fig. 7.4, the first committed step of the pathway. Aspartate carbamoyltransferase (also known as aspartate transcarbamoylase) is an allosteric enzyme with the pathway end-product, CTP, acting as a negative effector and the purine nucleotide, ATP, as a positive effector.

The regulation of the mammalian pyrimidine biosynthetic pathway occurs at step (1). This is the first committed step in eukaryotes since the cytosolic carbamoyl phosphate pool is dedicated to pyrimidine nucleotide formation. The enzyme involved, carbamoyl phosphate synthase II, shows feedback

See *Energy in Biological Systems*, Chapter 7

allosteric inhibition by UTP. Carbamoyl phosphate synthase II is also subject to protein phosphorylation by the cyclic AMP-dependent protein kinase. The consequence of phosphorylation is an increase in activity due to a decreased affinity of the enzyme for its allosteric inhibitor UTP. Orotate monophosphate (OMP) decarboxylase (step 6) is inhibited by its produce UMP and, less effectively, by CMP. Phosphoribosyl pyrophosphate (PRPP) serves both as a substrate (for step 5) and as an allosteric activator (for step 1). This ensures that the production of orotate keeps pace with the availability of PRPP as a co-substrate for OMP formation.

In prokaryotes six structural genes code for the first six enzymes of the UMP pathway, whereas in mammals only three genes are involved.

The precursors of the pyrimidine ring are shown in Fig. 7.5. Note that the origin of the aspartate is glutamine. The splitting of glutamine has been termed *glutaminolysis* and this probably accounts for the significant requirement for glutamine found for rapidly dividing cells.

FORMATION OF CTP AND TTP FROM UMP. UMP undergoes further metabolism to generate UTP and CTP for RNA production, and dCTP and dTTP for DNA formation, as shown in Fig. 7.6.

As well as allosteric feedback inhibition of CTP synthase by its product CTP, the synthesis of this pyrimidine nucleotide is also controlled by the purine nucleotide GTP, acting as an allosteric activator. The allosteric inhibition of bacterial aspartate carbamoyltransferase by CTP is counteracted by the purine nucleotide ATP acting as an allosteric activator. Thus both the purine nucleotides, ATP and GTP, promote pyrimidine nucleotide production. This ensures that the two classes of nucleotides, purines and pyrimidines, are produced in roughly equal quantities to each other as monomeric precursors for RNA and DNA.

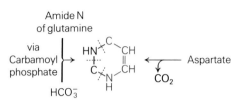

Fig. 7.5 The metabolic precursors of the pyrimidine ring are aspartate, a nitrogen atom derived from glutamine and a carbon atom derived from carbamoyl phosphate.

Box 7.2
Antimetabolites and cancer

Since cell division necessitates nucleic acid synthesis, the blocking of the latter stops the former. Glutamine serves as an amino donor in numerous reactions of nucleotide synthesis including the *de novo* synthesis of the purine ring (N-3 and N-9) and the conversion of IMP to GMP (the 2-amino group). For the pyrimidine ring there are carbamoyl phosphate formation and UTP conversion to CTP (4-amino). A naturally occurring glutamine analogue (azaserine, produced by *Streptomyces* spp.) and a synthetic analogue (Acivicin) have found use as anticancer drugs because of their ability to block nucleic acid synthesis by antagonizing the metabolic roles of glutamine (antimetabolite action).

Glutamine	Azaserine	Acivicin

Acivicin shows competitive inhibition of glutamine-utilizing enzymes but it efficacy as an anticancer drug is due to its reactivity (alkylation) at the substrate-binding site, making it an irreversible inhibitor. CTP synthase is the enzyme most affected *in vivo*, and CTP pools in liver tumours *in vivo* fall to 2% of normal levels within 2 hours after injection of the drug.

References McKeehan, W.L. (1986) in *Carbohydrate Metabolism in Cultured Cells*, ed. M.J. Morgan, Plenum Press, New York, pp. 111–50. Discusses the special role of glutamine in rapidly dividing cells. An explanation for this role of glutamine in terms of the sensitivity of pathway control is given in Newsholme, E.A., Crabtree, B. and Ardawi, M.S.M. (1985) 'The role of high rates of glycolysis and glutamine utilization in rapidly dividing cells. *Bioscience Reports* **5**, 393–400.

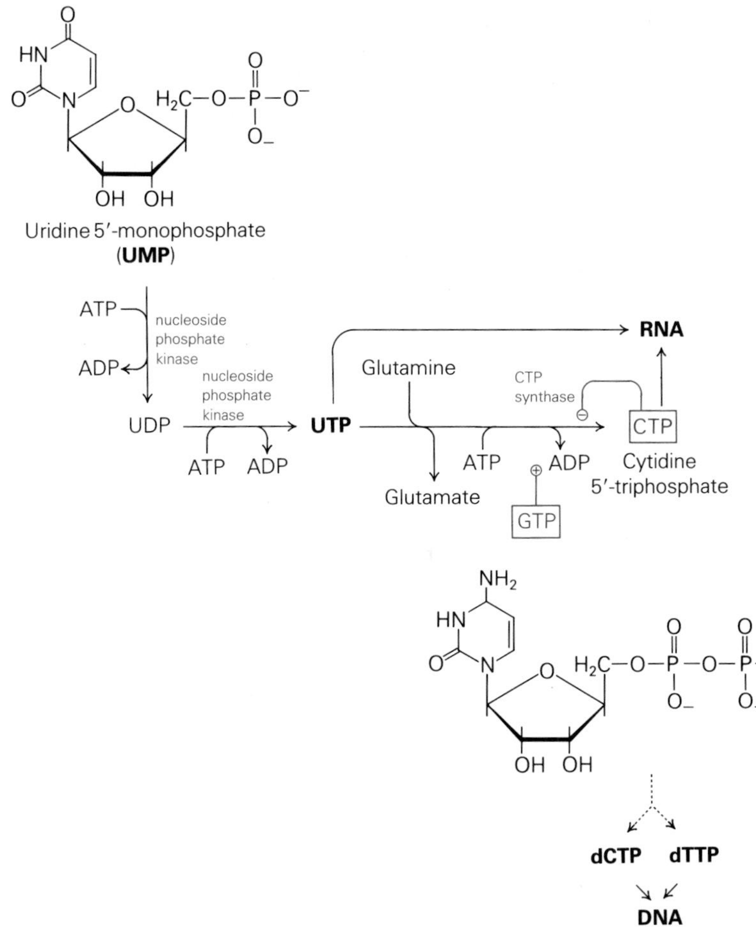

Uridine 5'-monophosphate
(**UMP**)

Fig. 7.6 Formation of pyrimidine nucleotides (from UMP) for nucleic acid synthesis. Allosteric activation (+) or inhibition (−) is shown in red. The pathway for dTTP formation is shown in detail in Fig. 7.11.

The details of the conversion to the deoxynucleoside triphosphates, dCTP and dTTP, are discussed later in this chapter.

Purine nucleotide biosynthesis de novo

Purine nucleotide biosynthesis involves assembly of the purine ring on the ribose sugar (as phosphoribosylpyrophosphate) from the outset. The amino acids aspartate and glutamine are again involved, as well as glycine, for the formation of **inosine monophosphate** (IMP) from which the purine nucleotides, AMP and GMP, are derived. **Serine** is important indirectly, both as a source of glycine and of the one-carbon units required by organisms, as formyltetrahydrofolate.

The ten-stage sequence for synthesis of IMP (Fig. 7.7) is an elegant example of building-up a molecule step-by-step in a metabolic pathway. As well as PRPP which utilizes two ATP molecules in its formation, four ATP molecules are hydrolysed to ADP and inorganic phosphate to drive this biosynthetic sequence. Steps 2, 4, 5 and 7 in Fig. 7.7, each involve ATP and are catalysed by synthase enzymes.

☐ The purine biosynthetic pathway is particularly active in **uricotelic** (see *Energy in Biological Systems*, Chapter 7) animals where the subsequent degradation of the purine base to uric acid is used as a means of nitrogen excretion. This realization of high activity led to the selection of pigeon liver as the most effective tissue for the elucidation of the purine pathway in vertebrates.

☐ Inosine is

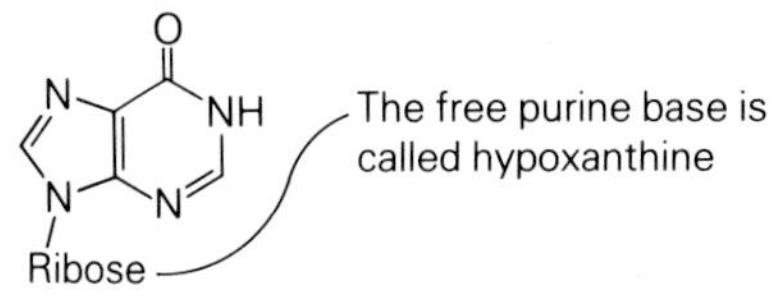

The free purine base is called hypoxanthine

References Snell, K. *et al.* (1988) Enzymic imbalance in serine metabolism in human colon carcinoma and rat sarcoma. *British Journal of Cancer* **57**, 87–90. Includes a brief discussion of the role of serine in purine and pyrimidine biosynthesis. For a more thorough discussion see Snell, K. (1984) Enzymes of serine metabolism in normal, developing and neoplastic rat tissues. *Advances in Enzyme Regulation* **22**, 325–400.

Box 7.3
Anticancer drugs and nucleotide biosynthesis

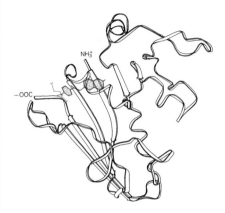

Structure of dihydrofolate reductase with bound methotrexate (in red). Redrawn from Stryer, L. (1988) *Biochemistry*, 3rd edn, p. 616, W.H. Freeman, New York.

The formation of deoxythymidylate is a strategic point at which to direct drugs to block DNA synthesis and thereby inhibit the proliferation of cancer cells .

The use of 5,10-methylenetetrahydrofolate in the thymidylate synthase (TS) reaction leads to the oxidation of the tetrahydrofolate carrier to dihydrofolate. Dihydrofolate reductase (DHFR) reduces dihydrofolate back to tetrahydrofolate, and serine hydroxymethyltransferase then donates the one-carbon unit to replenish methylenetetrahydrofolate. This latter reaction also generates the amino acid glycine which, with conversion of methylenetetrahydrofolate to 10-formyltetrahydrofolate, provides precursors for purine biosynthesis. The antifolate drug, methotrexate (Amethopterin), is a highly effective competitive inhibitor (active at 10^{-9} mol dm^{-3}) of dihydrofolate reductase and is used in cancer chemotherapy, particularly in the treatment of leukaemia. It is a synthetic analogue of folic acid.

Another useful drug is 5-fluorouracil which, by directly inhibiting thymidylate synthase, deprives growing cells of dTMP and thus dTTP for DNA synthesis and causes the phenomenon called 'thymineless cell death'. In fact, enzyme inhibition requires the conversion of 5-fluorouracil to 5-fluoro-2'-deoxyuridine 5'-phosphate by salvage pathway enzymes, which then competes at the dUMP binding site and forms an irreversible enzyme–substrate complex (so-called 'suicide-inhibition'). The cytotoxicity of 5-fluorouracil is also related to the formation of 5-fluoro-UTP and 5-fluoro-dUTP which become incorporated into RNA and DNA respectively and disrupt their normal function.

6-Mercaptopurine (6-μP) is an anticancer drug which again requires metabolic activation to be clinically effective, in this case by PRPP to form 6-thioIMP. This metabolite accumulates, acting as a negative effector of PRPP amidotransferase, the committed step of the *de novo* purine pathway. Further, it can inhibit the conversion of IMP to AMP at the adenylosuccinate synthase step, and IMP to GMP at the IMP to XMP (dehydrogenase) step. However, 6-thioIMP can also be a substrate and be converted to 6-thioGMP prior to incorporation into DNA and RNA. The former can lead to DNA strand breaks, the latter may inhibit RNA transcription and processing. 6-Mercaptopurine is a substrate for the enzyme xanthine oxidase and thus its cytotoxic actions can be potentiated by inhibiting this enzyme with allopurinol (the 'cocktail' approach to drug therapy called combination chemotherapy). A modified form of 6-mercaptopurine, azathioprine, is an important immunosuppressive drug for use in tissue transplantation.

The chemical modifications of sugars have also produced a range of drugs. For example, the replacement of the ribose moiety of nucleosides with arabinose has been used to produce anticancer drugs such as cytosine arabinoside (Cytarabine) and adenine arabinoside (Vidarabine).

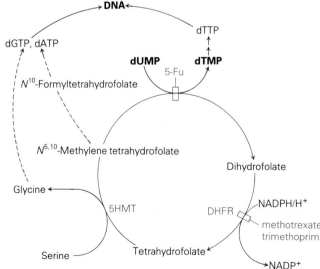

Inhibition of the thymidylate synthesis cycle. The cyclic formation of thymidylate involves the enzymes: TS, thymidylate synthase; DHFR, dihydrofolate reductase; and SHMT, serine hydroxymethyltransferase. Inhibitors of TS or DHFR can block DNA synthesis in actively dividing cells. Scheme adapted from Snell *et al.* (1988). Enzymic imbalance in serine metabolism in human colon carcinoma and rat sarcoma. *British Journal of Cancer*, **57**, 87–90.

References Weber, G. (1983) Biochemical strategy of cancer cells and the design of chemotherapy. (The Clowes Memorial Lecture). *Cancer Research* **43**, 3466–92. An excellent account of the rational use of biochemical principles to develop drugs against cancer, with particular emphasis on the pathways of purine and pyrimidine biosynthesis. This theme is also explored regularly in an annual series: *Advances in Enzyme Regulation*, edited by G. Weber, Pergamon Press, Oxford.

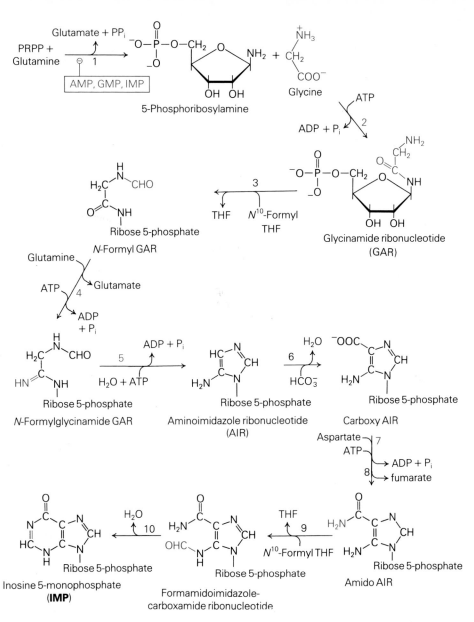

Fig. 7.7 *De novo* pathway for purine (IMP) biosynthesis. The enzymes involved in the ten steps are: (1) PRPP amidotransferase; (2) phosphoribosylamine-glycine ligase (or synthase); (3) phosphoribosylglycinamide *N*-formyltransferase; (4) phosphoribosylformylglycinamide synthase; (5) phosphoribosylaminoimidazole synthase (a cycloligase enzyme involved in ring closure through removal of H_2O); (6) phosphoribosylamino-imidazole carboxylase the enzyme involves neither biotin nor ATP for the incorporation of CO_2; (7) phospho-ribosylaminoimidazole – succinic carboxamide synthase; (8) adenylosuccinate lyase (or adenylosuccinase, the enzyme is also involved in a step after IMP formation; note the loss of the carbon of aspartate as fumarate (cf. argininosuccinate lyase in the urea cycle); (9) phosphoribosyl amidoimidazolecarboxamide *N*-formyl-transferase (as in step 3, one-carbon transfer is from N^{10}-formyltetrahydrofolate); (10) IMP cyclohydrolase (the ring closure involves loss of H_2O, but unlike that in step 5 it does not involve ATP hydrolysis). The naming of the intermediates in this pathway sequence is potentially confusing in that two alternative conventions are followed. In the figure itself the intermediates are depicted as ribonucleotide derivatives, whereas the enzymes in the legend are referred to by the alternative 'phosphoribosyl' – convention. Either system is legitimate, but not a mixture of the two! ATP-requiring steps in the pathway are numbered in red, the new atoms introduced at each step are shown in red, and allosteric control of the pathway is also shown in colour.

Exercise 3

Name and write the chemical structures of the nucleotide bases in DNA and RNA?

Inosine monophosphate is the first purine nucleotide formed, but it is not present at any great concentration under normal conditions either in the overall nucleotide pool or in nucleic acids. It is a branch-point intermediate and is rapidly and effectively converted along the route to adenosine 5'-monophosphate or along the alternative route to guanosine 5'-monophosphate (Fig. 7.8). As one might anticipate there is feedback control at this metabolic branch point; AMP is a competitive inhibitor, with respect to IMP, of adenylosuccinate synthase; and GMP inhibits IMP dehydrogenase. Thus the products of the branched pathways control their own synthesis (by *feedback inhibition*) at their respective branch-point enzymes. There is a second level of regulation called cross-over control) GTP serves as an energy source for the pathway to AMP, whereas ATP is required for GMP formation (see

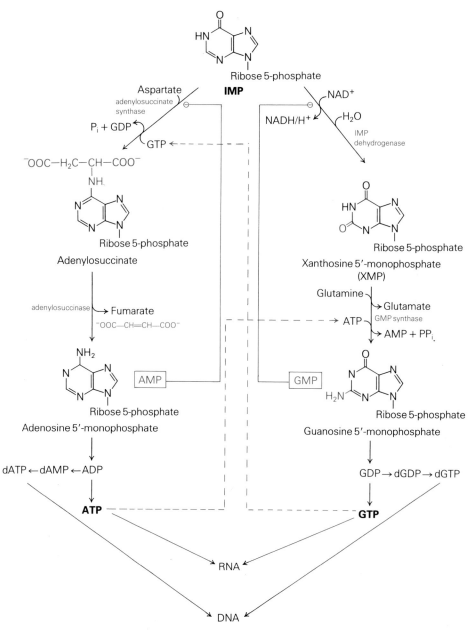

Fig. 7.8 Formation of purine nucleotides (from IMP) for nucleic acid synthesis. Allosteric feedback inhibition is shown in red.

Feedback inhibition: occurs when one
metabolite in a pathway inhibits the production of
another metabolite made earlier in that pathway.
This has the effect of not only turning off
production of the metabolite itself but also of all
subsequent metabolites in the pathway.

Reference Hirsch, M. and Kaplan, J. (1987)
Antiviral therapy. *Scientific American* **256** (4),
66–75. An informative and well-illustrated
article dealing with the development of drugs
to combat viral infection.

Fig. 7.8). Thus sufficient ATP leads to GMP being produced from IMP, and sufficient GTP leads to AMP production. In this way a balance of production of the ATP and GTP purine nucleotides is ensured.

As might also be expected the first committed step, catalysed by PRPP amidotransferase, of the overall purine synthetic sequence is another key point for regulation. In the presence of AMP, GMP or IMP the enzyme is converted to an *inactive* dimer, whereas PRPP promotes the conversion of the enzyme to the *active* monomer ($M_r = 133\,000$). The enzyme has separate, allosteric-effector binding sites for the oxopurine nucleotides (IMP and GMP) and for the aminopurine, AMP. The simultaneous binding of both an oxopurine and an aminopurine nucleotide results in a synergistic or co-operative feedback inhibition. The purine nucleotides, ATP and GTP (see Fig. 7.8), together with the pyrimidine nucleotides, UTP and CTP, can be used directly for the formation of RNA. Their use for DNA synthesis requires reduction of the ribose sugar to produce the corresponding deoxyribose nucleotide.

In summary, the precursors of the purine ring are shown in Fig. 7.9.

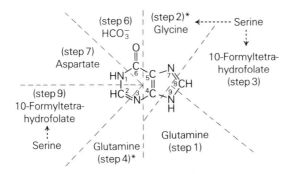

Fig. 7.9 The metabolic precursors of the purine ring system are shown in this diagram (compare with Fig. 7.5). The first five steps build up the imidazole ring, the subsequent five steps produce the second (pyrimidine) ring which is fused to the imidazole ring to form the nine membered purine ring. Steps 5* and 10 achieve ring closure, in both cases by removal of a water molecule forming a C–N bond. (* Indicates the four ATP utilizing steps). The IMP ring builds up: 9-457-83612, i.e. the purine telephone number!

7.4 Biosynthesis of deoxyribonucleotides

The production of DNA requires the provision of both purine and pyrimidine deoxyribonucleoside triphosphates. Analytical measurements in a variety of tissues and cell types indicate significant differences between the size of each of the four pools. For example the dCTP pool can be an order of magnitude greater than the dGTP pool in the S phase of the cell cycle (Fig. 7.10), and there is some evidence that disturbances to the normal balance can be mutagenic.

Deoxyribonucleotides are formed by reduction at the 2'-position of the ribose ring of the corresponding ribonucleoside diphosphate and is catalysed by the **ribonucleotide reductase system** (Fig. 7.11). The exception to this is dTMP, which is formed by methylation of the uridine pyrimidine base when it is already present as the deoxyribonucleotide. This reaction is catalysed by thymidylate synthase, an enzyme that is highly conserved evolutionarily (Fig. 7.12).

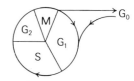

Fig 7.10 The cell cycle. Cells which are actively dividing pass through various metabolic phases (the cell cycle) between one mitotic division and the next. The phases are: M, mitosis; G_0, a resting or non-dividing state which is highly variable in length depending on both cell type and extracellular factors; G_1, the pre-synthetic DNA phase involving ribonucleotide biosynthesis and which, in turn, requires the production of the relevant enzymes (e.g. PRPP synthase); S, DNA synthesis phase involving the production of enzymes concerned with deoxyribo-nucleotide and DNA synthesis, including ribonucleotide reductase, dCMP deaminase, dTMP synthase, thymidylate kinase, the salvage enzyme thymidine kinase, and DNA polymerases; G_2, this is a stage of cellular organization for division, including the organization of the microtubular proteins for mitotic spindle formation and the condensation of interphase chromosomes. An increase in cellular deoxyribonucleotide pools during S-phase precedes the onset of cell division. The timing of the phases of the cell cycle can be determined by pulse labelling using radioactively-labelled precursors (such as ^{3}H-thymidine into DNA for the S-phase). In actively dividing mammalian cells (such as bone marrow) timings of about 8, 3 and 1 hours are observed for the S_1, G_2 and M phases, respectively; G_1 is variable (e.g. from 3–12 hours). Tissue culture cells that are depleted of a nutrient, or lack the appropriate growth factor stimulus, stop dividing and 'rest' in G_0.

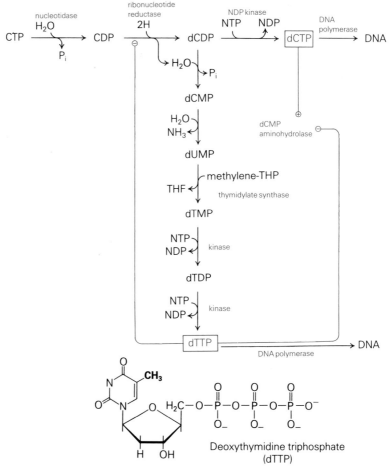

Fig. 7.11 Pathway for formation of dTTP from CTP. Allosteric feedback inhibition is shown in red.

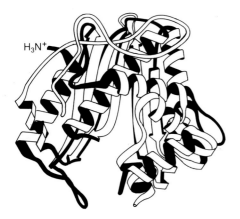

Fig. 7.12 Molecular model of part of the thymidylate synthetase from *Lactobacillus casei* showing one monomer. This enzyme is highly conserved evolutionarily and a high proportion of the conserved residues are in the central cleft thought to represent the active site. The protein is a homodimer. Redrawn from Mathews, C.K. and van Holde, K.E. (1990) *Biochemistry*, Benjamin/Cummings, Redwood City, CA.

Exercise 4

Explain the term 'promiscuity' as applied to deoxyribonucleotide biosynthesis.

The ribonucleotide reductase system

In animals, plants, and bacteria such as *E. coli*, the ribonucleotide reductase system can be represented as shown in Fig. 7.13. Note that the pentose phosphate pathway not only provides ribose 5-phosphate for ribonucleotide synthesis but also regenerates NADPH for use in the formation of deoxyribonucleotides. Thioredoxin, the hydrogen carrier to the reductase, is a small protein (M_r 12 000) with a –Cys–Gly–Pro–Cys– sequence at the active centre (Fig. 7.13). These two cysteine thiol groups participate in redox reactions with the reversible oxidation to an intramolecular disulphide bridge.

Enzyme conformation and activity is controlled by nucleoside triphosphate binding. It has a very complex, highly specific (and mechanistically poorly understood) regulation in which different types of nucleotide binding at one site can determine the particular nucleotide which serves as a substrate for the enzyme at the catalytic (substrate-binding) site. These complex control patterns ensure that the enzyme does not remain dedicated to the production of just one deoxyribonucleotide. This 'promiscuity' is essential to produce the variety of deoxyribonucleotides required for DNA replication.

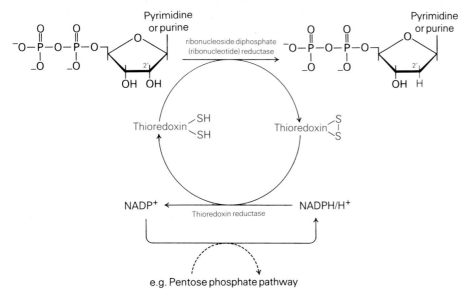

Fig. 7.13 Formation of deoxyribonucleotides by the ribonucleoside diphosphate (ribonucleotide) reductase system. The enzyme ribonucleotide reductase is shown interacting with NADPH/H$^+$ through reduction of the thioredoxin carrier by the flavoprotein, thioredoxin reductase.

7.5 Salvage pathways

The *de novo* pathways for synthesizing pyrimidines and purines are energy demanding. However, a considerable saving of energy can be achieved by the use of salvage pathways to rescue preformed purines and pyrimidines, either as the free bases or as nucleosides resulting from nucleic acid degradation (Fig. 7.14). In addition, certain tissues (erythrocytes, leukocytes, brain) are deficient in enzymes of *de novo* purine synthesis and therefore rely on salvage pathways for nucleotide formation. In fact mammalian liver is a major exporter of purine bases and nucleosides for salvage by those tissues which cannot synthesize purines *de novo*. Indeed, in the mammal as a whole, the purine salvage pathways utilize more phosphoribosyl pyrophosphate than does the *de novo* purine biosynthesis pathway.

The degradative enzyme adenosine deaminase (ADA) can use deoxyadenosine as well as adenosine as a substrate. Similarly, purine nucleoside phosphorylase can use deoxyadenosine or deoxyguanosine. There are, however, separate adenosine and deoxyadenosine kinases. The latter can phosphorylate deoxyadenosine, deoxyguanosine, deoxycytidine (and the synthetic anticancer analogue cytosine arabinoside) Hypoxanthine/guanine phosphoribosyltransferase (HGPRTase Fig. 7.14) can convert the synthetic compound 6-mercaptopurine (also used in cancer therapy) to the corresponding nucleotide.

The pyrimidine nucleotides can be degraded by the action of nucleotidase and deaminase and phosphorylase enzymes, to uracil or thymine. There is a very active dUTP pyrophosphatase which catalyses

$$dUTP \rightarrow dUMP + PPi.$$

This, together with the lack of production of dUDP by the ribonucleotide reductase system (referred to earlier), is all part of a cellular strategy to avoid mis-incorporation of dUTP into DNA. Uracil (cytosine) and thymine can be further broken down to β-alanine and 2-methyl-3-aminopropanoate respectively. This latter compound can be used to monitor DNA turnover, as it is

Box 7.4
Immunodeficiency

The growing interest in immunodeficiency (a feature of AIDS) has led to the finding that decreased adenosine deaminase (ADA) activity is an enzyme marker for severe combined immunodeficiency syndrome (an autosomal recessive inborn error of metabolism). In patients with ADA deficiency, deoxyadenosine builds up, which in turn leads to accumulation of dATP produced by the salvage pathway. T-cell lymphocytes do not mature and this is probably because dATP is an effective inhibitor of ribonucleotide reductase and thus prevents provision of dCTP, dGTP and dTTP for DNA synthesis. Purine nucleoside phosphorylase deficiency is also found in a different immunodeficiency disease affecting T lymphocytes. In this case, deoxyguanosine and thus dGTP accumulates, CDP reduction is inhibited and the lack of dCTP again impairs DNA synthesis. The infusion of deoxyadenosine, or deoxyguanosine or of deoxythymidine, can be cytotoxic although, in fact, they have been used in cancer treatment. In each case, salvage pathways enable the deoxyribonucleoside triphosphate to accumulate, thus inhibiting ribonucleotide reductase. With deoxyguanosine and deoxythymidine, cytotoxicity can be, at least in part, reversed by use of deoxycytidine as this replenishes the dCTP pool.

Acquired immunodeficiency syndrome (AIDS) is caused by HTLV, human T-cell lymphotrophic virus. Azidothymidine (AZT) is used in the treatment of AIDS and may prevent progression of the disease if given at an early stage.

Verma, I.M. (1990) Gene therapy. *Scientific American*, **263**(5), 34–41. Clinical trials of lymphocyte therapy for ADA deficiency are already underway.

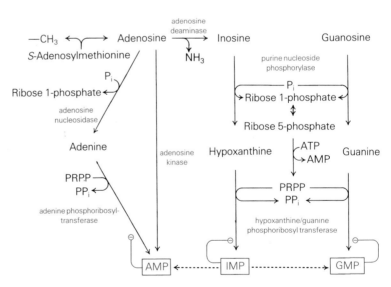

Fig. 7.14 Pathways of purine nucleotide salvage from the nucleosides or free bases. The key enzymes involved are: ADA, adenosine deaminase; purine nucleoside phosphorylase; APRTase, adenine phosphoribosyl-transferase; HGPRTase, hypoxanthine/guanine phosphoribosyltransferase. Product feedback inhibition (see also Fig. 7.8) is shown in red. Although PRPP can itself be salvaged from ribose 1-phosphate as indicated in the figure, this route is probably of only minor quantitative significance.

excreted without further metabolism in urine. Its concentration increases in urine after, for example, radiation damage.

Pyrimidine salvage involves the activity of a phosphoribosyltransferase, which seems to be a separate enzyme from the orotate phosphoribosyl-transferase of the *de novo* pathway. It can salvage uracil, and can also use orotate, thymine and 5-fluorouracil as substrates, but not cytosine, to produce ribonucleoside monophosphates. Pyrimidine-nucleoside phosphorylase and deoxythymidine phosphorylase can utilize ribose 1-phosphate and deoxyribose 1-phosphate, respectively, to salvage nucleosides. There are several distinct pyrimidine nucleoside kinases, e.g. for uridine/cytidine, for deoxycytidine and for deoxythymidine phosphorylation.

Exercise 5

Why would a nuclear safety officer advise that a person's urine samples be checked after a nuclear accident?

7.6 Overview

Purines and pyrimidines needed for nucleic acid biosynthesis can be synthesized *de novo* or by salvage pathways. For the maintenance of life the cell needs to have a full complement of these present as the nucleoside triphosphates. Purines and pyrimidines are required not only for the replication of DNA prior to cell division but also for RNA production and the maintenance of DNA. The enzymes that catalyse the synthesis of purine and pyrimidines are highly evolved and regulated, presumably as a result of evolutionary pressures.

Interference with the pathways of biosynthesis can be lethal to cells. For this reason, a wide variety of compounds which are analogues of nucleotides have been developed by the pharmaceutical industry with a view producing drugs that can kill actively dividing cells. Such drugs are used in tumour therapy.

Answers to Exercises

1. See Fig. 7.1.
2. The input in glucose (or glucose 6-phosphate) which may be completely oxidized with the generation of NADPH and CO_2. However, it is also possible to withdraw ribose 5-phosphate or an intermediate stage.
3. See Table 7.1 and Fig. 7.1(b).
4. In order to make DNA all four deoxynucleoside triphosphates must be present. The enzyme generates all four randomly.
5. To monitor increase in 2-methyl-3-aminopropanoate concentrations.

FILL IN THE BLANKS

1. AMP or adenosine 5'-monophosphate, contains the base adenine or _____ , the sugar _____ and a phosphate. The 6-amino group is derived from _____ , as is one of the ring nitrogens. Two of the purine ring carbon atoms originate from _____ . Amino acids contributing to the ring skeleton include _____ and _____ .

Choose from: glycine, glutamine, aspartate, ribose, 6-aminopurine, N^{10}-formyl tetrahydrofolate.

2. The _____ metabolic route for producing purine nucleotides involves assembly of the _____ ring on the _____ sugar. The necessary enzymes are located in the _____ of the cell. Several amino acids supply atoms for the _____ ring, and these include _____ , _____ and glycine. An initial product is _____ from which AMP and _____ are derived subsequently. The amino acid _____ supplies one-carbon units via _____ _____ .

Choose from: serine, GMP, IMP, aspartate, purine (2 occurrences) *de novo*, ribose, cytosol, glutamine, formyl tetrahydrofolate.

MULTIPLE-CHOICE QUESTIONS

3. CMP, cytidine 5'-monophosphate

A. Contains a pyrimidine base
B. Contains a cyclic phosphate
C. Contains deoxyribose
D. Contains nitrogen atoms derived from glutamine
E. Contains a C_3 unit derived from glutamate.

4. dTTP, deoxythymidine 5'-triphosphate

A. Contains a purine base
B. Contains a methyl group derived from methylene tetrahydrofolate
C. Is a precursor of DNA
D. Is derived from UTP
E. Is derived from CTP

SHORT-ANSWER QUESTIONS

5. The pyrimidine biosynthetic pathway is an energy demanding process. From the reactions of the pathway draw up a balance sheet to show the *net* utilization of ATP molecules required to synthesize *de novo* one molecule of CTP from aspartate, glutamine, bicarbonate and ribose-5-phosphate.

6. In Exercise 2 the net utilisation of ATP for the formation of CTP was requested. Repeat the exercise for GMP synthesis from ribose 5-phosphate, glycine, glutamine, aspartate, bicarbonate and 10-formyltetrahydrofolate.

7. Studies on cells in tissue culture included the introduction of exogenous cytidine into the medium. This nucleoside was radioactively [14C]-labelled in both the pyrimidine and ribose rings. After a suitable incubation period DNA was isolated from the cells and hydrolysed to release the constituent nucleosides. Deoxycytidine so isolated showed the same relative specific radioactivities of deoxyribose and base rings as in the exogenous cytidine. What does this suggest about the mechanism of deoxyribonucleotide biosynthesis?

8. The addition of hydroxyurea to animal cells in culture leads to an accumulation of cells at the G_1/S border. Washing and resuspending the cells in fresh medium relieves the inhibition and allows cells to enter DNA replication in synchrony. The addition of a high concentration ($mmol\ dm^{-2}$) of deoxythymidine can achieve the same effect. Comment on this achievement of synchrony by hydroxyurea or by deoxythymidine.

9. The production of monoclonal antibodies requires making a hybridoma, i.e. a myeloma cell immortal in tissue culture fused with the antibody-generating spleen cell (*Molecular Biology and Biotechnology*, Chapter 10). The fusion or hybridisation process produces a mixture of myeloma-spleen, myeloma-myeloma, and spleen-spleen hybrids as well as leaving unfused spleen and myeloma cells. The selection procedure must eliminate all but the first category and the strategy use is as follows: select the myeloma cells by growing them in medium containing 8-azaguanine,

thus only mutants lacking enzyme E survive. After fusion of these mutant myeloma cells with spleen cells, place in HAT (hypoxanthine-aminopterin-thymidine) selection medium. Hybridomas survive since enzyme E is made by the spleen cell partner; unfused myeloma cells and myeoma-myeloma hybrids die out because they both lack enzyme E; spleen and spleen-spleen hybrids have E but do not survive long in culture.

What is enzyme E and how do hybridomas survive in HAT medium whereas mutant myeloma-myeloma hybrids do not?

8

Lipid biosynthesis

Objectives

After reading this chapter, you should be able to:

☐ describe how a variety of fatty acids and their derivatives are synthesized by biological systems;

☐ explain how triacylglycerols are synthesized in a variety of tissues;

☐ outline some of the metabolic pathways involved in the biosynthesis of phosphoacylglycerols;

☐ appreciate how lipid molecules are transported between their sites of synthesis and utilization;

☐ describe how sphingolipids are synthesized.

8.1 Introduction

Lipids may be classified into three major groups: simple, complex, and polyisoprenoid lipids. A familiarity with their structures is essential before reading this chapter, which reviews the biosynthesis of lipids excluding the polyisoprenoid compounds whose functions and biosynthesis are dealt with in Chapter 9. The biosynthesis of triacylglycerols, phospholipids and glycolipids will be described here, fatty acids are molecular components of all three groups, and so their production will be outlined first.

Triacylglycerols are insoluble in the aqueous environment of the cell and fatty acids have a tendency to form micelles in water. Since fatty acid biosynthesis occurs in the aqueous cytoplasm and synthesis of lipids occurs on membrane surfaces in contact with water, enzyme mechanisms have developed to overcome the problems of lipid solubility.

See Biological Molecules, Chapter 6

☐ A fatty acid is a long-chain hydrocarbon with a terminal carboxyl group. The hydrocarbon chain is generally unbranched and the fatty acid usually has an even number of carbon atoms. For example, palmitic acid is $CH_3(CH_2)_{14}COOH$.

At the pH of the cell (7.0), fatty acids are largely dissociated so that palmitic acid, for example, exists mainly as palmitate, $CH_3(CH_2)_{14}COO^-$.

8.2 The biosynthesis of fatty acids

Fatty acids are components of the diet, but obviously the types of fatty acid obtained in this way will reflect the lipid composition of the foods. If other

Reference Gunstone, F.D., Harwood, J.L. and Padley, F.B. (eds) (1986) *The Lipid Handbook*, Chapman and Hall, London. An extremely good reference book; covers most aspects of the chemistry, biochemistry and industrial and medical uses of lipids.

types of fatty acid, in addition to those in the diet, are required, then the organism must synthesize them. Some organisms, including humans, cannot synthesize some fatty acids and these are the essential fatty acids (p. 162).

The biosynthesis of fatty acids can be thought of as occurring in two stages:

1. The production of fatty acids from simple precursors. **Palmitate** (Fig. 8.1) is the major product of these reactions in most cell types. These reactions are usually designated as the *de novo* pathway.
2. The subsequent modifications of these newly produced fatty acids. The modifications include chain shortening (retroconversion), chain elongation, and the introduction of one or more double bonds into the fatty acid.

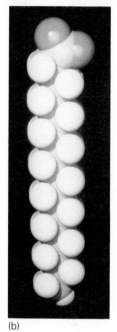

(a)

Palmitate

The biosynthesis of fatty acids has been studied in cell free systems obtained from *E. coli*, yeast and pigeon liver. Plants also synthesize palmitate. Initial experiments used acetate labelled with carbon-14 (^{14}C) and deuterium (^{2}H) as the source of the carbon and hydrogen atoms respectively. These experiments showed a pattern of labelling in the resulting fatty acid consistent with head-to-tail linking of the acetate, i.e.

$$8\ C_2{}^2H_3{}^{14}COO^- \rightarrow C^2H_3 - {}^{14}CH_2 - (C^2H_2 - {}^{14}CH_2)_{12} - C_2{}^2H_2{}^{14}COO^-$$

Since the mechanisms of β-oxidation had already been elucidated it was thought that synthesis of palmitate might be accomplished by a reversal of the degradative steps. It is now known that this occurs only to an insignificant extent, and that a separate biosynthetic pathway is involved (see Table 8.1). This separation of biosynthetic and degradative pathways is usual in biochemical systems, conferring the advantage of each process being controllable independently.

After it was shown that a simple reversal of β-oxidation was not the major pathway of fatty acid biosynthesis, it was demonstrated that *two-* and *three-*carbon molecules contributed carbons to the fatty acid skeleton and especially that carbon dioxide (or HCO_3^- in solution) stimulated this process. The CO_2 is required to carboxylate the C_2 molecule, **acetyl CoA** producing the three carbon compound **malonyl CoA** (Fig. 8.2). This reaction is irreversible under physiological conditions and the enzyme responsible, **acetyl CoA carboxylase**, requires **biotin** (a member of the B group of vitamins, see also Section 3.2) as a prosthetic group.

(b)

Fig. 8.1 (a) Crystals of palmitic acid viewed using polarized light. Courtesy G. Oppermans, Buxton Micrarium, UK. (b) Molecular model of palmitate. Courtesy Dr C. Freeman, Polygen, University of York, UK.

☐ The biosynthesis of fatty acids using a cell-free system did not work in phosphate buffers! No radioactively-labelled acetate was incorporated into long-chain fatty acids. However, if a bicarbonate buffer were substituted for the phosphate buffer then long-chain fatty acids were synthesized. This implies the involvement of carbon dioxide in fatty acid biosynthesis. However, if $H^{14}CO_3^-$ is used as a source of CO_2 for fatty acid biosynthesis, in the presence of non-radioactive acetate, no radioactivity is found in the fatty acids. Thus $^{14}CO_2$ is not incorporated but has a catalytic effect. In fact, the precursor of fatty acids is malonate, a C_3 compound, which then loses CO_2. Malonate is formed from an acetyl unit and CO_2.

Table 8.1 *Principal features of the metabolic pathways of fatty acid synthesis and degradation*

Synthesis	β-Oxidation
Occurs in *cytosol*	Occurs in *mitochondria*
Intermediates linked to acyl carrier protein	Intermediates linked to coenzyme A
Reductant is NADPH + H$^+$	Oxidants are NAD$^+$ and FAD
Enzymes organized into a multifunctional protein	Enzymes not associated in a complex
Citrate necessary for maximum rate of synthesis	Citrate not required
L-3-hydroxyacyl CoA intermediate	D-3-hydroxybutyryl ACP intermediate

Reference Gurr, M.I. and Harwood, J.L. (1991) *Lipid Biochemistry*, 4th edn, Chapman and Hall, London. A good, quite detailed coverage of most aspects of general lipid biochemistry.

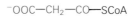

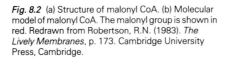

(a)

Fig. 8.2 (a) Structure of malonyl CoA. (b) Molecular model of malonyl CoA. The malonyl group is shown in red. Redrawn from Robertson, R.N. (1983). *The Lively Membranes*, p. 173. Cambridge University Press, Cambridge.

(b)

Production of malonyl CoA

The production of malonyl CoA is an essential first step in fatty acid biosynthesis and requires **biotin carboxyl carrier protein** (BCCP), a dimeric protein of two identical subunits of M_r 22 500; **biotin carboxylase** (BC), also a dimer of two identical subunits (M_r 51 000 each) and **carboxyl transferase** (CT) a tetramer of two pairs of identical subunits with M_r of 30 000 and 35 000 respectively.

The carboxylation of acetyl CoA to give malonyl CoA occurs in two stages. Firstly the biotin prosthetic group of the BCCP is carboxylated. The source of carbon is HCO_3^-. Because the formation of a new C–C bond requires an input of energy the reaction is coupled to the hydrolysis of ATP, so that the overall reaction has a negative free energy change.

$$\text{BC}$$
$$\text{BCCP} + HCO_3^- + \text{ATP} \rightarrow \text{BCCP–COO}^- + \text{ADP} + P_i$$

The second stage in the sequence is the transfer of the carboxyl group from the BCCP–COO$^-$ to the acceptor, acetyl CoA:

$$\text{BCCP–COO}^- + CH_3\text{CO-CoA} \rightarrow \text{BCCP} + {}^-OOCCH_2\,\text{CO-CoA}$$

In animals, the enzymes required for carboxylation are isolated as a tightly bound multienzyme complex of M_r $4–8 \times 10^6$. **Citrate** is required to initiate the polymerization of the monomers (M_r 400 000). This explains why the biosynthesis of fatty acids occurs only in the presence of citrate.

The reactions of fatty acid synthesis

Once malonyl CoA has been generated from acetyl CoA, a sequence of six enzyme-catalysed reactions is repeated in cyclic fashion as many times as is necessary to synthesize the required fatty acid. These reactions are illustrated in Fig. 8.3 and are catalysed by a **fatty acid synthase** (FAS) complex of enzymes. Each turn of the spiral of reactions increases the length of the

□ Multienzyme complexes are highly organized, physically associated collections of enzymes. The enzymes are arranged so that the product of one enzyme is the substrate for the next, thus limiting the distance substrate molecules have to diffuse during a series of metabolic reactions. Multienzyme complexes increase metabolic efficiency and can also provide a hydrophobic environment for the reactions which may be very important when dealing with lipid molecules.

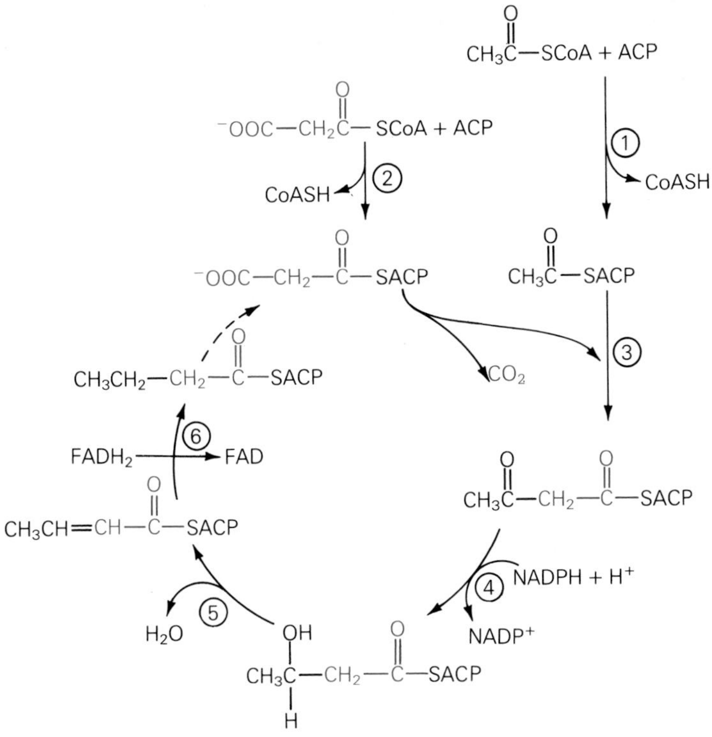

Fig. 8.3 Reactions of fatty acid biosynthesis. Step 1 is catalysed by acetyl transacylase and step 2 by malonyl transacylase. Step 3 is catalysed by β-ketoacyl-ACP synthase and 4 by β ketoacyl ACP-reductase, while steps 5 and 6 are catalysed by enoyl ACP-dehydratase and crotonyl APC-reductase respectively.

growing fatty acid chain by two carbons. The source of this carbon is malonyl CoA, although as already discussed (marginal note on p. 155), the carbon derived from bicarbonate is not added to the fatty acid but is lost as CO_2. The reactions of the cycle are a condensation, a reduction, a dehydration and a second reduction.

A small protein, called the **acyl carrier protein** (ACP), is associated with the FAS, and acts as a carrier of the intermediates of fatty acid synthesis. In *E. coli* the ACP has an M_r 10 000. It contains a sulphydryl group (–SH) attached to a prosthetic group with similar features to coenzyme A (Fig. 8.4). Fatty acid

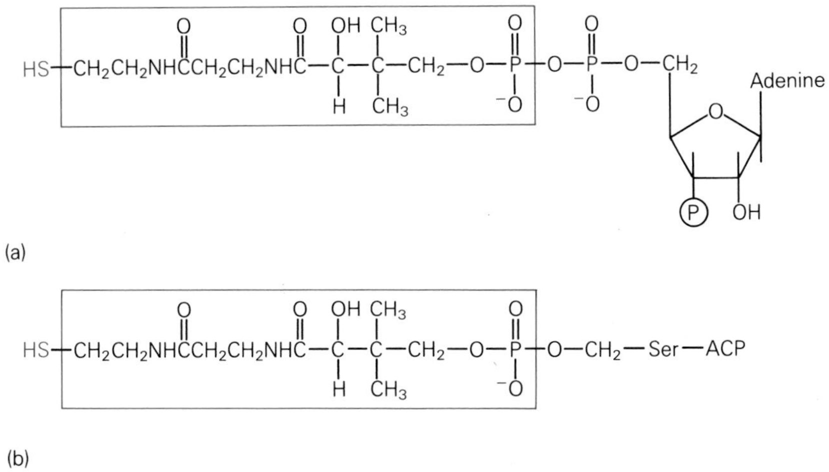

(a)

(b)

Fig. 8.4 Similarity in structure between (a) coenzyme A, and (b) the acyl carrier protein prosthetic group.

Exercise 1

Why does the feeding of double-labelled acetate, $C^2H_3{}^{14}COO^-$, result in head-to-tail labelling of biosynthesized fatty acids

$(C^2H_3{}^{14}CH_2(C^2H_2{}^{14}CH_2)_n C^2H_2{}^{14}COO^-)$?

synthesis is primed when acetyl CoA reacts with ACP, the reaction being catalysed by **ACP acyl transferase**. The acetyl group is subsequently transferred to a second enzyme β-ketoacyl ACP synthase.

Malonyl CoA also reacts with an ACP, the reaction being catalysed by **ACP malonyl transferase**. A condensation reaction with loss of CO_2, now takes place between the bound acetyl and malonyl groups producing **acetoacetyl-ACP**. This reaction is catalysed by β-ketoacyl-ACP synthase. The acetoacetyl-ACP is reduced by NADPH to give **β-hydroxybutyryl-ACP** in a reaction catalysed by **β-ketoacyl-ACP reductase**.

The next reaction of the biosynthetic cycle involves a dehydration of the β-hydroxybutyryl-ACP, catalysed by **β-hydroxybutyryl-ACP dehydratase**, to give **crotonyl S-ACP**. A second reduction step, involving NADPH, catalysed by **enoyl-ACP** reductase converts the crotonyl S-ACP to **butyryl-ACP**.

The cycle of reactions is now complete, and essentially results in two acetyl (C_2) molecules joining together to give a butyryl (C_4) group. The butyryl-ACP can now react with a further malonyl CoA and enter the same sequence of reactions to extend the growing fatty acid by a further *two* carbons. The sequence of reactions is repeated until the required chain length is reached. The completed fatty acid is then cleaved from the FAS complex by a **thioesterase**. The thioesterase has a high activity with C_{16} and C_{18} substrates, thus **palmitate** and **stearate** are the major products of the FAS.

Structure of the fatty acid synthase complex

The FAS of prokaryotes and plants dissociates into individual enzymes when the cell is disrupted, allowing study of the individual steps of fatty acid synthesis. In yeast, avian and mammalian liver, and mammary gland the enzymes are much more tightly bound together and are difficult to separate. They are thought to occur as dimers of multifunctional subunits. The animal enzyme is composed of two identical subunits of M_r 250 000. The yeast enzyme has two different subunits, an α subunit of M_r 213 000 and a β subunit of M_r 203 000. The yeast FAS has a structure $\alpha_6\beta_6$ with an M_r of 2.4×10^6. Figure 8.5 shows an electron micrograph of the FAS complex from the fungus *Neurospora*.

The multifunctional proteins may be arranged as globular domains connected by polypeptide bridges (Fig. 8.6). These globular domains are the sites of the enzymic reactions. In animals it is thought that the two subunits of

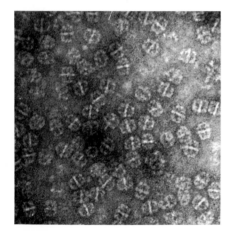

Fig. 8.5 Electron micrograph of fatty acid synthase complex (FAS) from *Neurospora* (× 44 000). Courtesy of Dr N.M. Packter, Department of Biochemistry and Molecular Biology, University of Leeds, UK.

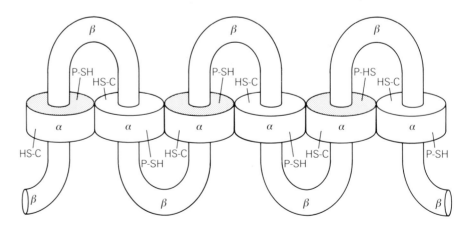

Fig. 8.6 Arrangement of the globular domains of FAS. Redrawn from Wakil, S., Stoops, J. and Joshi, V. (1983) *Annual Review of Biochemistry* **52**, 537–79.

Reference Harwood, J.L. (1988) Fatty acid metabolism. *Annual Review of Plant Physiology and Plant Molecular Biology*, **39**, 101–38. An extensive review of fatty acid synthesis and oxidation. Naturally, given the journal, the emphasis is on activities in plants.

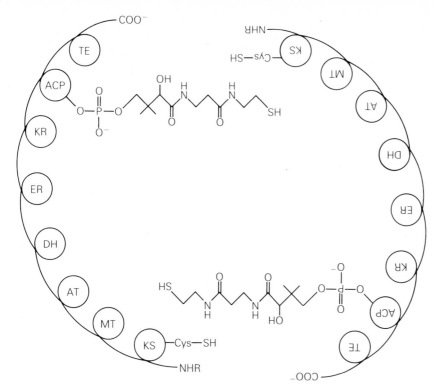

Fig. 8.7 Possible arrangement of the dimers of animal fatty acid synthase (FAS). The abbreviations used are AT, acetyl transacylase; MT, malonyl transacylase; KS, β ketoacyl synthase; KR, β ketoacyl reductase; DH dehydratase; EH, enoyl reductase; TE, thioesterase (see Fig. 8.2). Redrawn from Wakil, S., Stoops, J. and Joshi, V. (1963) *Annual Review of Biochemistry*, **52**, 537–79.

the FAS dimer are arranged in a head-to-tail fashion (Fig. 8.7). Within the FAS complex, the intermediates of fatty acid synthesis are passed from enzyme to enzyme, the product growing in length during the process. The proteins in the subunits have a high proportion of hydrophobic amino acids residues. This presumably provides a non-aqueous environment for the synthesis of the fatty acids.

The source of acetyl CoA and NADPH for fatty acid biosynthesis

The synthesis of palmitate by the fatty acid synthase complex may be summarized by the equation:

$$\text{Acetyl CoA} + 7\,\text{Malonyl CoA} + 14\,\text{NADPH} + 7\text{H}^+ \rightarrow$$
$$\text{Palmitate} + 7\text{CO}_2 + 14\,\text{NADP}^+ + 8\,\text{CoA} + 6\text{H}_2\text{O}$$

Since malonyl CoA is synthesized from acetyl CoA a supply of acetyl CoA and NADPH is necessary for the production of fatty acids.

Fatty acids are synthesized in the cytosol whereas acetyl CoA is produced in the mitochondrion from pyruvate by the pyruvate dehydrogenase complex. The mitochondrion is impermeable to acetyl CoA and a shuttle mechanism is necessary to move it to the cytosol. This mechanism (Fig. 8.8) involves the synthesis of citrate from oxaloacetate and acetyl CoA. The citrate is exported from the mitochondria to the cytosol. Here it is cleaved by the enzyme citrate lyase generating acetyl CoA which may be used in fatty acid synthesis:

See *Energy in Biological Systems*, Chapter 4

$$\text{Citrate} + \text{ATP} + \text{CoA} \rightarrow \text{Acetyl CoA} + \text{ADP} + \text{P}_i + \text{Oxaloacetate}$$

Acetyl CoA is therefore transported from the mitochondria to the cytosol at the expense of ATP.

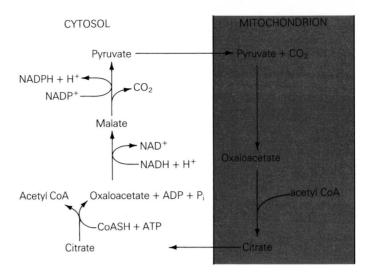

Fig. 8.8 The transfer of acetyl CoA from the mitochondrion to the cytosol as citrate. The reducing power of NADH is converted to NADPH at the same time.

NADPH required for fatty acid biosynthesis can be produced from the excess oxaloacetate accumulated in the cytosol from the bypass. It has been estimated that about 40% of the NADPH required for palmitate biosynthesis is produced by the mechanism; the remainder is provided by the pentose phosphate pathway. Oxaloacetate is reduced to malate (Fig. 8.8) using NADH generated by glycolysis. The reaction is catalysed by malate dehydrogenase.

$$\text{Oxaloacetate} + \text{NADH} + \text{H}^+ \rightarrow \text{malate} + \text{NAD}^+$$

Malate is decarboxylated to pyruvate by malic enzyme generating NADPH:

$$\text{Malate} + \text{NADP}^+ \rightarrow \text{Pyruvate} + \text{NADPH} + \text{H}^+ + \text{CO}_2$$

The pyruvate can diffuse back into the mitochondrion where it is carboxylated to regenerate oxaloacetate.

Regulation of fatty acid biosynthesis

Acetyl CoA carboxylase catalyses the formation of malonyl CoA, an irreversible reaction which commits the cell to fatty acid synthesis. In animals the major source of acetyl CoA is citrate, while in *E. coli* it is derived from pyruvate. In the absence of citrate there is very little carboxylation of acetyl CoA (see earlier) and synthesis of fatty acids ceases. In the long term, the amount of acetyl CoA carboxylase in the cell is affected by diet, and by the hormones **insulin** and **thyroxine**. In all cases regulation is achieved by regulating the rate of enzyme (protein) synthesis. For example, fasting animals show elevated levels of acetyl CoA carboxylase shortly after being fed a carbohydrate-rich meal. This allows the excess carbohydrate to be converted to fatty acids and subsequently stored as triacylglycerols.

Little is known about the short-term regulation of the FAS complex of animals. Long-chain fatty acyl CoA derivatives are known to inhibit the FAS, while fructose 1,6-bisphosphate and NADPH stimulate activity, increasing the rate of biosynthesis of fatty acids.

Reference Wakil, S. *et al.* (1983) Fatty acid synthesis and its regulation, *Annual Reviews of Biochemistry*, **52**, 537–79. Rather old, but good stuff!

Elongation of fatty acids

Palmitate is the principal product of the FAS (see earlier) and a major component of triacylglycerols. However, some long-chain (C_{20}, C_{24}) fatty acids are required by the body. These are not produced by the FAS but by the elongation of pre-existing fatty acids. There are two elongation systems. One is located in the endoplasmic reticulum, the other in mitochondria. The two systems differ in a number of respects as listed in Table 8.2. The endoplasmic reticulum elongation system does not contain ACP but otherwise the reactions are the same as those of the cytosolic FAS. The mitochondrial elongation system does not require CO_2, but simply reverses the final reactions of β-oxidation (*Energy in Biological Systems*, Chapter 6). Both elongation systems also work with unsaturated fatty acids.

☐ The role of the endoplasmic reticulum in lipid biosynthesis has largely been elucidated by studies on microsomes. Microsomes are small membrane-bound structures formed when the cell is homogenized. They are artefacts derived largely from the endoplasmic reticulum and Golgi apparatus membranes when subjected to mechanical breakage (*Cell Biology*, Chapter 1).

Table 8.2 *Fatty acids may be elongated by a mitochondrial or endoplasmic reticulum pathway. The mitochondrial system uses acetyl CoA as the carbon source and NADH. The endoplasmic reticulum pathway uses malonyl CoA and NADPH*

Site	Substrate	Cofactor	Product
Mitochondria	Saturated acyl CoA + acetyl CoA	NADH	Fatty acid increased in chain length by two carbons
	Unsaturated acyl CoA	NADH	Unsaturated fatty acid increased in chain length by two carbons
Endoplasmic reticulum	Saturated acyl CoA + malonyl CoA	NADPH	Fatty acid increased in chain length by two carbons
	Unsaturated acyl CoA	NADPH	Unsaturated fatty acid increased in chain length by two carbons

Biosynthesis of unsaturated fatty acids

In animals, unsaturated fatty acids are produced by desaturation of preformed saturated fatty acids. Unsaturated fatty acids are required to provide triacylglycerols with low melting points. These are necessary since triacylglycerols containing only saturated fatty acids are solids at body temperature, are poorly incorporated into transport lipoproteins and hydrolysed by lipases at relatively slow rates. This may seem an unusually complex method of synthesizing unsaturated fatty acids given that the FAS complex introduces a double bond during the stage of the cycle catalysed by β-hydroxybutyryl S-ACP dehydratase. It must be remembered, however, that the product of this enzyme contains a *trans* double bond while naturally-occurring unsaturated fatty acids have a *cis* configuration. Thus a separate pathway is required to synthesize unsaturated fatty acids with the *cis* structure.

Mono-unsaturated (containing *one* double bond) fatty acids are produced in the endoplasmic reticulum of the cell by an enzyme complex whose mode of action is not fully understood. This **desaturase complex** contains cytochrome b_5 and two enzymes **cytochrome b_5 reductase**, and a so called **cyanide sensitive factor** (CSF). The CSF may be the enzyme responsible for introducing the double bond. A possible arrangement of these factors is shown below:

☐ Δ is the symbol used to represent a double bond. Δ^9 indicates a double bond between carbons 9–10 of the compounds, *c* indicates *cis* configuration, while *t* indicates a *trans* isomer.

Reference Jeffocoat, R. (1979) The biosynthesis of unsaturated fatty acids and its control in mammalian liver, in *Essays in Biochemistry*, **15**, 1–36, Campbell, P.N. and Marshall, R.D. (eds), Academic Press, New York. A good review of the biosynthesis of unsaturated fatty acids.

□ Mixed-function oxygenases catalyse the insertion of one atom of oxygen into an organic substrate; the other atom from the oxygen molecule is reduced to water. This can be summarized:

$$AH + XH_2 + O_2 \rightarrow AOH + H_2O + X$$

where AH is the substrate, XH_2 is an electron donor and AOH is the oxidized substrate. An example of a mixed-function oxidase is phenylalanine hydroxylase which converts phenylalanine to tyrosine (Section 1.4).

Four desaturases are known in animal tissues introducing double bonds at the Δ^4, Δ^5, Δ^6 and Δ^9 positions.

These enzymes require O_2 and are examples of mixed function oxidases. The Δ^9 is introduced preferentially, producing oleate ($9cC_{18+1}$) from stearate ($C_{18:0}$). If a second double bond is introduced then animal enzymes can only introduce this bond between the carboxyl end of the fatty acid and the Δ^9 bond. Animals cannot introduce double bonds between the Δ^9 and the terminal methyl carbon, although plants can.

Skipped methylene double bonds

Polyunsaturated fatty acids have their double bonds arranged in such a way as to maintain a 'skipped methylene' or unconjugated arrangement, i.e.:

$$-CH=CH-CH_2-CH=CH-$$

Using a combination of desaturation and elongation it is possible to build up *four* families of polyenoic fatty acids. They are named after the parent fatty acid from which they are derived (Table 8.3).

As can be seen from Fig. 8.9 **linolenic** and **linoleic acids** contain double bonds between the methyl end of the molecule and Δ^9. Since these bonds cannot be introduced by animal desaturases they must be obtained in the diet, and are examples of essential fatty acids. Although these compounds are similar to vitamins from a nutritional point of view, quite large amounts are required in the diet. The essential fatty acids are required to produce a series of lipid derivatives which produce local physiological effects. These derivatives are the **prostaglandins, thromboxanes and leukotrienes** which are collectively termed the **eicosanoids**.

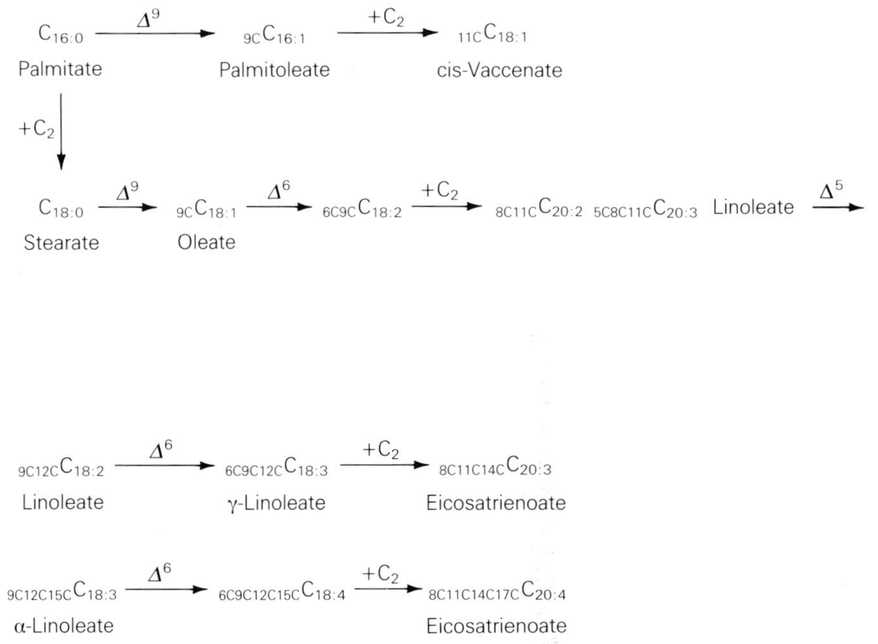

Fig. 8.9 The biosynthesis of polyunsaturated fatty acids by elongation and desaturation.

Exercise 2

Draw the structures of oleate and linolenate.

Prostaglandins: *originally classified as ether-soluble (the E series), or phosphate-soluble (the F series, from the German* fosphat*). This is a peculiar nomenclature because it is not based on the structure of the compound.*

α-β Unsaturated ketone: *the structure contains a ketone functional group immediately adjacent to a carbon–carbon double bond.*

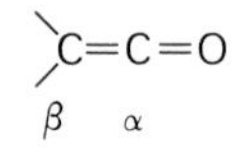

$$\begin{array}{c} \diagdown \\ C=C=O \\ \diagup \\ \beta \quad\ \alpha \end{array}$$

Table 8.3 *Families of polyunsaturated fatty acids. Each family will contain polyenolate formed by elongation and desaturation of the parent fatty acid*

Family name	Chain structure at methyl end
Palmitoleate	$CH_3(CH_2)_5$—C=C— (with H, H above the double-bonded carbons)
Oleate	$CH_3(CH_2)_7$—C=C— (with H, H above the double-bonded carbons)
Linoleate	$CH_3(CH_2)_4$—C=C— (with H, H above the double-bonded carbons)
Linolenate	CH_3CH_2C=C— (with H, H above the double-bonded carbons)

* Since C_2 addition or desaturation between the carboxyl end and the first double bond will not affect the structures shown above, all members of the family will have the same methyl-terminal structure.

8.3 The eicosanoids

Eicosanoids are local acting, short-lived, pharmacologically-active compounds having powerful effects on the cells which produce them and on adjoining cells.

PROSTAGLANDINS are produced in trace amounts by all animal cells. The precursors of **prostaglandins** are formed from 20-carbon fatty acids containing three, four or five double bonds (Fig. 8.10). There are three series of prostaglandins (named because they were originally isolated from the seminal fluid produced by the prostate gland) called the A, E, and F series. These groups are distinguished by the functional group in the cyclopentane ring. The A series have an α,β-**unsaturated ketone**, the E series a β-**hydroxy ketone** and the F series a **1,3-diol**. The subscript numeral indicates the number of double bonds in the side-chain and the α indicates the configuration of the C-9 hydroxyl group.

The prostaglandins derived from arachidonate, C_{20}-5,8,11,14, are biologically the most important in humans. Linoleate ($C_{18:2}$) is the principal source of the C_{20} fatty acid and is converted by elongation and desaturation into the prostaglandin precursors (see Fig. 8.9). The cyclopentane ring is introduced into the prostaglandin by a process of oxidative cyclization catalysed by a **prostaglandin synthetase complex** (prostaglandin endoperoxidase synthetase or 'cycloxygenase'). The ring system is formed by joining carbons 8 and 13 (Fig. 8.11). The reaction requires molecular oxygen and is inhibited by a number of drugs. Figure 8.12 shows the mechanism of oxygen addition.

Prostaglandins have diverse physiological effects, affecting the cardiovascular, gastrointestinal and the reproductive systems as well as modulating renal functions.

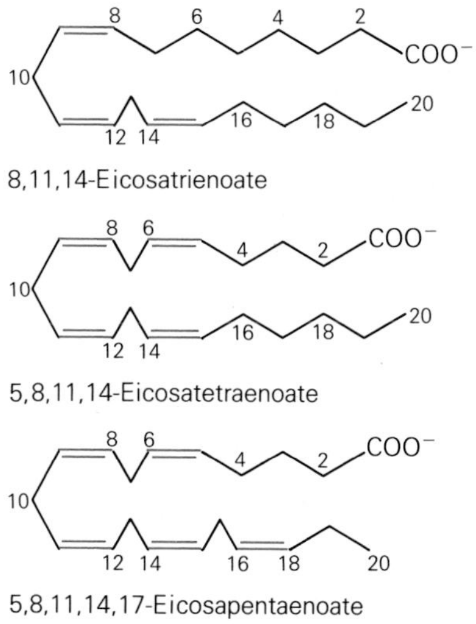

THROMBOXANES are very short-lived derivatives of prostaglandins with a six-membered oxygen-containing ring (oxane) instead of the cyclopentane ring of the prostaglandins (Fig. 8.13). **Thromboxane synthetase**, the enzyme responsible for thromboxane biosynthesis, is particularly active in blood platelets and lung tissue. It is also found in spleen, kidney, leucocytes and

8,11,14-Eicosatrienoate

5,8,11,14-Eicosatetraenoate

5,8,11,14,17-Eicosapentaenoate

Fig. 8.10 The twenty-carbon fatty acid precursors of prostaglandins.

β-hydroxyketone: *the structure has a ketone and alcohol functional groups separated by a methylene group.*

OH   OH
| |
—CH—CH₂—CH—

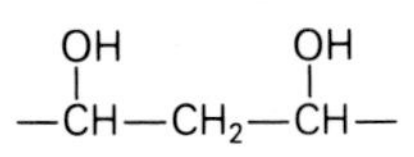

1,3-Diol: *the structure has two alcohol groups separated by a methylene group.*

OH   O
| ||
—CH—CH₂—C—H
   β

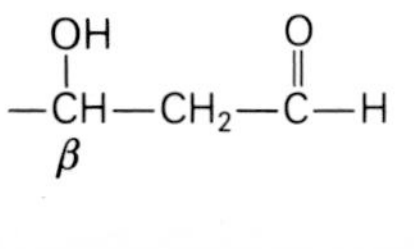

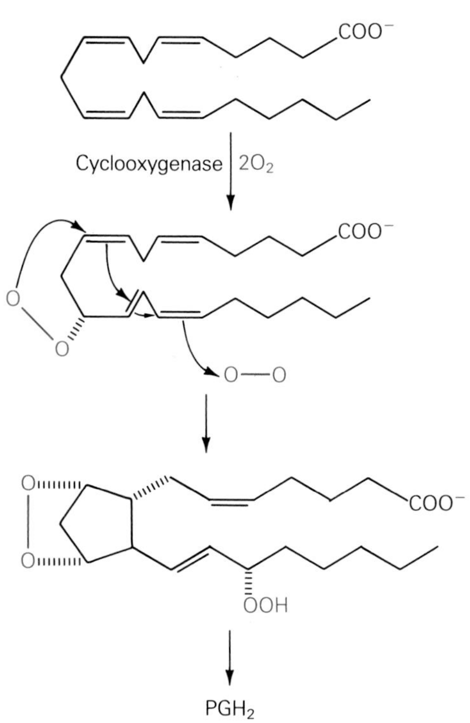

8,11,14-Eicosatrienoate
(Diheme-γ-linoleiate)

PGE$_1$

PGF$_{1\alpha}$

Cyclooxygenase | 2O$_2$

5,8,11,14-Eicosatrienate chrachidenate
(Arachidonate)

PGE$_2$

PGF$_{2\alpha}$

5,8,11,14 17-Eicosapentaenoate

PGE$_3$

PGF$_{3\alpha}$

PGH$_2$

Fig. 8.12 The mechanism of cyclo-oxygenase.

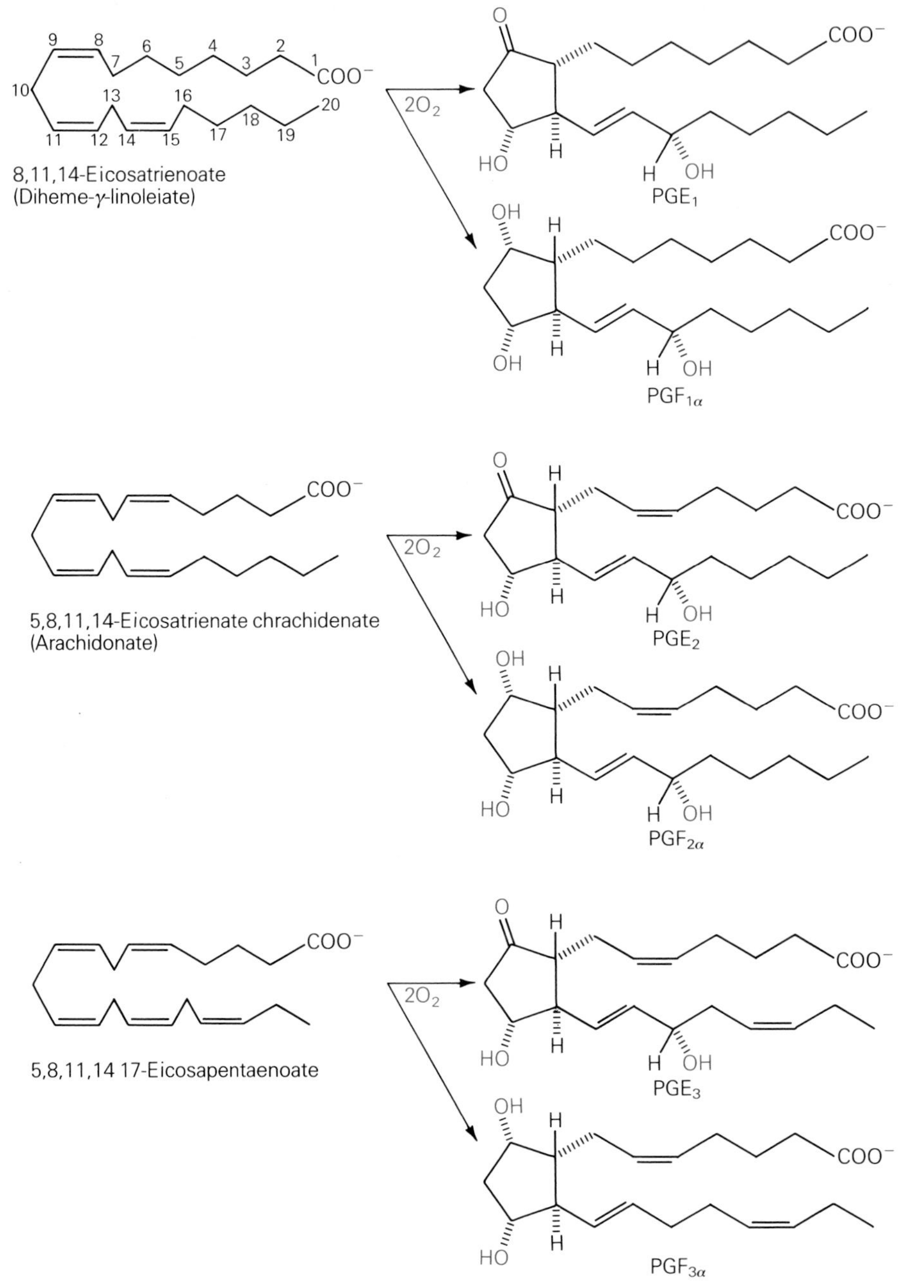

Fig. 8.11 The biosynthesis of prostaglandins.

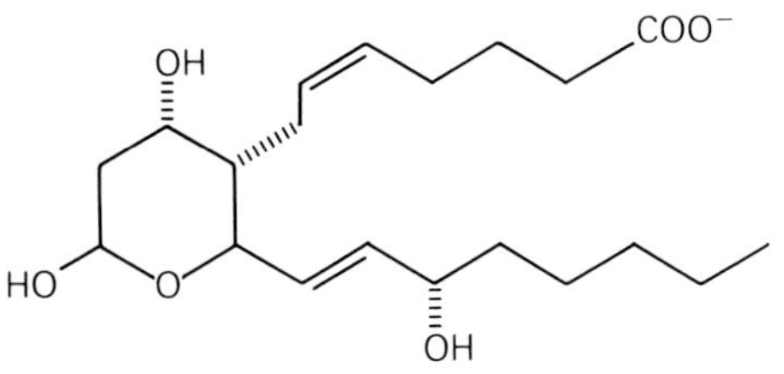

Fig. 8.13 The structure of thromboxane A$_2$.

some human blood vessels. The conversion of 5, 8, 11, 14-eicosatetraenoate into prostaglandin H$_2$, an intermediate common to the biosynthesis of thromboxanes and prostaglandins, is shown in Fig. 8.14.

It is difficult to study the physiological effects of thromboxanes because of their short life in the body, but the use of isolated organs indicates they stimulate vasoconstriction. They also promote platelet aggregation and thromboxane A$_2$ is produced by platelets in response to a stimulus for example by a 'foreign surface'. Thromboxane A$_2$ promotes broncho-constriction in the lungs and constriction of the pulmonary vascular bed.

Reference Johnson, M., Carey, F. and McMillan, R.M. (1983) Alternative pathways of arachidonate metabolism; prostaglandins, thromboxane and leukotrienes, in *Essays in* *Biochemistry*, **19**, 40–141, Campbell, P.N. and Marshall, R.D., eds. Academic Press, New York. A comprehensive review of eicosanoid synthesis.

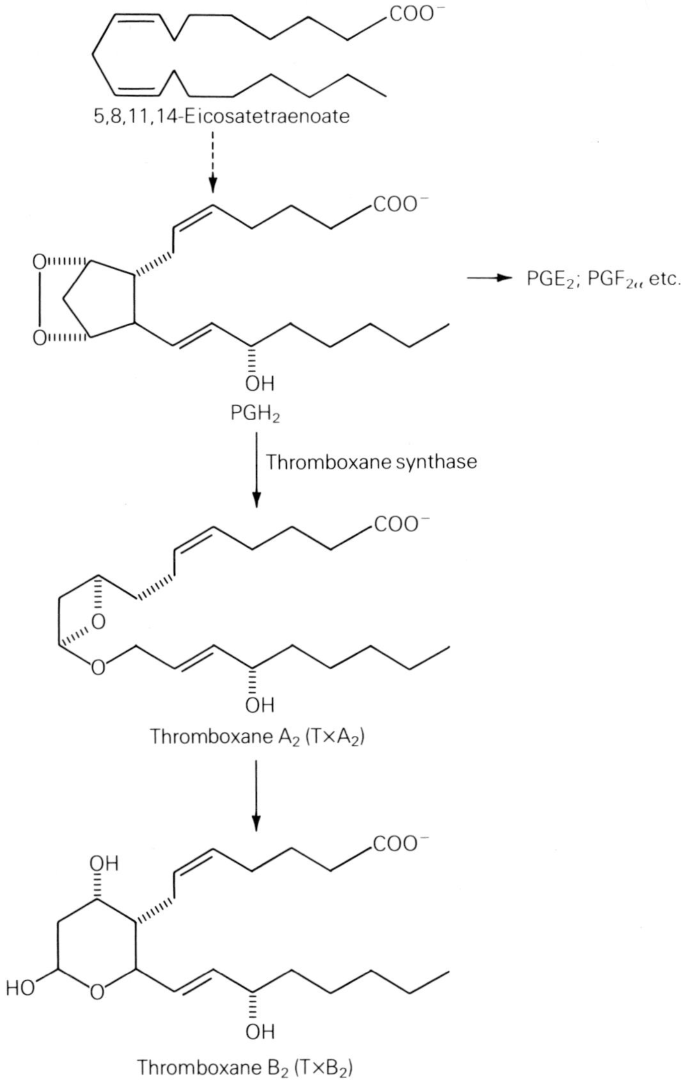

Fig. 8.14 The biosynthesis of thromboxane A₂ from arachidonate.

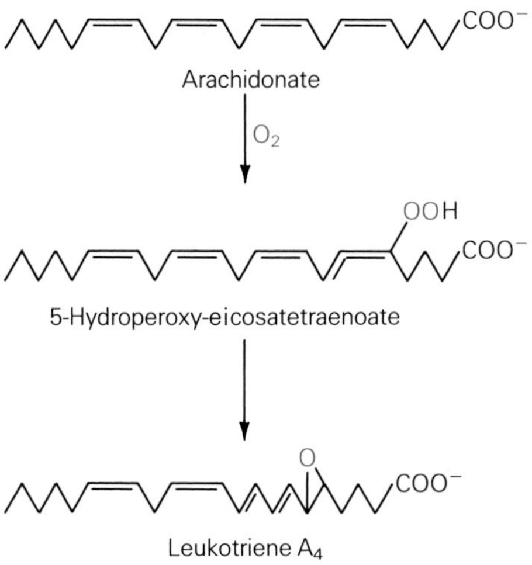

Fig. 8.15 Biosynthesis of leukotriene A₄ by C_5 oxidation of arachidonate. Redrawn from Johnson, M. *et al.* (1983) *Essays in Biochemistry* **19**, 85.

LEUKOTRIENES are produced by the action of the enzyme **5-lipoxygenase** which introduces oxygen into carbon-5 of arachidonate (Fig. 8.15). Leukotrienes are potent vasoconstrictors and thus reduce coronary blood flow.

PROSTACYCLINS were found during the screening of a variety of tissues for their ability to synthesize thromboxanes. It was found that vascular tissues converted prostaglandin H_2 into an unstable metabolite with biological properties antagonistic to those of thromboxane A_2. This metabolite was later given the name prostacyclin or PGI₂ (Fig. 8.16).

Prostacyclin synthetase is concentrated principally in vascular tissue. PGI₂ has a half-life of a few minutes in the blood. It is a potent vasodilator, stimulating a lowering of blood pressure by causing vasodilation in the kidney, gut, mesentery and skeletal muscle.

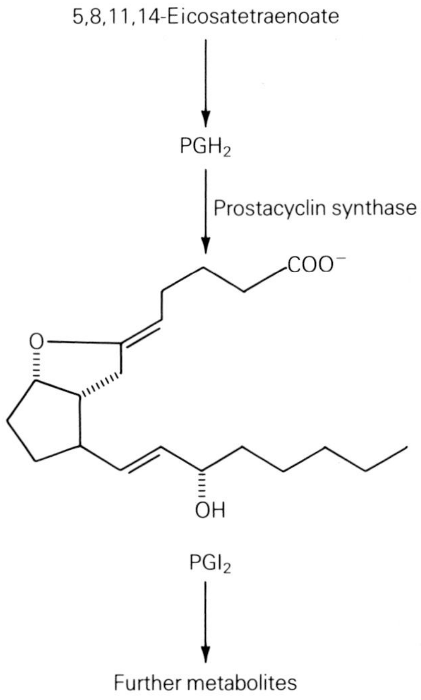

Fig. 8.16 Structure of prostacyclin PGI₂.

A number of non-steriodal, antiinflammatory agents (NSAID) operate through blocking prostanoid biosynthesis. Examples are aspirin, phenylbutazone and indomethacin. These drugs inhibit prostaglandin biosynthesis by acting on the cyclo-oxygenase enzyme, competing with arachidonate for the active site of the enzyme. The acid groups (–COO⁻–) of the NSAID binds to the same site as the –COO⁻ of arachidonate, while the hydrophobic groups of the NSAID bind to a hydrophobic binding site on the enzyme.

Aspirin irreversibly acetylates a serine residue of the enzyme by binding to a group at the active site:

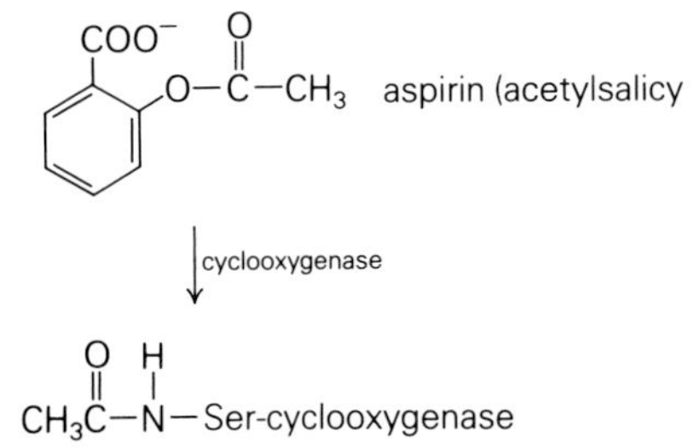

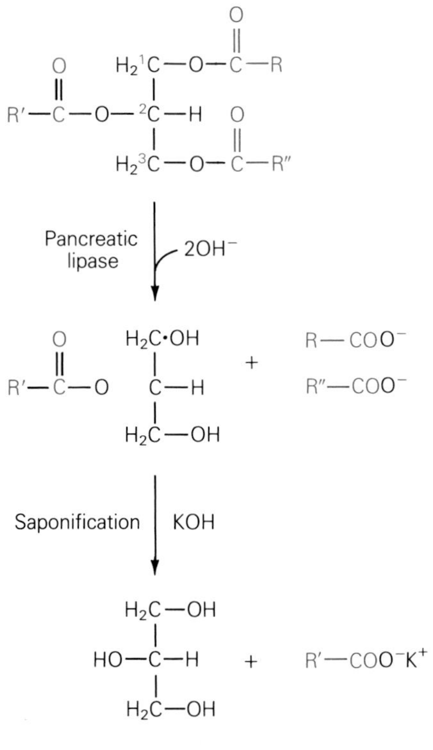

$$COO^- \quad O$$
aspirin (acetylsalicy

$$O-\overset{\overset{O}{\|}}{C}-CH_3$$

cyclooxygenase

$$CH_3\overset{\overset{O}{\|}}{C}-\overset{\overset{H}{|}}{N}-Ser\text{-}cyclooxygenase$$

8.4 The biosynthesis of triacylglycerols

Triacylglycerols are highly concentrated stores of metabolic energy and serve as storage lipids. They are a major reserve of energy, mobilized when the organism is energy deficient. Triacylglycerols are esters of glycerol with three molecules of fatty acid. They are synthesized principally in the liver, gut and adipocytes of animals. The three fatty acids attached to glycerol are usually different. It is possible to study the distribution of fatty acids in triacyl-glycerols using the enzyme **pancreatic lipase**. This enzyme selectively hydrolyses the esters linkages at carbons 1 and 3 of the glyceride. Subsequent saponification would then release the central fatty acid at carbon 2 (Fig. 8.17).

The stepwise release of the fatty acids can be followed by either thin layer chromatography or gas–liquid chromatography of the products (Fig. 8.18). Analyses of this sort have indicated that stored triacylglycerols contain a high proportion of unsaturated fatty acids such as oleate, at position 2 of the glycerol, while carbons 1 and 3 are often esterified with palmitate. The identities of the fatty acids at the various positions depends, of course, on the specificities of the enzymes involved in the biosynthesis of the triacylglycerols. Two metabolic pathways are involved in the production of triacylglycerols: the *de novo* pathway and the dihydroxyacetone phosphate pathway.

The *de novo* pathway for triacylglycerol biosynthesis utilizes glycerol 3-phosphate as the immediate precursor of the triacylglycerol (Fig. 8.19).

Fig. 8.17 Hydrolysis of a typical triacylglycerol by pancreatic lipase, followed by saponification.

Reference Kuksis, A. (ed.) (1978) *Fatty Acids and Glycerides*, Plenum Press, New York. An old but very informative account of many aspects of the structure, metabolism, synthesis and separation of fatty acids and glycerides.

Reference Macrae, A.R. (1989) The versatility of lipases for industrial uses. *Trends in Biochemical Sciences*, **14**, 125–6. A two-page synopsis covering the applications of lipase enzymes in industry.

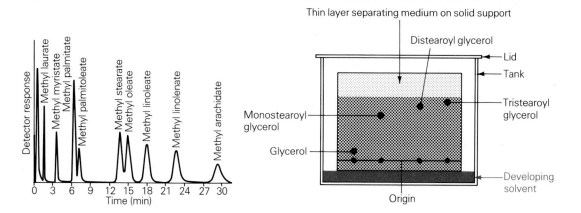

Fig. 8.18 (a) Separation of methyl esters of fatty acids by gas–liquid chromatography. Redrawn from Bohinski, R.C. (1987) *Modern Concepts in Biochemistry*, 5th edn, p. 420. Allyn and Bacon, Boston. (b) Separation of acylglycerols by thin-layer chromatography. Redrawn from Yudkin, M. and Offord, R. (1975) *Comprehensible Biochemistry*, p. 543. Longman, London.

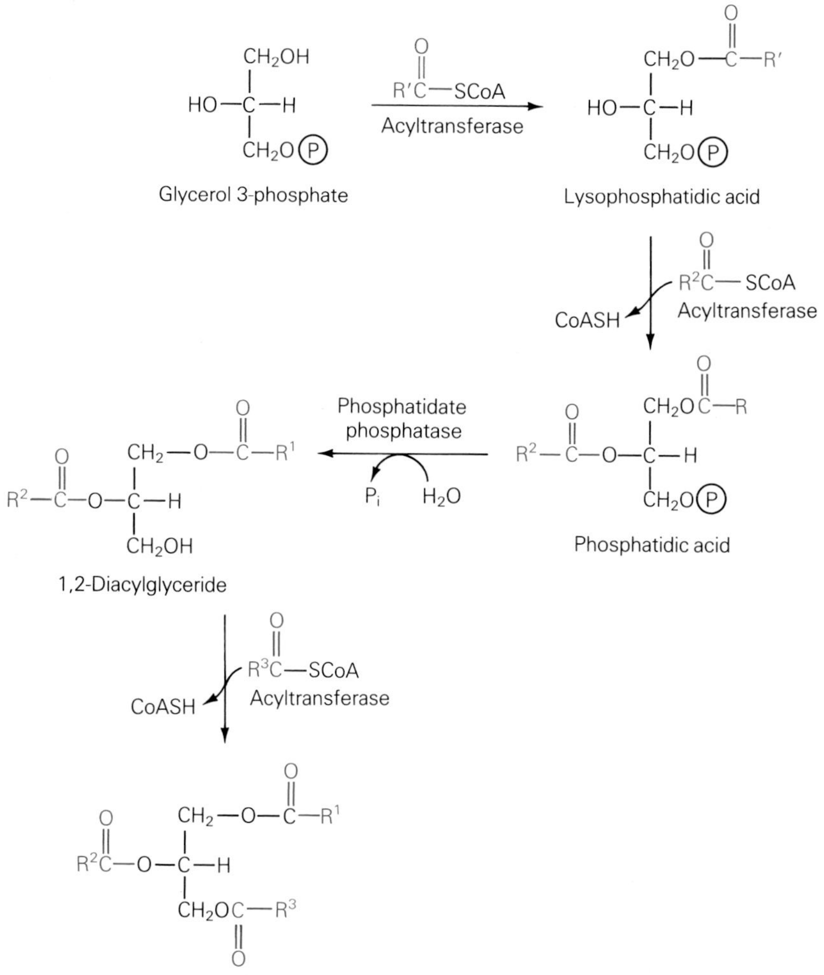

Fig. 8.19 The *de novo* pathway for the biosynthesis of a triacylglycerol.

Exercise 3

Draw one structure of a triacylglycerol containing two molecules of palmitate and one of stearate. It is possible to construct other triacylglycerols using the same fatty acid molecules?

The glycerol 3-phosphate is sequentially esterified with two fatty acyl CoA molecules to give **phosphatidate**. Phosphatidate is also an intermediate in the biosynthesis of complex lipids. Removal of the phosphate by a **phosphatase** generates a diacylglycerol. Subsequent esterification of the diacylglycerol gives the complete triacylglycerol. These reactions are essentially irreversible because glycerol 3-phosphate and fatty acyl CoA are metabolically active forms of glycerol and fatty acids, respectively.

Glycerol 3-phosphate may be produced directly by the phosphorylation of glycerol by the enzyme **glycerol kinase** in the presence of ATP.

$$\text{Glycerol + ATP} \xrightarrow[\text{Mg}^{2+}]{\text{glycerol kinase}} \text{Glycerol 3-phosphate + ADP}$$

However, it may also be produced from dihydroxyacetone phosphate, an intermediate in glycolysis.

$$\text{Dihydroxyacetone-phosphate} + \text{NADH} + \text{H}^+ \xrightarrow[\text{dehydrogenase}]{\text{glycerol 3-phosphate}}$$

$$\text{Glycerol 3-phosphate} + \text{NAD}^+$$

White adipose tissue, however, has only low levels of glycerol kinase, and therefore most of the glycerol 3-phosphate must be obtained from glycolysis by the second mechanism.

The dihydroxyacetone phosphosphate pathway is found in rat intestinal cells and, indeed, the endoplasmic reticulum of these cells lacks glycerol 3-phosphate dehydrogenase. Dihydroxyacetone phosphate is therefore used directly for the formation of acylglycerols. The essential features of this pathway are shown in Fig. 8.20.

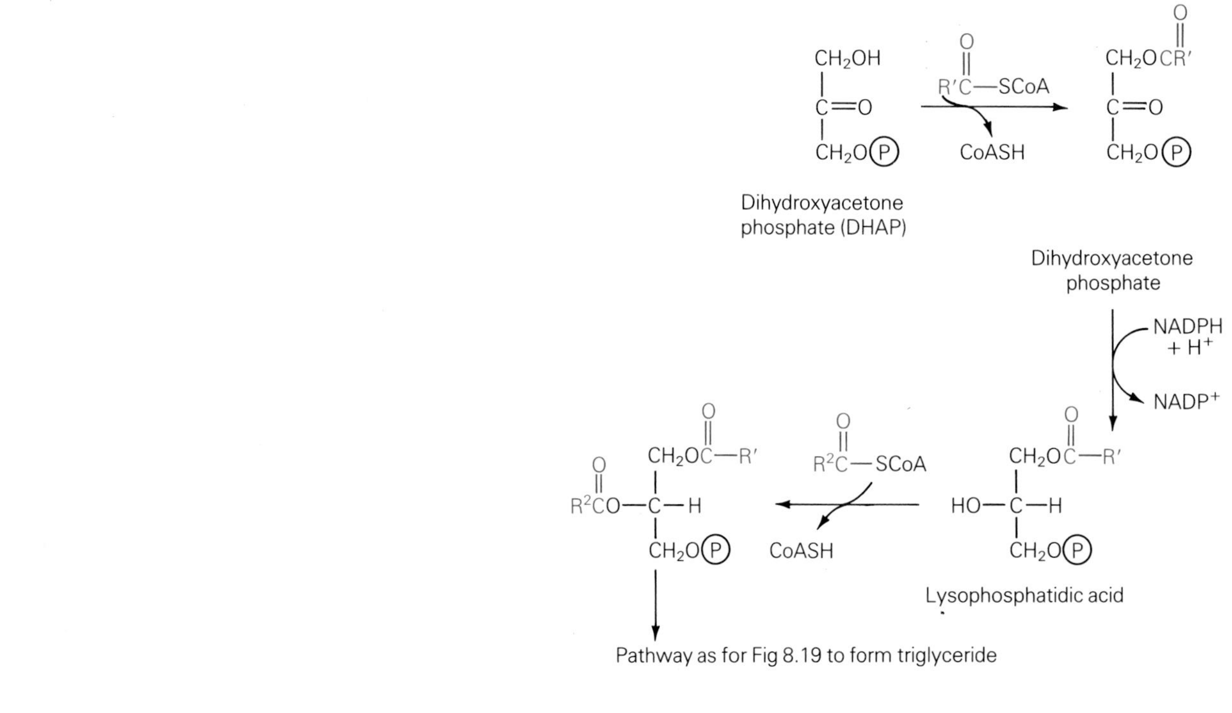

Fig. 8.20 The dihydroxyacetone phosphate pathway for the biosynthesis of a triacylglycerol.

The regulation of triacylglycerol biosynthesis

The control of the biosynthesis of triacylglycerols is largely achieved by regulating the levels of glycerol 3-phosphate within the cells. The major pathways involved in producing this intermediate are glycolysis and the pentose phosphate pathway which also produces NADPH. Glycerol 3-phosphate can also be produced, of course, by direct phosphorylation of glycerol. Thus the concentration of glycerol 3-phosphate in the cell is influenced by the activities of enzymes involved in the above pathways, and also those regulating gluconeogenesis and lipogenesis.

8.5 The biosynthesis of phosphoacylglycerols

Phosphoacylglycerols are major components of all biological membranes (Section 1.6) as well as being components of serum lipoproteins (Section 8.6) and bile and lung surfactants. The major **phosphoacylglycerols** found in eukaryotic membranes are **phosphatidylethanolamine** (cephalin), **phosphatidylcholine** (lecithin) and **phosphatidylserine** (Fig. 8.21). The diacylglycerol portion of the phosphoacylglycerols is derived from phosphatidate by the action of phosphatidate phosphatase: this enzyme cleaves the phosphate from the terminal alcohol group. The enzymes which catalyse the synthesis of phosphoacylglycerols are found in membranes of the endoplasmic reticulum or of mitochondria. Phosphoacylglycerol synthesis occurs on the cytosolic side of these membranes.

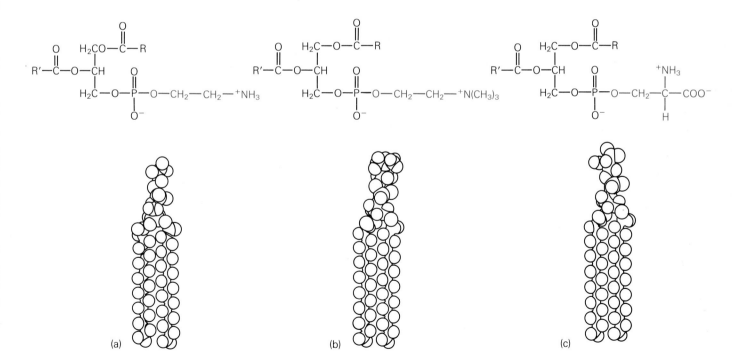

Fig. 8.21 Structures and molecular models of (a) phosphatidylethanolamine, (b) phosphatidylcholine, and (c) phosphatidylserine. All molecular models are shown with saturated acyl chains and are redrawn from Robertson, R.N. (1983) *The Lively Membranes*, p. 25. Cambridge University Press, Cambridge.

The biosynthesis of phosphatidylethanolamine

Ethanolamine is phosphorylated by the enzyme **ethanolamine kinase** (Fig. 8.22). The product, **phosphorylethanolamine**, then reacts with **cytosine triphosphate** (CTP). The resulting **CDP–ethanolamine** acts as a carrier of the ethanolamine for transfer to a diacylglycerol, giving the completed phosphatidylethanolamine. This sequence of reactions occurs on the endoplasmic reticulum membrane. In addition, a minor pathway involving the biosynthesis of phosphatidylethanolamine by the decarboxylation of phosphatidylserine occurs in the mitochondria of liver (Fig. 8.23).

The biosynthesis of phosphatidylcholine

There are two pathways for the biosynthesis of phosphatidylcholine. One occurs in the liver and involves the methylation of phosphatidylethanolamine (Fig. 8.24). The methyl group donor is *S*-**adenylosylmethionine**, a general contributor of methyl groups in biosynthetic reactions (Section 6.7).

The second pathway for phosphatidylcholine biosynthesis mimics that of phosphatidylethanolamine (Fig 8.22). **Choline** is metabolically activated by phosphorylation (Fig. 8.25). The **phosphorylcholine** reacts with CTP to give **CDP–choline**. Subsequent transfer of the choline to a diacylglycerol gives the required phosphatidylcholine.

The biosynthesis of phosphatidylserine

The major pathway for the biosynthesis of phosphatidylserine involves base exchange of the ethanolamine moiety of phosphatidylethanolamine for serine:

$$\text{Phosphatidylethanolamine} + \text{Serine} \rightarrow \text{Phosphatidylserine} + \text{ethanolamine}$$

However, serine can also react *directly* with a CDP–diacylglycerol to give phosphatidylserine (Fig. 8.26).

Phosphoacylglycerols, like triacylglycerols, are synthesized by enzymes embedded within the endoplasmic reticulum membrane of the cell. Polar lipids associate with specific apoproteins forming **lipoprotein particles**. The best characterized lipoproteins are the transport lipoproteins of blood plasma.

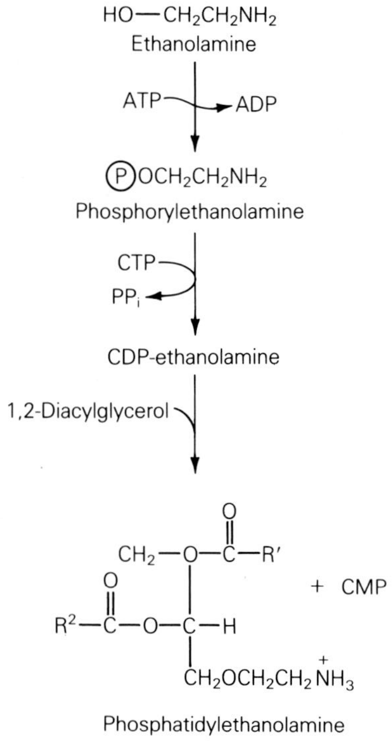

☐ *S*-Adenosylmethionine is the donor of metabolically active methyl groups in most biosynthetic reactions involving methylation (Section 6.7).

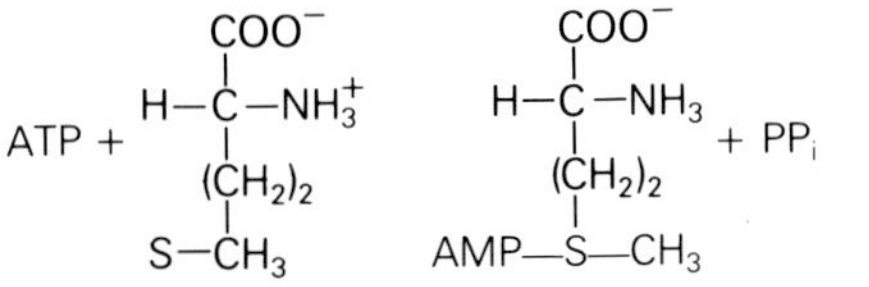

Phosphoadenosinephosphosulphate (PAPS) is a general donor of metabolically active sulphate groups (Section 4.4). PAPS is synthesized in two steps:

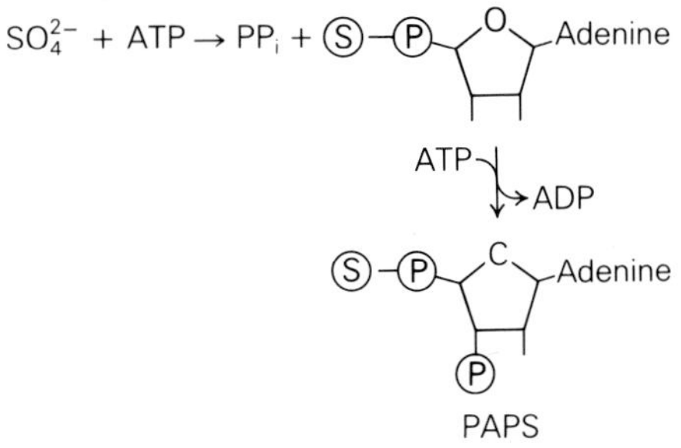

$$SO_4^{2-} + ATP \rightarrow PP_i + \text{(S)}-\text{(P)}\text{—Adenine}$$

PAPS

Fig. 8.22 Biosynthesis of phosphatidylethanolamine.

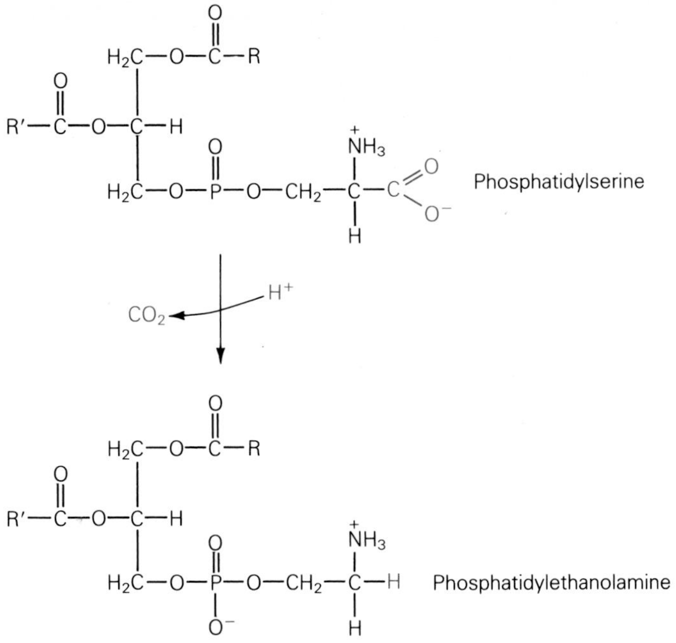

Fig. 8.23 Decarboxylation of phosphatidylserine to give phosphatidylethanolamine.

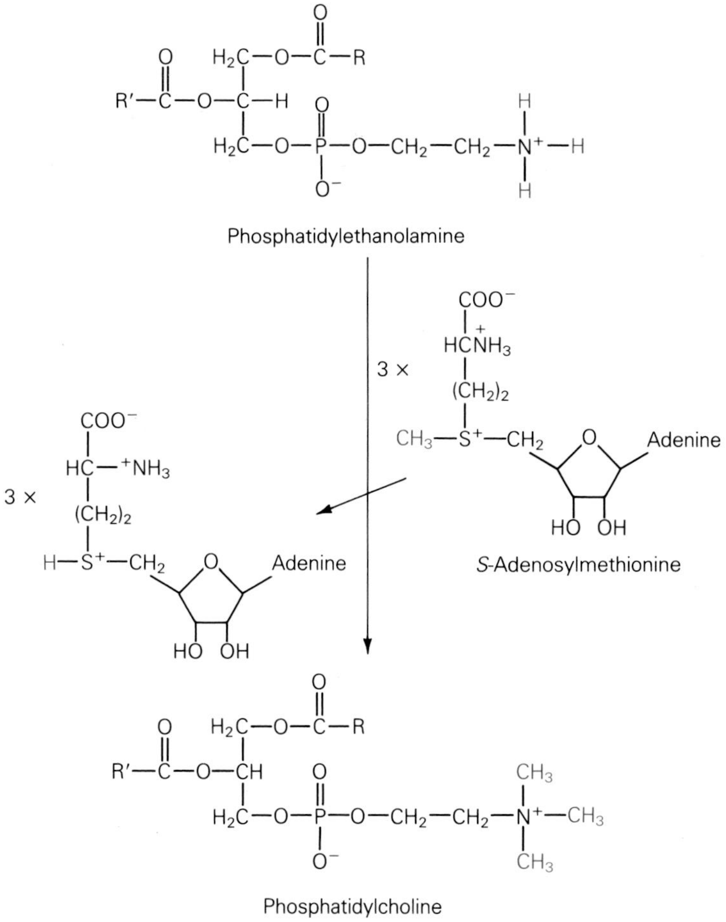

Fig. 8.24 Methylation of phosphatidylethanolamine to give phosphatidylcholine.

Box 8.2
Lung surfactant and respiratory distress syndrome

The pulmonary lung surfactant is a complex lipoprotein material, which coats the internal surface of the lung alveoli. It functions by lowering the surface tension at the alveolar lining–air interface to about $10 \, \text{nNm}^{-1}$, reducing the work associated with inflating the lungs. The composition of surfactant varies. In general it consists of about 10% protein and roughly 90% lipid. Phospholipids predominate, forming about 85% of the lipids, with phosphatidylcholine being particularly abundant. The remaining 15% is made up of neutral lipids, cholesterol and fatty acids.

Respiratory distress syndrome is a developmental disorder when the lungs do not produce sufficient pulmonary surfactant to allow normal respiration following birth. This results in morphological and physiological alterations to the lungs which are a major cause of death in premature babies. The syndrome affects about 50% of premature babies in the UK. About a quarter of these babies die, with a portion of the remainder left handicapped. The pathology is largely due to the difficulty in expanding the lungs which tend to collapse on expiration. This greatly increases the work associated with breathing, resulting in respiratory failure and death unless ventilatory support is given.

Medical treatment of the syndrome has been to hasten development of the immature lungs by hormone treatment and to infuse lipid mixtures into the lungs to mimic the effects of the natural surfactant. Both types of treatment have had some success.

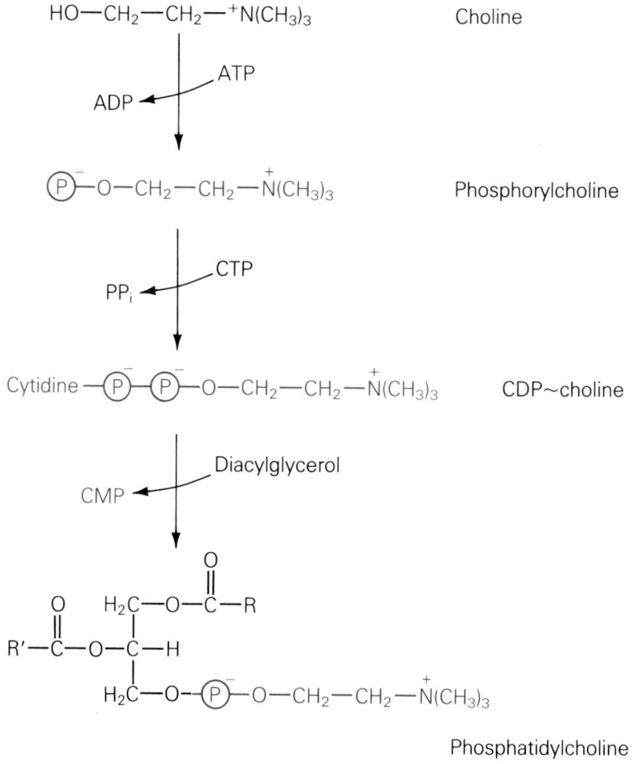

Fig. 8.25 Biosynthesis of phosphatidylcholine involving CDP–choline.

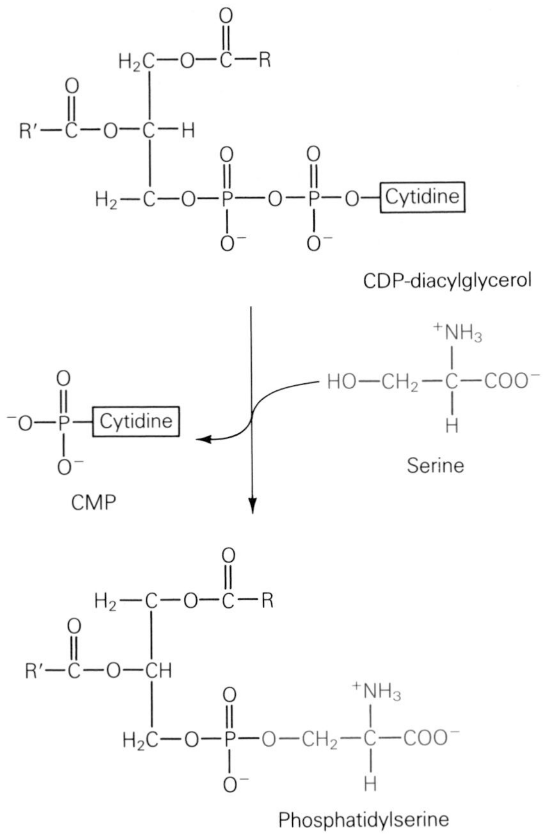

Fig. 8.26 Biosynthesis of phosphatidylserine using a CDP-diacylglycerol.

8.6 Lipid transport

Since most lipids are insoluble in water, the transport of these compounds in blood is accomplished by associating them with specific proteins which presumably have hydrophobic areas on their surfaces, to which lipids can bind. Triacylglycerols, phospholipids and **cholesterol** (see also Section 9.3) are transported in body fluids as particles called **lipoproteins**. Lipoproteins are classified in groups according to their densities. The three major groups are: (i) very low density (VLDL), (ii) low density (LDL), and (iii) high density lipoproteins (HDL).

A fourth group, consisting of particles called **chylomicrons** is also found in lymph and blood following absorption of dietary fats. Some of the characteristics of these lipoproteins are summarized in Table 8.4.

The protein part of the lipoproteins (apoprotein), are designed to solubilize hydrophobic lipids allowing them to be transported around organisms. All lipoproteins share a common basic structural arrangement consisting of a central core of hydrophobic lipids. The core is surrounded by polar lipids, which in turn is covered by a coat of proteins (Fig. 8.27). A variety of diseases, the hyperlipoproteinaemias, are associated with elevated levels of serum lipoproteins. Five major types have now been identified, depending upon which lipoproteins show increased levels in the plasma (Table 8.5).

Cholesterol: *derived from the Greek words* chole, *meaning bile (appreciable amounts of cholesterol are found in bile), and* stereos, *solid.*

Chylomicrons: *Chyle is the milky fluid appearing in the lacteals of the lymphatic vessels of the small intestine when fat is being absorbed from the gut. From the Greek,* chyles, *juice. The milky appearance results from the presence of microscopic globules of lipid,* micro, *small.*

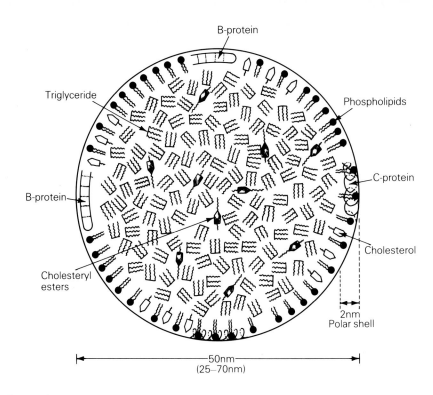

B-protein

Triglyceride

Phospholipids

C-protein

B-protein

Cholesterol

Cholesteryl esters

2nm
Polar shell

50nm
(25–70nm)

Fig. 8.27 Structure of a lipoprotein particle (VLDL). Redrawn from Morrisett, J.D. *et al.* (1977) *Biochimica et Biophysica Acta*, **472**, 125.

While most lipoproteins are synthesized from endogenous fatty materials, chylomicrons are derived from dietary fat (Fig. 8.28). Dietary triacylglycerols are emulsified by bile salts, and hydrolysed by digestive lipases to give fatty acids and 2-monoacylglycerols. These are absorbed by intestinal muscosal cells and used by the cell to give form triacylglycerols. The triacylglycerols are combined with cholesterol, phospholipids, and proteins to form

Table 8.4 *Some major features of the lipoproteins: VLDL (very low density lipoproteins), LDL (low density), HDL (high density) and chylomicrons*

	VLDL	LDL	HDL	Chylomicrons
Diameter (nm)	30–70	15–25	7.5–10.0	100–1000
Density (g cm^{-3})	0.95–1.006	1.006–1.063	1.0063–1.21	<0.95
Approximate composition (%)				
protein	10	20	50	1
phospholipid	19	24	30	4
cholesterol	19	45	18	6
triacylglycerol	50	10	5	90
Sites of synthesis	Gut Liver	Liver (from VLDL)	Liver (plasma)	Gut
Function	Transport TGA from liver	Transport of cholesterol	Transport of cholesterol	Transport of dietary TGA

TGA, triacylglycerols.

Reference Brown, M.S. and Goldstein, J.C. (1983) Lipoprotein metabolism in the macrophage: implications for cholesterol deposition in atherosclerosis, in *Annual Review of Biochemistry*, **52**, 223–61. A comprehensive review of lipoprotein metabolism and its implications for cardiovascular disease.

Box 8.3
Hyperlipoproteinaemias

Hyperlipoproteinaemias are a set of five metabolic disorders which are, genetically based and lead to an overproduction of one of the types of lipoprotein and its consequent increased concentration in plasma.

Type I hyperlipoproteinaemia is a rare disorder in which the chylomicron fraction is increased. The disorder is due to a deficiency of the enzyme lipoprotein lipase in adipose tissue and a consequent inability to hydrolyse the triacylglycerols in the chylomicrons. The result of the disorder is an inability to metabolize triacylycerols.

Type II hyperlipoproteinaema (or hypercholesterolaemia) leads to an increase in either LDL (type IIA) or LDL and VLDL (Type IIB). Both of these types are found as an acquired form, or as hereditary defects. Since sufferers have increased plasma cholesterol levels they are at an increased risk of atherosclerosis.

Type III hyperlipoproteinemia is characterized by an abnormal lipoprotein, floating β-lipoprotein, appearing in the plasma. Sufferers have an increased risk of atherosclerosis.

Type IV hyperlipoproteinemia is characterized by an increased amount of VLDL in the plasma. It is an hereditary condition associated with obesity and diabetes mellitus.

Type V hypolipoproteinemia is rare and characterized by increased chylomicron levels. It is often associated with diabetes mellitus and liver and kidney diseases.

See also Box 9.1

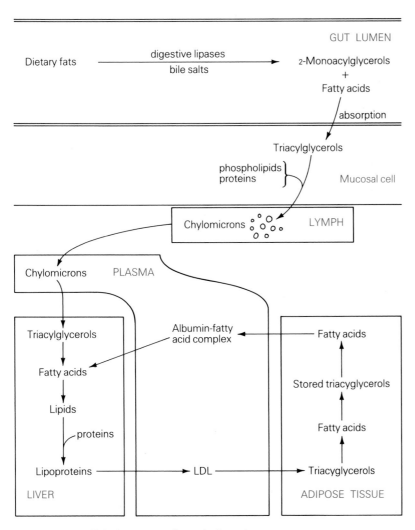

Fig. 8.28 The transport of lipids between gut, liver and adipose tissue.

Reference Nilsson-Ehle, P., Garfinkel, A.S. and Karnovsky, M.C. (1980) Lipolytic enzymes and plasma lipoprotein metabolism, in *Annual Review of Biochemistry*, **49**, 667–93. A detailed review of lipoprotein metabolism.

Table 8.5 *Types of hyperlipidaemias. From Campbell, P.N. and Smith, A.D. (1982)* Biochemistry Illustrated, *Churchill Livingstone, Edinburgh.*

Type	Lipoprotein elevated	Cholesterol level	Triacylglyceride level
I	Chylomicrons (also possibly VLDL)	+	+++
IIa	LDL	+++	±
IIb	LDL and VLDL	++	++
III	'Floating' LDL	+++	+++
IV	VLDL	±	++
V	VLDL and chylomicrons	+	+++

± normal to slightly increased. ++ moderately increased. +++ greatly increased.

chylomicrons which are extruded from the cell into the lymphatic system, eventually reaching the blood system via the thoracic duct. Chylomicrons are removed from the circulation by the liver. Here, the triacylglycerols are hydrolysed to glycerol and fatty acids. The fatty acids may be oxidized to provide metabolic energy but, more importantly, are used in the synthesis of other triacylglycerols for lipoprotein formation.

Very low density lipoproteins transport triacylglycerols to adipose tissue where they form an energy reserve. When these reserves are required, the triacylglycerols are hydrolysed releasing fatty acids which are then bound to albumin and transported in the plasma to the liver for oxidation.

See *Energy in Biological Systems*, Chapter 6

CH₃
|
(CH₂)₁₂
|
CH
‖
CH
|
H—C—H
|
⁺NH₃—C—H
|
CH₂OH

Fig. 8.29 Sphingosine.

8.7 The biosynthesis of sphingolipids

Sphingolipids are components of animal cell membranes and may have an important role in cell recognition and in the regulation of differentiation and embryogenesis. They are continually being synthesized by many animal cells. Unlike the lipids mentioned so far in this chapter, the sphingolipids are based on the alcohol **sphingosine** (Fig. 8.29). Essentially there are two classes of sphingolipids: the **sphingomyelins** and the **glycolipids** (Fig. 8.30). In general they are found only in eukaryotic cells.

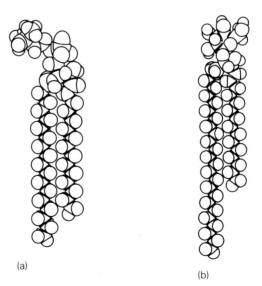

(a)

(b)

Fig. 8.30 Molecular model of (a) a sphingomyelin and (b) a glycolipid (specifically a galactocerebroside). Both models redrawn from Robertson, R.N. (1983) *The Lively Membranes*, pp. 26 and 27. Cambridge University Press, Cambridge.

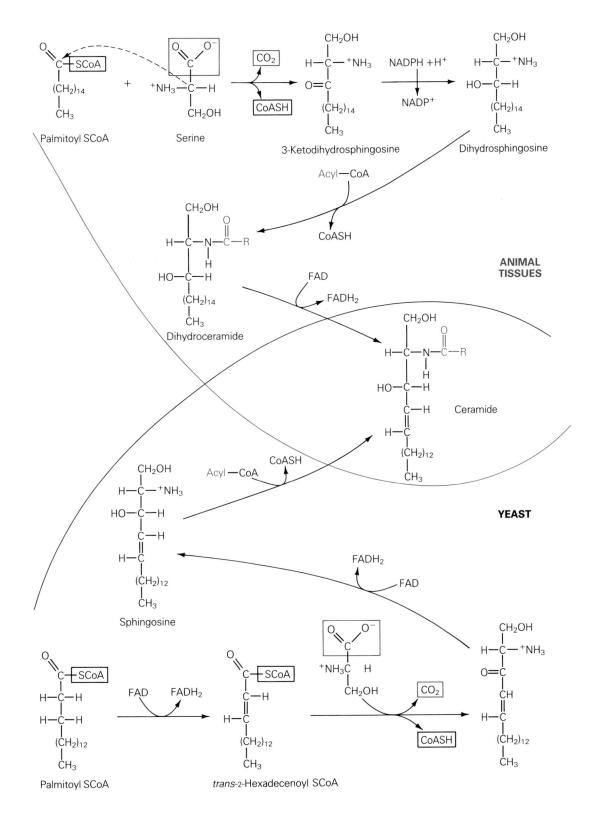

Fig. 8.31 The biosynthesis of a ceramide by animal and yeast systems.

SPHINGOSINE is synthesized in the endoplasmic reticulum of animal cells from palmitoyl CoA and serine (Fig. 8.31). This reaction is catalysed by the enzyme **3-ketosphinganine synthase** and produces **3-ketodihydrosphingosine**. The condensation of a fatty acid and an amino acid would not generally be thermodynamically favourable. However, because the reaction is accompanied by a decarboxylation and because the 'activated' coenzyme A derivative of palmitate is involved, the reaction goes to completion. The 3-ketodihydrosphingosine is then reduced in an NADPH-linked reaction to give **dihydrosphingosine**. This is acylated, and the subsequent unsaturation involving an FAD-linked enzyme generates a **ceramide**. This sequence of reactions means that *free* sphingosine does not occur in animal tissues. In yeast the situation is somewhat different: palmitoyl CoA can be desaturated to yield ***trans*-2-hexadecanoyl CoA**, which can condense with serine (Fig. 8.31). The subsequent reduction of the keto group yields sphingosine directly. Acylation by an acyl CoA gives the ceramide.

SPHINGOMYELINS can be synthesized by the addition of a phosphorylcholine group from CDP-choline to the ceramide (Fig. 8.32). This reaction recalls that involved in the biosynthesis of phosphatidylethanolamine (see Section 8.5). The importance of this particular pathway is now the subject of some dispute. It now appears to be generally accepted that the biosynthesis of sphingomyelin occurs by the transfer of phosphorylcholine from phosphatidylcholine to a ceramide, rather than from CDP-choline:

Ceramide + Phosphatidylcholine → Sphingomyelin + Diacylglycerol

GLYCOLIPIDS are also formed using a ceramide backbone. The two groups of glycolipids, **cerebrosides** and **gangliosides**, are formed by the sequential addition of sugars to the ceramide. The sugars are generally donated from **nucleoside diphosphate sugars** (Fig. 8.33, see also Section 4.2): examples

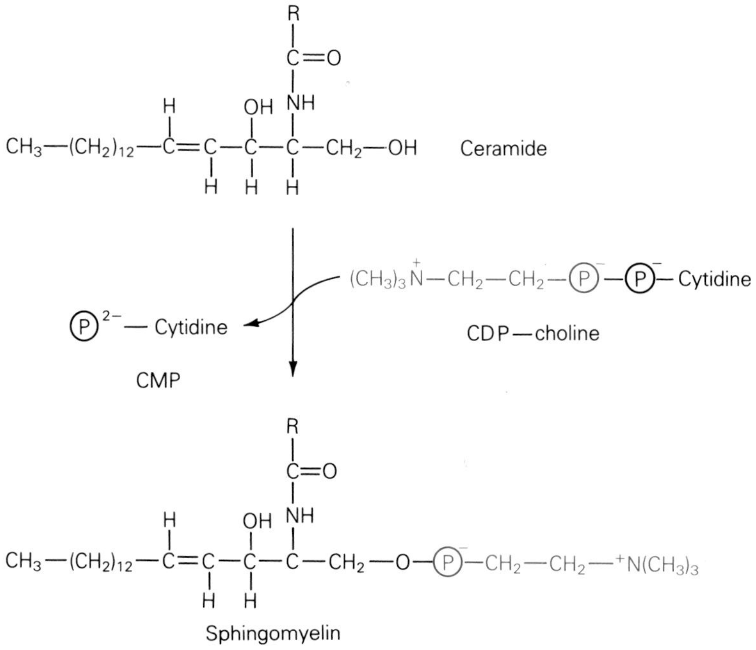

Exercise 4

Suggest an experiment which could distinguish whether CDP-choline or phosphatidylcholine is the immediate donor of the phosphorylcholine group in sphingomyelin biosynthesis.

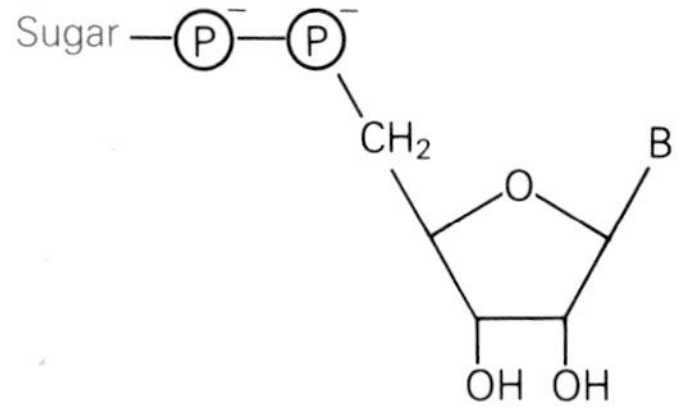

Fig. 8.33 Generalized nucleoside diphosphate sugar (NDP–sugar). B stands for base, thus uracil would give a UDP–sugar derivative.

Fig. 8.32 Formation of a sphingomyelin.

Reference Kanfer, J.N. and Hakomori, S-i (1983) *Sphingolipid Biochemistry*, Plenum Press, New York. A very comprehensive text, covering most aspects of the chemistry and biochemistry of sphingolipids.

are UDP-glucose, UDP-galactose, GDP-fucose. For example, **galacto-cerebrosides** are formed by the addition of galactose to the ceramide from UDP-galactose by a **galactosyltransferase**.

$$\text{UDP-galactose} + \text{Ceramide} \xrightarrow[\text{transferase}]{\text{galactosyl}} \text{Galactocerebroside} + \text{UDP}$$

Other sugars are added by the action of other, specific, **glycosyltransferases**.

Sulphatides are formed by the sulphation of the galactocerebroside (Fig. 8.34). The sulphate is donated from **phosphoadenosinephosphosulphate (PAPS)** (Section 4.4).

Gangliosides are formed by similar mechanisms to cerebrosides. The difference is that **sialic acid** residues (sialate residues) are added from **CMP-sialate** (Fig. 8.35) by sialyltransferases.

The glycosyltransferases involved in glycolipid biosyntheses are found in the Golgi apparatus. The Golgi apparatus is thus the site of glycosylation of

Exercise 5

Write a series of equations to represent the synthesis of a glucocerebroside.

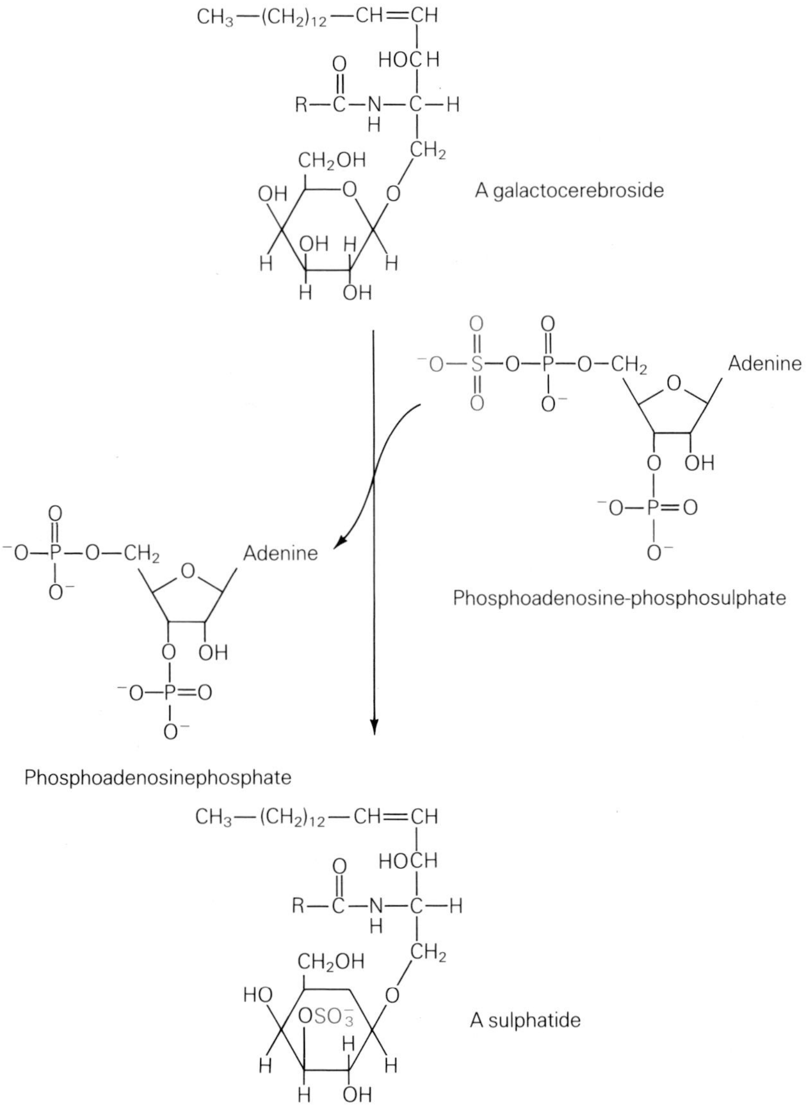

Fig. 8.34 Conversion of a galactoceramide to a sulphatide.

Exercise 6

Which class of enzymes is involved in transferring fucose and glucose from donor molecules to form glycolipids?

Glycosyltransferases: *are enzymes involved in transferring a sugar molecule from a donor (normally a nucleoside diphosphate sugar) to an acceptor molecule:*

NDP-sugar + acceptor →

NDP + acceptor-sugar

See *Molecular Biology and Biotechnology,*
Chapter 5

glycolipids. Glycosylation of glycoproteins also occurs to a large extent in the Golgi apparatus. However, it appears that two sets of glycosyltransferases operate, one for lipid substrates, the other specialized for glycoprotein biosynthesis. The biosynthesis of fatty acids and lipids therefore involves the co-ordinated activities of many enzymes, distributed in several compartments of the cell.

Exercise 7

List the intracellular compartments involved in the biosynthesis of the molecular constituents of a ganglioside, and the cellular sites where the intermediate stages are assembled to give the complete structure.

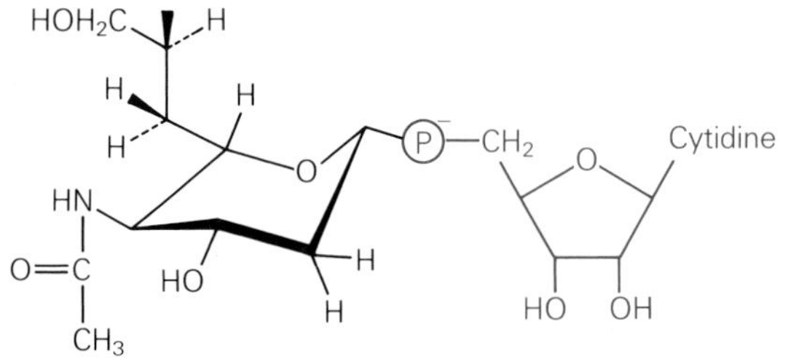

Fig. 8.35 Structure of CMP-sialate.

8.8 Overview

The processes involved in the biosyntheses of lipids are complicated and highly diverse mirroring the variety of types of lipids found in organisms. The metabolic steps involved in these biosynthetic reactions are endergonic and require an input of free energy, usually made available from the hydrolysis of ATP, in the formation of intermediates activated as coenzyme A derivatives or as nucleoside diphosphate derivatives.

Many enzymes are involved in these processes, and are distributed in differing intracellular compartments of the cell, emphasizing that the functional economy of the cell depends upon a full integration of its activities.

Answers to exercises

1. C-1 of acetyl CoA (as malonyl CoA) is attached to the methyl carbon of a second acetyl CoA producing head-to-tail labelling. The resulting four-carbon compound will go through a cycle of reductions and is then available for further additions of acetyl CoA molecules.

2. Linoleate.

$CH_3(CH_2)_7CH\!\!=\!\!CH(CH_2)_7COO^-$

Linolenate.

$CH_3(CH_2)_4CH\!\!=\!\!CHCH_2CH\!\!=\!\!CH(CH_2)_7COO^-$

3.
$$H_2-C-O-\overset{O}{\overset{\|}{C}}-(CH_2)_{14}CH_3$$
$$H-C-O-\overset{O}{\overset{\|}{C}}-(CH_2)_{14}CH_3$$
$$H_2-C-O-\overset{O}{\overset{\|}{C}}-(CH_2)_{16}CH_3$$

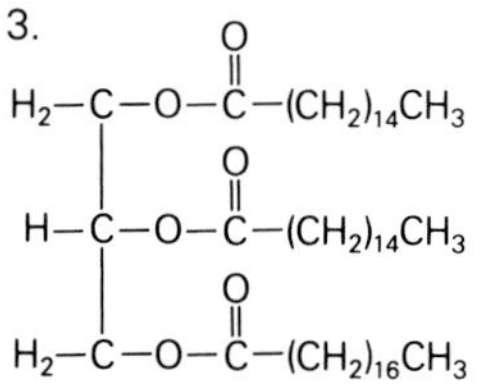

The possible arrangements of palmitate (P) and stearate (S) in a triacylglycerol are:

```
C—P        C—P
|          |
C—P        C—S
|          |
C—S        C—P
```

4. Follow the fate of radioactively labelled precursors, e.g. CDP –(^{14}C) choline and phosphatidyl choline labelled with ^{32}P, and analyse the sphingomyelin to see which label predominates.

5. Equations as for ceramide synthesis in yeast (see Fig. 8.31) then

UDP-Glc + ceramide → glucocerebroside

6. Fucosyltransferase; glucosyl-transferase.

7. Cytosol; SER membrane; Golgi apparatus
Cytosol: fatty acid synthesis
Cytosol/ER: ceramide
Golgi apparatus: glycosylation.

FILL IN THE BLANKS

1. Fatty acids are produced by the _____ _____ _____ complex in the cytosol. The major product of this enzyme is _____ . The production of _____ _____ is stimulated by _____ _____ and requires _____ . The function of the latter is to activate _____ _____ . Long-chain, C_{20-24}, fatty acids are produced by _____ reactions in _____ or _____ _____ . These two elongation systems differ in their cofactor requirements, which are _____ and _____ , and sources of carbon, _____ _____ and _____ _____ respectively. A further series of reactions produces _____ fatty acids. There are _____ families of these fatty acids. Each contain one or more _____ bonds. The _____ families of acids are called _____ , _____ , _____ and _____ .

Choose from: acetyl CoA (2 occurrences), biotin, carbon dioxide, double, elongation, endoplasmic reticulum, fatty acid synthase, four (2 occurrences), linoleate, linolenate, malonyl CoA (2 occurrences), mitochondria, NADH, NADPH, oleate palmitae, palmitoleate, unsaturated.

MULTIPLE-CHOICE QUESTIONS

2. Which of the following is/are correct:

A. The biosynthesis of palmitate requires $FADH_2$; the breakdown requires FAD.
B. The biosynthesis of palmitate requires NADH; the breakdown requires FAD.
C. The biosynthesis of palmitate requires NADH; the breakdown requires NAD^+.
D. The biosynthesis of palmitate requires $FADH_2$; the breakdown requires NAD^+.
E. The biosynthesis of palmitate occurs in the cytosol; the breakdown in the mitochondrion.
F. The biosynthesis and metabolism of palmitate occur in the mitochondrion.
G. The carrier of the intermediates of palmitate biosynthesis is ACP and of the catabolism is coenzyme A.
H. The carrier of the intermediates of palmitate biosynthesis and of the catabolism is coenzyme A.

3. Which of the following is/are correct? The biosynthesis of triacylglycerols involves:

A. sphingosine
B. pancreatic lipase
C. glycerol 3-phosphate
D. dihydroxyacetone phosphate

4. Which of the following is/are correct? The biosynthesis of phosphoacylglycerols involves:

A. phosphatidate
B. decarboxylation of phosphatidyl serine
C. S-adenylosylmethionine
D. galactosyltransferase

5. Which of the following is/are correct? The biosynthesis of sphingomyelins involves:

A. fatty acyl CoA
B. glycerol kinase
C. NADPH-linked reduction
D. CDP–choline

6. Which of the following is/are correct? The biosynthesis of glycolipids involves:

A. GDP–galactose
B. UDP–glucose
C. phosphatidate
D. CDP–choline

SHORT-ANSWER QUESTIONS

7. In an experiment to study the biosynthesis of fatty acids in a cell-free system, $C^2H_3COSCoA$ and $^{14}CO_2$ were supplied as the precursors. What would be the ratio of $^2H : {}^{14}C$ in the palmitate produced?
If the experiment were repeated in the absence of CO_2 but with C^2H_3CO SCoA and $^-OO^{14}C\text{-}CH_2CO$ SCoA what would be the $^2H : {}^{14}C$ ratio?
In a final experiment C^2H_3CO CoA and $^-OOC\text{--}^{14}CH_2COCOA$ were supplied. What would be the expected $^2H : {}^{14}C$ ratio? What do these results indicate about the mechanism of palmitate biosyntheses? How would the amounts of isotopes be measured?

8. How many molecules of ATP, NADPH and acetyl CoA would be required for the biosynthesis of one molecule of (a) palmitate, and (b) palmitoleate?

9. Describe how the essential fatty acid linoleate can be converted to prostaglandins.

10.   List the principal carriers of the following molecules in the plasma. (a) fatty acids, (b) triacylglycerols, (c) cholesterol.

9
Polyisoprenoids and porphyrins

Objectives

After reading this chapter you should be able to:

☐ describe the role played by cholesterol as a precursor of many materials in organisms;

☐ outline the biosynthesis of cholesterol in vertebrate tissues and its regulation;

☐ appreciate how this pathway relates to similar reactions in intermediary metabolism;

☐ discuss the diverse roles of polyisoprenoids in nature and their biosynthesis by a common metabolic pathway;

☐ describe how porphyrins are formed and discuss their importance in living systems.

9.1 Introduction

This chapter deals with the biosynthesis of two groups of compounds, the polyisoprenoids and the porphyrins. The two groups of compounds are unrelated chemically and yet there are many features of their structure, biosynthesis and biological role where one can discern similarities. The polyisoprenoids include such vitally important compounds as vitamins A, D, E and K, cholesterol, and the hydrophobic tail of chlorophyll. The porphyrins include the haem groups of haemoglobin and the cytochromes as well as the ring structure of chlorophyll. Clearly, life as we know it is very much dependent on these compounds. Parallels may also be seen in the way the compounds are built up. Although the final products of biosynthesis are fairly large and are formed by a polymerization-like process from smaller units, they are not huge polymers in the sense that polysaccharides and proteins are. Furthermore, a large number of modifications occur, once the basic units are joined together, to give the final, functional molecule. Finally, because many of the compounds have conjugated double bond systems, they tend to be coloured or at least to absorb light in the ultra-violet region of the spectrum.

See *Biological Molecules*, Chapter 7

9.2 Polyisoprenoid biosynthesis

In addition to the fatty acids (Chapter 8), a tremendous variety of products in nature are derived from acetyl CoA. They are formed from acetyl CoA via multiple condensation of branched-chain C_5 compounds called isoprenoid intermediates and the term **polyisoprenoid** is used to describe such products.

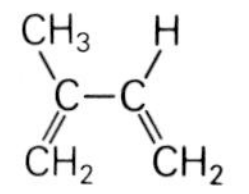

Isoprene

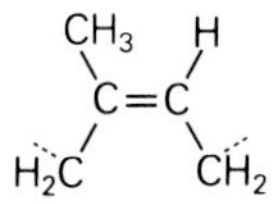

Isoprenoid unit

See Chapter 4

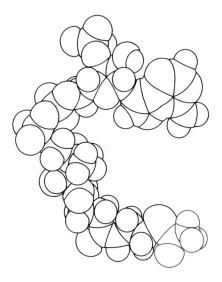

Fig. 9.1 Space-filled model of acetyl CoA. The active acetate group is shown in red.

☐ *α and β Configurations.* A *β* configuration indicates that the substituent lies *above* the plane of the ring system. This form is usually denoted in the literature by a wedge (◀). The *α* configuration refers to a substituent *below* the plane of the rings and is designated by dashes (– – –) or (≡). In the case of cholesterol (Fig. 9.2a), the hydroxyl group at C-3, the methyl groups (C-18 and C-19), and the side-chain attached to C-17 are all *β* and lie above the rings.

☐ *Monooxygenases.* A monooxygenase enzyme catalyses the incorporation of one atom of oxygen, derived from molecular oxygen, into an organic substrate. The remaining oxygen atom is converted to water with the aid of a co-reductant, frequently NADPH. Most substrates become hydroxylated in the course of this reaction:

$$RH + O_2 + NADPH + H^+ \rightarrow$$
$$ROH + H_2O + NADP^+$$

They include several classes of metabolites of diverse chemical structure, with a wide range of essential functions in animals, plants and microorganisms. Examples include the sterols, and the many substances derived from them, the highly-coloured carotenoids and the related vitamin A, rubber and the hydrocarbon portions of compounds such as chlorophyll, plastoquinone and ubiquinone, and vitamins E and K. The isoprenoid tail portion of these molecules makes them hydrophobic, that is, water-hating or lipophilic (lipid loving) enabling them to be inserted and orientated in biomembranes.

All polyisoprenoid compounds are assembled entirely from the two carbon atoms of the acetyl units of acetyl CoA (Fig. 9.1) via a complex series of reactions in the cytoplasm of the cell. Isoprenoid derivatives are the intermediates that act as their building blocks. The carbon skeleton becomes highly reduced in the process, especially in the case of sterols. The reducing equivalents required for the appropriate steps are invariably supplied by NADPH (Section 1.4).

9.3 *Importance of cholesterol*

The isoprenoid of greatest general importance in mammalian metabolism is **cholesterol**. This is the major *sterol* in vertebrate tissues and has a complex cyclic structure comprising 27 carbon atoms with a single hydroxyl group. This group is essential for its activity in membranes and its role in physiological functions. Cholesterol consists of four fused rings with an eight-carbon side-chain and a secondary alcohol group in the *β*-configuration at C-3 (Fig. 9.2).

Membranes

Despite its appearance when written on paper, the cholesterol molecule is rather long and thin. The hydroxyl group generates hydrophilic properties at one end of the molecule and permits it to be positioned in membranes alongside the phospholipids. Its chair–boat–chair form is the preferred conformation (Fig. 9.2). This structure has the same overall dimensions as phospholipids but in contrast to these it is rigid and flat. Cholesterol is incorporated into the lipid bilayer of membranes, especially the external plasma membranes of cells. Its major function in eukaryotic membranes is to modify their fluidity. Cholesterol is particularly abundant in the myelin component of brain and nervous tissue; the myelin sheath consists of concentric layers of plasma membrane. Sterols are characteristically not found in the membranes of prokaryotic organisms.

Precursor of important metabolites

Cholesterol plays a key role in many tissues of higher animals as the starting point for a number of vital and potent substances. These include bile acids, sex hormones and the hormones of the adrenal cortex (illustrated briefly in Table 9.1). In these cases the catabolic pathways concerned are initiated by hydroxylation reactions. They are catalysed by monooxygenase enzymes, which utilize cytochrome P450 systems, and side-chain cleavage.

Sterol: derived from the Greek word for 'solid' (stereos) alcohol Cholesterol indicates its origin in bile.

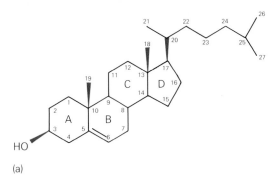

(a)

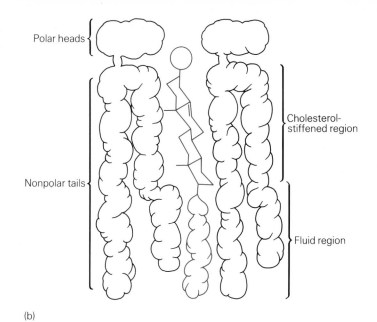

Polar heads

Nonpolar tails

Cholesterol-stiffened region

Fluid region

(b)

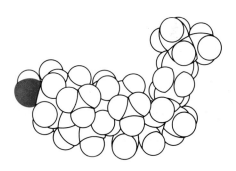

HO

(c)

Fig. 9.2 (a) Structure of cholesterol (cholestan-3β-ol) showing its numbering system. ▼ denotes group above plane of rings: the β configuration.
(b) Position of cholesterol in membranes.
(c) Conformation of cholesterol in membranes.
(d) Space-filled model of cholesterol.

Table 9.1 *Functions of products of cholesterol catabolism*

Tissue	Product	Function
Liver	Bile acids	Emulsifying agents in digestion of lipids
Adrenal cortex	Adrenal cortical hormones (e.g. cortisol)	Ion balance and carbohydrate metabolism
Testes	Testosterone	Male secondary sex characteristics
Ovaries and placenta	Oestrogens	Female secondary sex characteristics; ovarian cycle
	Progesterone	Maintenance of pregnancy
Skin	Vitamin D_3 (cholecalciferol)	Calcification of bone; prevention of rickets

Clinical problems

Men, from middle age onwards, become increasingly likely to suffer heart attacks or strokes. Everybody is surely aware that high levels of cholesterol in the plasma (for example, above 200 mg/100 cm^3; 5.2 mmol dm^{-3}) predispose individuals to such serious medical consequences and other pathologies. If high levels persist, free cholesterol may tend to be deposited within plaques on the artery wall leading to blockage of the intima and restriction of blood flow (Fig. 9.3b); if a clot forms there may be fatal consequences. These atherosclerotic conditions therefore lead to coronary artery disease, a major cause of death in Western countries, and stroke.

Another important clinical condition associated with an imbalance of cholesterol metabolism is its deposition in the bile duct or gall bladder. This gives rise to gall stones in which cholesterol forms the major, and sometimes the only, constituent. Indeed, cholesterol was originally isolated from human gallstones and only later identified as a component of membranes.

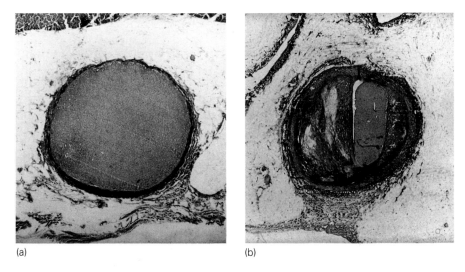

(a) (b)

Fig. 9.3 (a) Cross-section of a 'clean' artery from a healthy individual. (b) Cross-section of an atherosclerotic artery. This partially occluded artery shows a proliferation of cells (on the left-hand side) together with accumulation of cholesterol and collagen.

Transport of cholesterol and its esters

Cholesterol esters are solid, waxy substances, strongly hydrophobic in nature and insoluble in water. They therefore have to be transported from tissue to tissue in the plasma in a soluble, packaged form called **lipoprotein particles** (Section 8.6). The cholesterol esters are found in the core of the particle. This core is surrounded by a surface layer comprising proteins, phospholipids and free cholesterol in contact with the aqueous phase (see Fig. 8.27).

Lipoproteins are classified according to their densities which relate to their compositions. The principal ones, as far as cholesterol metabolism is concerned, are **HDLs** (high density lipoproteins) which are rich in protein, and **LDLs** (low density lipoproteins) which are larger particles rich in **cholesteryl ester**. LDLs are important in transporting cholesterol to non-hepatic (peripheral) tissues and cells and also to regulate synthesis. Characteristics of these and other lipoproteins are given in Tables 9.2 and 8.4.

Table 9.2 *Characteristics of human lipoproteins*

Lipoprotein	Origin	Density (g cm⁻³)	Composition (%)					Polypeptides
			Triacylglycerol	Cholesteryl ester	Cholesterol	Phospholipid	Protein	
Chylomicrons	Intestines (dietary origin)	0.94	90	4		5	1	Apo A, B, C
VLDL	Liver (dietary CHO)	0.94–1.01	50–70	10	5	15	10–15	Apo B, C, E
LDL	VLDL in plasma (after uptake of triacylglycerol)	1.01–1.06	6	39	8	25	22	Apo B
HDL	Liver	1.07–1.21	3	22	3	22	50	Apo A

Biosynthesis of cholesterol and other isoprenoid compounds

Extensive and elaborate investigations have been pursued for many years in order to elucidate the pathway by which cholesterol is synthesized in various tissues. One of the first metabolic pathways to be examined using isotopically labelled compounds (both stable and radioactive isotopes were used) was that concerned with cholesterol biosynthesis. Initially, it could be shown that both ^{2}H- and ^{13}C-labelled acetate were incorporated into cholesterol after feeding these compounds to rats and mice. These 'heavy' stable isotopes were detected in the product by mass spectrometry. Subsequently, the results were confirmed using radioactive [^{14}C]acetate. It was then shown by detailed degradative studies that every carbon atom in the cholesterol molecule was derived from either a methyl or a carboxyl carbon of acetate. The pattern of labelling in the sterol skeleton and in the side-chain was such that a repeating five-carbon (C_5) isoprenoid unit was implicated as an intermediate.

In a series of experiments from many laboratories over a long period, details of each individual step leading from acetyl CoA, through many intermediates, were clarified and the entire sequence is now well established. A brief outline of the pathway is shown in Fig. 9.4. This reveals the involvement of C_5 isoprenoid intermediates, of a C_{30} hydrocarbon (**squalene**) found in liver and certain fish oils, and of the first sterol to be produced, **lanosterol**. Over 30 enzyme-catalyzed reactions take place in the formation of cholesterol. They have all been studied in some depth and appear to be common throughout nature.

All mammalian tissues and cells contain cholesterol within their membranes and most can synthesize it. However, the liver is by far the most important site in this respect in mammals since over 70% of the total capacity for synthesis takes place here. Typically, about 1 g a day is made in adult humans. Other organs, with the exception of intestines, have much lower rates and contribute about 10% of the total. Some dietary cholesterol is excreted in the faeces, especially when intake is high.

□ Bloch, Cornforth and Popjak are names frequently associated with this early period. Cornforth was awarded the Nobel Prize in Chemistry in 1975 for his role in this work. Popjak is a distinguished Czech scientist who performed these experiments in the United Kingdom. Bloch carried out key experiments in 1950 in which the doubly-labelled acetate, ^{13}CH$_3$^{14}CO$_2^-$, was used and their positions in cholesterol determined. Bloch received the Nobel Prize in Physiology or Medicine in 1964.

3 Acetate (C_2) ⟶ Isoprenoid unit (C_5) —6×→ Squalene (C_{30}) ⟶

Lanosterol (C_{30}) ⟶ Cholesterol (C_{27})

Fig. 9.4 Outline of the pathway of cholesterol biosynthesis.

Exercise 1

The C_8 side-chain of cholesterol is derived from [1-^{14}C]acetate (carboxyl or 'c'-labelled) and [2-^{14}C]acetate (methyl or 'm'-labelled), via an isoprenoid intermediate, as shown

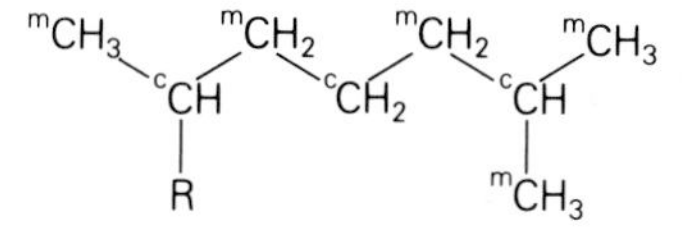

^{m}CH$_3$ ^{m}CH$_2$ ^{m}CH$_2$ ^{m}CH$_3$
^{c}CH ^{c}CH$_2$ ^{c}CH
R ^{m}CH$_3$

Identify the reaction pathway from these substrates leading to (a) the C_{30} squalene and (b) the side-chain of cholesterol. Give some indication of how these experiments may have proceeded.

References Schoepfer, G.J. (1981) Sterol biosynthesis. *Annual Review of Biochemistry* **50**, 585–621. Schoepfer, G.J. (1982) Sterol biosynthesis. *Annual Review of Biochemistry* **51**, 555–85. Two comprehensive articles, which together form a detailed review of the biosynthesis of cholesterol and related polyisoprenoid compounds. Largely restricted to studies on animal cells. Kleinig, H. (1989) The role of plastids in isoprenoid biosynthesis. *Annual Review of Plant Physiology and Molecular Biology*, **40**, 39–59. High level overview of isoprenoid production in plant plastids.

Formation of mevalonate

The sequence of reactions leading to the production of cholesterol and other isoprenoids became much clearer after **mevalonate** had been identified as an intermediate. This C_6 metabolite (Fig. 9.5) has a number of important structural features which permit it to react in an appropriate way. For example, it contains a carboxylic acid group which may be lost by decarboxylation to generate a branched-chain C_5 product (the basic isoprenoid unit), as well as primary and tertiary alcohol groups. Only the (R)-stereoisomer is metabolically active. Its formation results from two consecutive condensation steps which yield **3-hydroxy-3-methylglutaryl CoA (HMG CoA)** and which then undergoes reduction to mevalonate (Fig. 9.5).

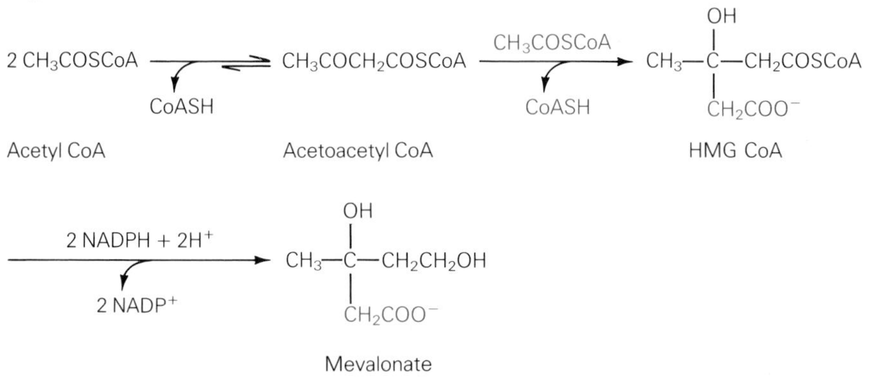

Fig. 9.5 Formation of mevalonate.

The initial reactions involve cytosolic enzymes, thiolase and HMG CoA synthase, to give in turn acetoacetyl CoA and HMG CoA (the formation of which are favoured by loss of coenzyme A). These enzymes perform similar actions to, but are distinct from, the mitochondrial enzymes responsible for ketone body formation. Acetyl CoA for cholesterol biosynthesis, as well as for fatty acid formation, is generated in the cytosol by cleavage of citrate, a reaction catalysed by ATP citrate lyase. This was established by showing that both fatty acid and cholesterol biosynthesis in the liver were inhibited by addition of an antimetabolite, hydroxycitrate (not to be confused with isocitrate). This compound specifically inhibits the cytosolic cleavage of citrate.

HMG CoA reductase (and indeed all the later enzymes involving 'lipid' substrates) is present in membranes of the smooth endoplasmic reticulum. It is in integral membrane protein. Two reductive steps are required, each using a molecule of NADPH as reducing equivalent. In the process the acyl thioester is completely reduced via an enzyme-bound aldehyde derivative (mevaldate CoA), to the primary alcohol:

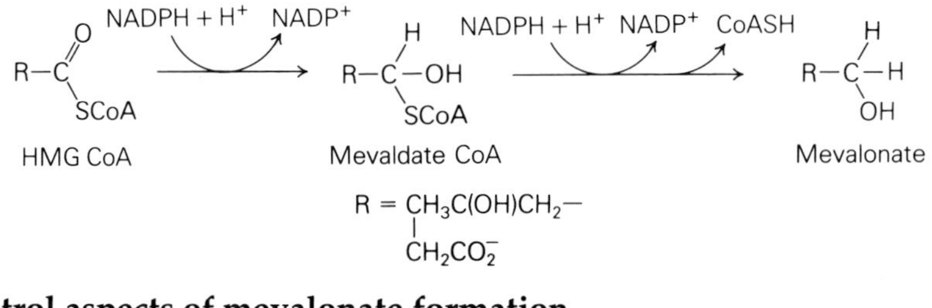

Control aspects of mevalonate formation

The pathway to mevalonate is favoured by the subsequent reduction of HMG CoA with resulting loss of coenzyme A. This is catalysed by HMG CoA reductase in a unidirectional manner under physiological conditions. The reduction is the major rate-limiting step in the complex process of cholesterol

□ Pharmaceutical firms have taken an interest in the use of (−)**hydroxycitrate** as a possible weight-reducing agent, since lipid biosynthesis should be blocked after its administration. Both fatty acid and cholesterol formation could potentially be controlled. Unfortunately, however, this compound proved too toxic for serious development.

Exercise 2

Devise an experiment, making use of a liver homogenate, a suitable ^{14}C-labelled substrate and hydroxy-citrate to show that cytosolic citrate generates acetyl CoA for both fatty acid and cholesterol biosynthesis.

See *Energy in Biological Systems*, Chapter 6

Reference Endo, A. (1981) Biological and pharmacological activity of inhibitors of 3-hydroxy-3-methylglutaryl coenzyme A reductase. *Trends in Biological Sciences* **6**, 10–13. A brief review of competitive inhibitors of HMG CoA reductase and their role in controlling hypercholesterolaemia.

formation. Furthermore, mevalonate does not generally have any alternative metabolic routes it can follow, apart from phosphorylation, which leads to isoprenoid compounds. Mevalonate is therefore committed to sterol biosynthesis and, hence, it is appropriate that control be exerted at this point.

The activity of HMG CoA reductase in liver and other organs and cells is under stringent control. It is carefully regulated to match the cholesterol requirements of the body, which in turn depends on dietary intake and the need of peripheral tissues for this metabolite. Generally, about 50% of the requirement for cholesterol in adult humans is derived from *de novo* synthesis, the remainder being obtained from the diet.

Fluctuations of cholesterol and carbohydrate content in the diet, with consequent changes in hormone level, affect HMG CoA reductase severely. Sterol synthesis, on both a long- and short-term basis, is therefore similarly affected. Novel inhibitors of this enzyme are currently being used as a means of lowering plasma cholesterol (together, of course, with restriction of dietary intake). This important aspect will be developed below and in Boxes 9.1 and 9.2.

See Chapter 3

HORMONAL EFFECTS. One of the experimental situations which has been examined is the effect of lack of food. Starvation results in a decrease in the production of **insulin** and an enhancement of **glucagon** secretion. These responses may arise from reduced intake of carbohydrate or they may be a consequence of the diabetic state. They give rise, in a complementary manner, to a diminished activity and amount of HMG CoA reductase. The effect is similar to that found with other key anabolic enzymes. Many enzymes concerned with fatty acid biosynthesis, such as those involved in the oxidative stages of the pentose phosphate pathway, are similarly depressed. This pathway is essential for the provision of NADPH for reductive biosynthesis.

Insulin exerts its effect at the transcriptional level and enhances production of the mRNA for HMG CoA reductase in the fed state. In the starved state, glucagon may cause phosphorylation and inactivation of the reduced amount of HMG CoA reductase *via* the cAMP/protein kinase cascade. An AMP-activated protein kinase has recently been implicated in a similar phosphorylation process, initiated by glucagon. Thus, long- and short-term responses are apparent (Fig. 9.6) acting through different mechanisms on HMG CoA reductase.

See *Cell Biology*, Chapter 8

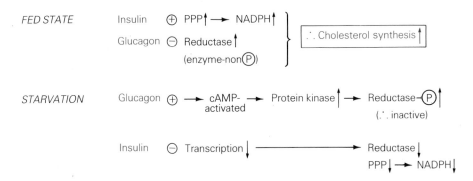

Fig. 9.6 Roles of insulin and glucagon on availability of active enzymes. Reductase refers to the HMG CoA reductase: the phosphorylated form of this enzyme is inactive; PPP, the pentose phosphate pathway.

EFFECT OF SURPLUS DIETARY CHOLESTEROL. In the fed state, dietary cholesterol is taken up by liver cells from chylomicron remnants after removal of much of the triacylglycerol component as fatty acids. This is effected by means of a receptor-mediated mechanism, associated with one of the apo-proteins on the surface of the chylomicron. The process is similar to uptake by non-hepatic cells (see below). When dietary levels of cholesterol are high this

7-α-Hydroxycholesterol

procedure meets all the requirements of the cell for membrane formation. It also exceeds the ability of catabolic pathways within the cell for its degradation and in this situation another means of controlling cholesterol synthesis is switched on. This form of inhibition also acts at the level of HMG CoA reductase but now suppression of gene transcription is in addition to the hormonal effects mentioned above.

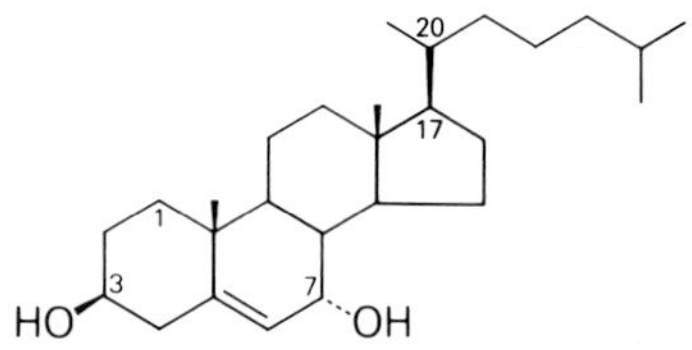

It is probable that the true inhibitory substance causing this effect in the liver and elsewhere is not cholesterol itself, but rather a hydroxylated derivative. Substances such as **7α-hydroxycholesterol** (in liver) or **20-hydroxycholesterol**, for example, are about 100-fold more potent than cholesterol in inhibiting gene transcription for HMG CoA reductase. However, these hydroxylated derivatives are much more water-soluble. Thus they inhibit the synthesis of HMG CoA reductase, possibly because they are more water-soluble, switching off cholesterol synthesis. The rationale for this effect is that these hydroxyl-ated products, in their appropriate tissue of origin, act as feedback inhibitors. They are natural catabolites of cholesterol which arise, for example, during bile acid formation in liver, or sterol hormone formation in, say, the adrenal cortex (see Fig. 9.16). Since they are the first intermediates in such pathways, they are ideally suited to signal an excess of cholesterol within the cell.

USE OF ANTIMETABOLITES. Recently a series of fungal metabolites has been discovered which act as potent inhibitors of HMG CoA reductase by competing with HMG CoA. They were identified after an intensive screening programme by pharmaceutical companies and isolated from the culture media of *Penicillium* spp. among others. They act as competitive inhibitors and have low toxicity for humans. Compactin and mevinolin have structures which are analogous to part of the HMG CoA molecule (Fig. 9.7). They have K_i values in the nanomolar region or less, and thus bind extremely tightly. These metabolites compete so well with HMG CoA that they block the formation of mevalonate almost entirely and hence suppress cholesterol biosynthesis. They have been tested in many species including humans and new drugs have recently been developed based on their structures. These drugs proved effective in reducing high plasma cholesterol and LDL-cholesterol (hypercholesterolaemia) and are now available in this country for clinical use in suitable patients.

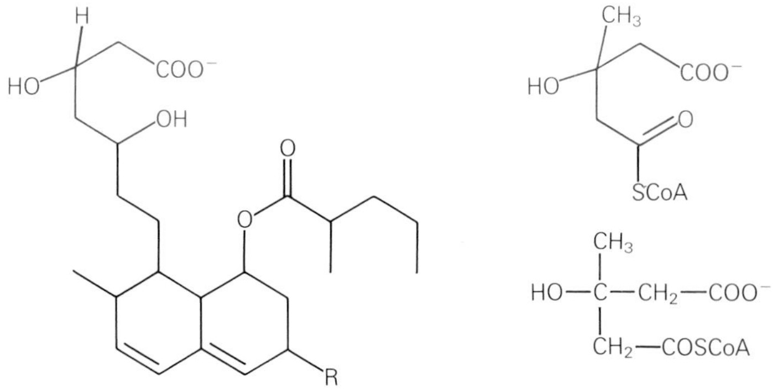

R = H, Ia; R = CH₃, Ib   Ic

Fig. 9.7 Resemblance of structures of fungal metabolites to HMG CoA.

ROLE OF THE LDL RECEPTOR. The LDL-cholesterol receptor controls the entry of cholesterol into many cells such as fibroblasts, lymphocytes, muscle cells and adipocytes, as well as liver cells. Delivery to cells is effected by a process known as receptor-mediated endocytosis which is initiated by the binding of the **apoprotein B-100** on the surface of the LDL (Fig. 9.8) to a receptor on the plasma membrane of the cell. Most cells of the body, despite their capacity for synthesis, make little of the cholesterol which is needed for their membranes. In adrenal glands, ovaries and testes, cholesterol is also required for steroid hormone production. These organs receive their cholesterol in general from the circulating plasma as LDL-cholesterol, mainly in ester form as cholesterol linoleate ($C_{18:2}$). This, in turn, is derived from VLDL (very low density lipoprotein) and chylomicron particles of liver and intestinal origin respectively.

☐ Apoprotein or apoenzyme is the term used to describe a protein without its cofactor or prosthetic group. **Apoprotein B-100** is a large protein (M_r 400 000) characteristic of LDL particles. One such molecule is associated in a monolayer with about 800 molecules of phospholipid and 500 molecules of free cholesterol. The hydrophobic core of the particle contains approximately 1500 molecules of esterified cholesterol.

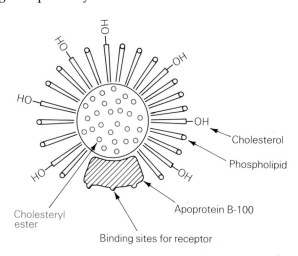

Fig. 9.8 Structure of LDL particle showing its central core of cholesterol ester (mainly linoleate) and outer layer comprising phospholipid, cholesterol and a single molecule of apoprotein B-100 (see also Fig. 8.27).

The process of LDL binding to cell surfaces, internalization, vesicle hydrolysis and redistribution of cholesterol was elegantly studied and described by Brown and Goldstein. The sequence of events which take place and the mechanisms which operate may be outlined as follows:

1. The LDL particle first binds to the cell surface because of the strong affinity of its apoprotein B-100 to a receptor located on a coated pit within the plasma membrane.
2. The LDL-receptor complex is then internalized by endocytosis, the membrane protein clathrin playing a prominent role in this action. The plasma membrane in this region invaginates and forms first a coated vesicle, and then an endosome, which fuses with a lysosome. Lysosomes are organelles responsible for degrading cell debris by hydrolysis and contain a battery of appropriate hydrolytic enzymes.
3. The receptor is then recycled to the cell surface. The LDL-protein is hydrolysed to amino acids and the sterol ester to cholesterol plus fatty acids.
4. The liberated cholesterol is then available to participate in a variety of structural, biosynthetic and regulatory events (Fig. 9.9). In part, it is used for membrane assembly upon cell division and, in appropriate cells, to make steroid hormones. Moreover, it activates an acyl CoA cholesterol acyl transferase (ACAT) which removes some excess free cholesterol by esterification, for storage within droplets in the cell.

☐ These experiments were performed by Brown and Goldstein in Texas and spanned a decade. They used mainly cultured human fibroblasts and other cells as source material. It is for this work and the important information gained from it that they were rewarded in 1985 with the Nobel Prize for Physiology or Medicine. The study of cholesterol in all its facets has been the subject of no less than 13 Nobel prizes over the years.

Reference Brown, M.S. and Goldstein, J.L. (1984) How LDL receptors influence cholesterol and atherosclerosis. *Scientific American*, **251(5)**, 58–66. An excellent, readable account of the role of these receptors in cholesterol uptake into cells.

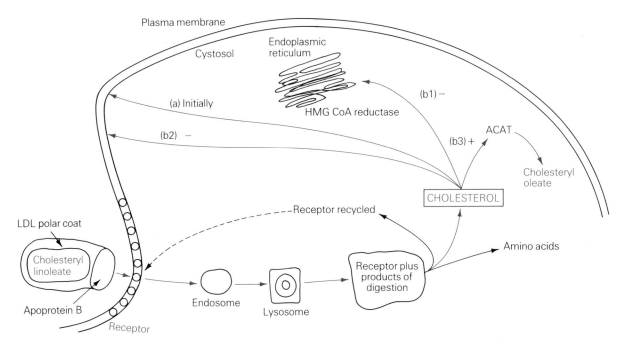

Fig. 9.9 Internalization of an LDL particle by a fibroblast. The subsequent activities caused by the cholesterol released are also shown.

5. Surplus cholesterol, beyond the requirements of the cell, suppresses the synthesis of the receptor for apoprotein B and also controls the formation of HMG CoA reductase by feedback mechanisms. These mechanisms operate at the level of production of the appropriate mRNAs. Thus, both *de novo* synthesis of cholesterol in the cell and its uptake from the plasma are cut off. In addition, excess cholesterol increases the rate of degradation of the enzyme, which normally turns over with a very short half-life of approximately 2–4 hours.

In moderate cases of hypercholesterolaemia, the underlying problem may be related to a reduction in number of LDL receptor proteins, brought about by a high cholesterol diet. This situation may be alleviated by drugs or bile acid-binding resins and, of course, by reducing dietary intake. These regimens would all stimulate, in various ways, increased synthesis of the receptor. This, in turn, would lead to a greater uptake of cholesterol from the plasma into the liver and enhanced secretion *via* bile acids. Conversely, a diet high in saturated (animal) fats tends to aggravate the situation, but the reason behind this is not yet clear.

Interrelationships between pathways

Acetyl CoA and its condensation products, acetoacetyl CoA and HMG CoA, are intermediates for the early stages of a number of major pathways of intermediary metabolism in the cytosol and/or mitochondria of liver cells. Fatty acid and sterol biosynthetic pathways possess a number of similarities. Both take place in the fed state and are initiated by cytoplasmic acetyl CoA, derived from citrate and, initially, glucose. Both involve considerable requirement for NADPH at various reductive steps. The pathways may be differentiated from one another since palmitate formation requires ATP to generate malonyl CoA (Chapter 8). This molecule which takes part in the initial condensation is formed from acetyl CoA in a reaction catalysed by

□ NADPH is invariably the reducing agent used in the cytoplasm for biosynthetic pathways. This cofactor originates from the oxidative reactions of the pentose phosphate pathway, and malate enzyme (*Energy in Biological Systems*, Chapter 5).

When the LDL receptor is absent or non-functioning, a condition called familial hypercholesterolaemia results. This is an inherited disorder where the faulty gene shows itself clinically in a grossly elevated plasma LDL-cholesterol content. LDL-derived cholesterol is deposited in tendons (xanthomas) and in arteries (atheromas). Such a situation leads to an accelerated development of atherosclerotic conditions in these individuals.

The prevalence of heterozygotes in European, American and Japanese populations is about 1 in 500 persons, whereas homozygotes number 1 in 10^6 in the United States. In heterozygotes, the coronary atherosclerosis usually develops after the age of 30 while in homozygotes coronary heart disease tends to begin in childhood and frequently causes death from myocardial infarction before the age of 20.

The primary genetic defect in familial hypercholesterolaemia results from one of several mutations in the gene specifying the receptor for plasma LDL. Entry of LDL-cholesterol into liver and other cells cannot occur by the endocytic route and this material therefore remains in large amounts in the circulation. Homozygotes for this disease have cholesterol levels in excess of $700\,mg/100\,cm^3$. Drugs have recently become available, however, which lower plasma cholesterol levels and may prevent atherosclerosis in these and other vulnerable patients. They are based on products related to inhibitors of HMG CoA reductase.

See also Box 8.3

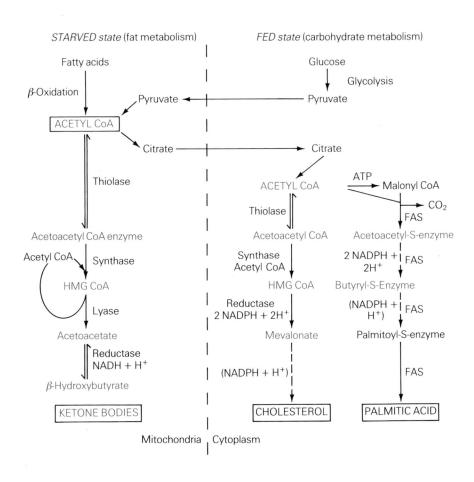

Fig. 9.10 Formation of cholesterol, palmitic acid and ketone bodies in liver cells. FAS, fatty acid synthase.

Box 9.2
Tumour cells

Control of activity of HMG CoA reductase is lost in cancer cells such as **hepatomas**. The enzyme in these liver cells is not affected by a high dietary intake of cholesterol, possibly owing to a defect in cholesterol transport across their plasma membranes. No constraint then applies to the production of this component for the plasma membrane and growth of new liver cells takes place rapidly.

Exercise 3

The reactions described in Fig. 9.10 are all directed towards synthesis of a final product or group of compounds. Describe the important factors which effectively make these pathways unidirectional.

acetyl CoA carboxylase, the key enzyme in this sequence, and requires the expenditure of ATP.

Liver mitochondria, in contrast, degrade fatty acids through β-oxidation to acetyl CoA and then ketone bodies when oxaloacetate becomes limiting. These products may be utilized in the starved or diabetic state for oxidative metabolism after transport to non-hepatic cells. The overall relationship between the key intermediates for these pathways is illustrated in Figure 9.10.

Conversion of mevalonate to squalene

The synthesis of squalene and cholesterol involves the conversion of mevalonate to the C_{15} compound **farnesyl-pyrophosphate** (farnesyl-PP). Indeed, isoprenoid compounds, including C_{10} products, are made from mevalonate. Mevalonate initially undergoes two successive phosphorylation steps at the primary alcohol group to give the pyrophosphate derivative, mevalonate 5-PP. This activated product is decarboxylated to the C_5 **isopentenyl-PP**. A concerted reaction takes place which also involves hydrolysis of a further molecule of ATP and loss of the tertiary alcohol group

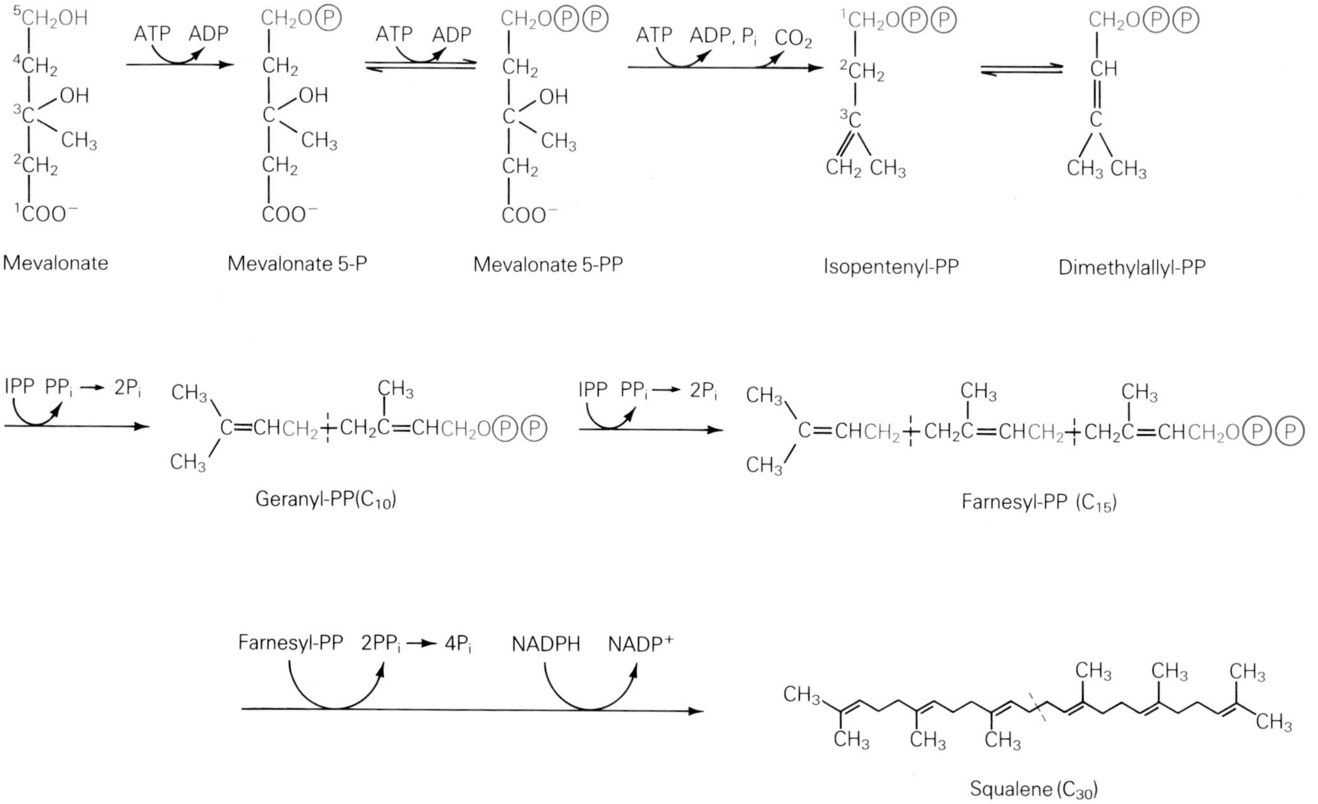

Fig. 9.11 Conversion of mevalonate to farnesyl pyrophosphate and subsequently to squalene. P$_i$, $H_2PO_4^-$; PP$_i$, $HP_2O_7^{3-}$; Ⓟ, HPO_3^-; ⓅⓅ, $HP_2O_6^{2-}$.

(Fig. 9.11). Isopentenyl-PP contains a nucleophilic methylene group and may be looked upon as the biologically active isoprene unit.

For polyisoprenoid synthesis to occur, 'head-to-tail' condensation between C_5 molecules must take place. Isopentenyl-PP is the 'tail' unit. Another active isomer, dimethylallyl-PP, is formed from isopentenyl-PP to form the 'head' unit. This reaction is catalysed by an isomerase. Dimethylallyl-PP is capable of an electrophilic reaction with the nucleophilic group of isopentenyl-PP.

Chain elongation is achieved in two stages through condensations catalysed by transferase enzymes to make the C_{10} compound **geranyl-PP** (leading to monoterpenes) and the C_{15} compound farnesyl-PP (Fig. 9.11). Dimethylallyl-PP reacts as the 'head' unit initially. The product of the first transferase reaction, geranyl-PP, contains a similar *allyl* group which allows it to condense further with another molecule of isopentenyl-PP. Pyrophosphate is released in the course of these reactions and is immediately hydrolysed to orthophosphate and therefore the sequence is driven in the direction of synthesis. In addition, a proton is lost sterospecifically at these steps to generate a *trans* double bond in the allylic product. *trans*-Geranyl-PP and *trans,trans*-farnesyl-PP are therefore formed.

The cytosolic enzyme in liver stops at the farnesyl-PP stage and two such C_{15} residues undergo reductive condensation in an uncommon type of reaction. Both pyrophosphate groups are eliminated in a 'head-to-head' condensation to give the C_{30} hydrocarbon product, **squalene**. The reaction proceeds in a complex manner through a C_{30} cyclopropane intermediate (presqualene-PP) which retains one of the pyrophosphate residues. In the final step, the 'linear' squalene is formed with the release of the second pyrophosphate; there is a requirement for NADPH at this point. The overall reaction may therefore be represented:

$$\text{R—CH}_2\text{OPP + PPOCH}_2\text{—R + NADPH} \xrightarrow{\text{Squalene synthase}}$$
$$\text{2 Farnesyl-PP}$$
$$\text{R—CH}_2\text{—CH}_2\text{—R + NADP}^+ + \text{H}^+ + \text{2 PP}_i$$
$$\text{Squalene}$$

Synthesis of longer-chain products

Many other transferase enzymes with wider specificity are found in both prokaryotes and eukaryotes. These transferases are capable of forming longer-chain products and always have important roles in the producing organism. Such compounds include chlorophyll, the electron transport intermediates ubiquinone and plastoquinone, the carotenoids, the polyprenols, and the very high M_r polymers rubber and gutta percha (Fig. 9.12) and many other products of wide-ranging structure and importance. As a general rule, C_{10} products are termed monoterpenes, C_{20} diterpenes (for example, gibberellins and phytol derivatives), C_{30} triterpenes (squalene and sterols) and C_{40} tetraterpenes (carotenoids).

How is the sterol ring formed?

Although squalene does accumulate in certain fish oils and in sebum, its *raison d'être* as an intermediate of the sterol ring system appears to be its structure and shape. It is converted into the sterol, lanosterol, in two steps, each catalysed by a separate enzyme. First a monooxygenase and then a cyclase. The double bond at either end of the molecule (adjacent to the *gem*-dimethyl groups) is attacked by molecular oxygen to generate squalene 2,3-epoxide

Allyl: the grouping CH_2=CH—CH_2 as in allyl chloride: CH_2=CH—CH_2Cl.

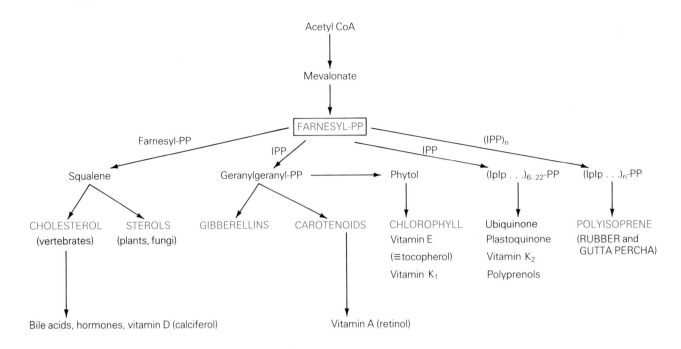

Acetyl CoA
Mevalonate
FARNESYL-PP
Farnesyl-PP
IPP
IPP
(IPP)n
Squalene
Geranylgeranyl-PP
Phytol
(Iplp . . .)6–22-PP
(Iplp . . .)n-PP
CHOLESTEROL
(vertebrates)
STEROLS
(plants, fungi)
GIBBERELLINS
CAROTENOIDS
CHLOROPHYLL
Vitamin E
(≡tocopherol)
Vitamin K1
Ubiquinone
Plastoquinone
Vitamin K2
Polyprenols
POLYISOPRENE
(RUBBER and
GUTTA PERCHA)
Bile acids, hormones, vitamin D (calciferol)
Vitamin A (retinol)

Fig. 9.12 Formation of important polyisoprenoids in nature.

(Fig. 9.13). This reaction is catalysed by a monooxygenase (epoxidase) enzyme which, typically, requires NADPH. The product now contains an active oxygen function sensitive to attack by an enzyme-derived proton from the cyclase. This opens the epoxide ring (Fig. 9.13) and initiates a series of electron shifts, mediated by a concerted movement of methyl groups and hydride ions (H^-).

The tetracyclic sterol nucleus is thus formed. **Lanosterol** is formed in animal and fungal cells and a similar intermediate occurs in plant cells. Because of the requirement for molecular oxygen, which forms the 3-β-hydroxyl group, cholesterol and other sterols can only be made by aerobic organisms.

Monooxygenase
(squalene epoxidase)
+ O2 + NADPH + H+
Squalene
H+
O
CH3
CH3
+ NADP+ + H2O
Squalene 2,3-epoxide
Cyclase
(squalene
oxidocyclase)
H+
H3C
H3C
H3C
HO
H3C
CH3
CH3
Lanosterol

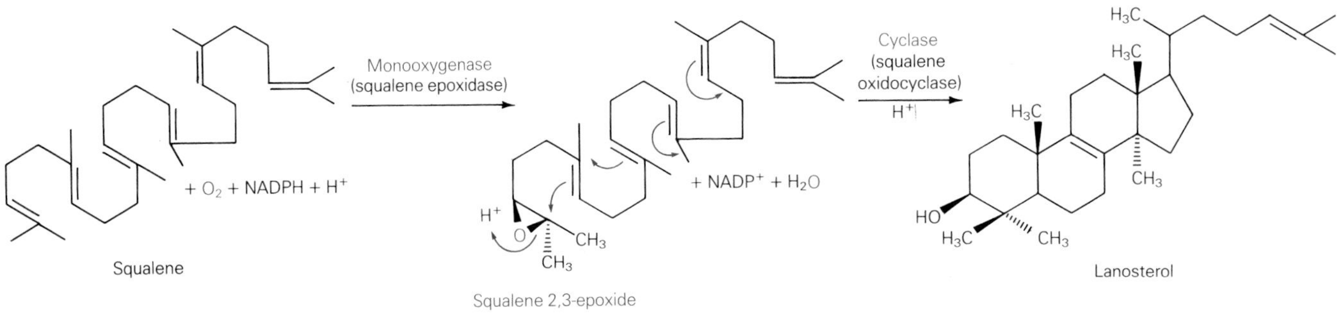

Fig. 9.13 Conversion of squalene to lanosterol.

Many separate stages are required to convert the C_{30} lanosterol into the C_{27} cholesterol and the reactions take place in the endoplasmic reticulum. The steps include:

- oxidation of three methyl groups after initial attack by monooxygenase enzymes and their subsequent elimination. This converts C_{30} to C_{27}. The C atoms are all eventually eliminated as CO_2 after oxidation to carboxyl groups;
- reduction of the double bond in the side-chain by means of NADPH; and
- migration of the double bond at C-8 to C-5 (in cholesterol).

7-Dehydrocholesterol, a precursor of vitamin D_3 in the skin, is formed in the course of the migration of the double bond from C-8 to C-5. The intermediate is then reduced with NADPH at the C-7 double bond to generate cholesterol.

The formation of cholesterol from acetyl CoA is summarized in Fig. 9.14, which highlights the input of various cofactors for the reactions concerned.

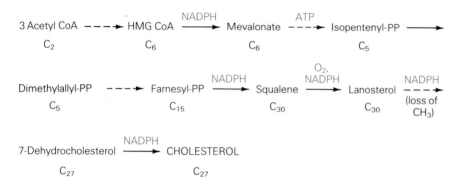

Fig. 9.14 Formation of cholesterol from acetyl CoA.

Exercise 5

Indicate the metabolic routes used to achieve the conversion of lanosterol into cholesterol. Why does exposure to ultraviolet light enhance the production of vitamin D in the skin?

Fate of cholesterol

After synthesis in the liver, cholesterol is packaged, together with phospholipid and apoproteins, into small disc-like particles. These are secreted into the circulation and, together with similar precursor species from the intestines, pick up more cholesterol from the surface of peripheral cells and become HDLs. Some of the cholesterol obtained in this way from the turnover of cell membranes or dying cells is esterified through the agency of the enzyme *lecithin* cholesterol acyl transferase (LCAT for short). The cholesteryl ester, thus formed, finds its way back into tissues after transfer into IDLs (intermediate density lipoproteins) and LDLs. Both of these particles may enter the liver (where the cholesterol is degraded to bile acids) and also non-liver cells. Surplus cholesterol is then eventually taken up again by 'nascent' HDL particles, and the cycle is repeated.

Conversion of cholesterol in the liver to **bile acids** represents the only catabolic route available for its degradation and eventual excretion from the body. Over 70% of the cholesterol is broken down in this manner. Bile acids are made after initial hydroxylation at the 7α-position when the product undergoes further reactions and partial cleavage of the side-chain. Cholate is formed. This material is activated to its coenzyme A derivative and then converted to **bile salts** after reaction with the amino group of glycine or

See *Biological Molecules*, Chapter 7

Lecithin: the old name for the phospholipid now called phosphatidylcholine. This lipid is an important component of most biological membranes.

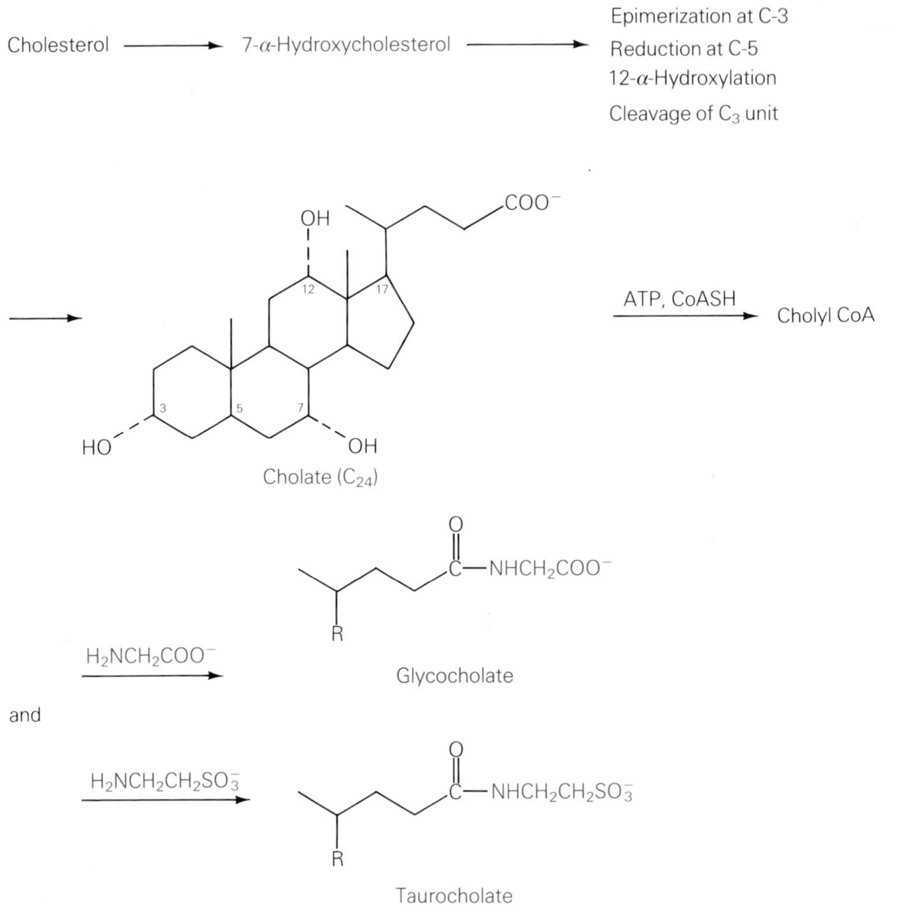

Cholesterol $\longrightarrow$ 7-α-Hydroxycholesterol $\longrightarrow$ Epimerization at C-3
Reduction at C-5
12-α-Hydroxylation
Cleavage of C_3 unit

Cholate (C_{24})

ATP, CoASH $\longrightarrow$ Cholyl CoA

$H_2NCH_2COO^-$ $\longrightarrow$

Glycocholate

and

$H_2NCH_2CH_2SO_3^-$ $\longrightarrow$

Taurocholate

Fig. 9.15 Formation of bile salts in mammalian liver. R, ring system of cholate with side-chain attached at C-17.

□ Reactions leading to the steroid hormones take place at the side-chain of cholesterol. Initial hydroxylations are followed by cleavage and sometimes total loss of the side-chain.

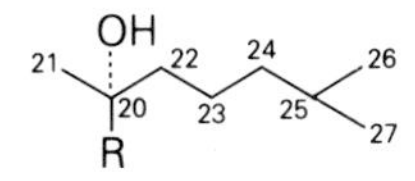

20α-Hydroxycholesterol (C_{27})

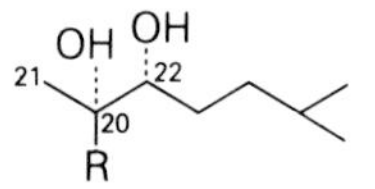

20α,22α-Dihydroxycholesterol (C_{27})

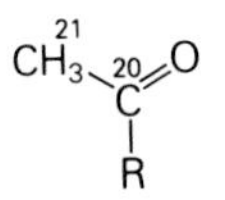

Pregnenolone (C_{21})

R represents the ring system in cholesterol

taurine. The reaction sequence yields the corresponding glycocholate and taurocholate as products (Fig. 9.15).

Bile salts have detergent properties and are secreted into the small intestine. Here they act as emulsifying agents to aid solubilization and digestion of dietary lipids (mainly triacylglycerols). Small droplets about 1–5 μm in diameter are produced which become more susceptible to hydrolytic attack by lipases.

Another metabolic fate for cholesterol is its conversion after attack by monooxygenase enzymes into **steroid hormones** (Fig. 9.16). These hormones (androgens, oestrogens, progestogens and the corticosteroids: glucocorticoids and mineralocorticoids) are made in testes, ovaries, placenta and the adrenal cortex. Cholesterol is stored in these tissues as its fatty acid ester within lipid droplets, and released when required. Steroid hormones form a major class of messengers that act in a variety of ways in their target tissues.

The reaction pathway always proceeds via hydroxylated derivatives of cholesterol and the C_{21} intermediate, pregnenolone. The hydroxylation steps require NADPH and molecular O_2. The reactions are catalysed by monooxygenases which contain cytochrome P450 as the terminal oxidase. In all these products, the C_6 terminus of the side-chain is cleaved initially between two adjacent hydroxyl groups to leave a residual C_2 unit. These carbon atoms may also be eliminated altogether as occurs in testosterone and oestrogen formation.

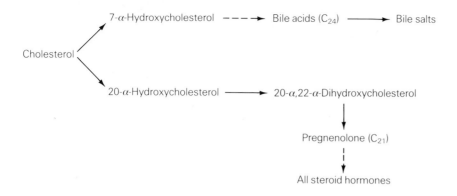

Fig. 9.16 Reaction pathways from cholesterol leading to bile salts and steroid hormones.

9.4 Further variations on the polyisoprenoid theme

Sterols (triterpenes) are quantitatively extremely important in all eukaryotic systems. They may therefore be considered, together with bile acids, major end products of the isoprenoid pathway. Likewise, **rubber** and **gutta percha** are produced as 'secondary' metabolites (see below) and are effectively secreted by the plants concerned in large amounts. **Phytol** derivatives, too, are abundant in nature when found as the side-chain of **chlorophyll**. Electron transport components, although biochemically vital, make use of a less major flux through the metabolic pathways involved. Steroid hormones and fat-soluble vitamins also come into this category.

Fat soluble vitamins

The best known fat soluble vitamins are A and D.

VITAMIN A (RETINOL) is essential for healthy skin and mucous membranes and the growth of cells. It is also needed for the synthesis of rhodopsin, a conjugate of the aldehyde, 11-*cis*-retinal, and protein. This material plays a key role in vision and is located in the retina. Exposure to light brings about the conversion of 11-*cis*-retinal to the all-*trans*-isomer, with a resultant change in geometry of the molecule and the generation of a sensory nervous impulse.

Vitamin A is found abundantly in fish liver oils and egg yolks. It can also be formed in the intestinal mucosa by the action of a dioxygenase which cleaves dietary β-**carotene** (a C_{40} tetraterpene, Fig. 9.17) at its central double bond into two molecules of **all-*trans*-retinal** (vitamin A aldehyde):

Fig. 9.17 Space-filled model of β-carotene.

β-Carotene

$+ O_2 \longrightarrow$

2

CHO

Retinal

The aldehyde may be reduced using NADH in an enzyme-catalysed reaction to give retinol. This alcohol is transported to the liver for storage mainly as retinyl palmitate after esterification with palmitic acid or other long-chain fatty acids.

□ **Dioxygenase** enzymes catalyse reactions where both atoms from a molecule of oxygen are incorporated into the substrate. Two molecules of the same monooxygenated derivative may be produced, as in retinal formation. Alternatively, both oxygen atoms may be incorporated into a single product. Contrast with monooxygenases described earlier in the marginal note on p. 194.

VITAMIN D (CALCIFEROL) is required for the healthy development of bones and teeth; deficiency gives rise to rickets when the bones are poorly calcified and become soft and pliable. **Vitamin D₃ (cholecalciferol)** is found in fish liver oils and also dairy products and egg yolks. It may be formed, however, in the skin by irradiation of 7-dehydrocholesterol with sunlight. 7-Dehydrocholesterol, a $\Delta^{5,7}$-unsaturated sterol (Fig. 9.18), is an immediate precursor of cholesterol in animal cells. The related ergosterol, found in yeasts and fungi, possesses the same ring system and behaves in a similar manner to generate ergocalciferol (vitamin D₂).

The presence of the conjugated double bond system in ring B of the 7-dehydrocholesterol molecule renders it susceptible to attack by ultraviolet light. Various electron shifts take place and the bond between C-9 and C-10 is cleaved to form precalciferol. This product undergoes isomerization to cholecalciferol in a second UV-mediated reaction. The process is normally adequate to provide sufficient vitamin D for our needs, even in the British climate, provided that the skin is partially exposed.

The active hormone arises after further hydroxylation reactions in liver and kidneys to give mainly the 1,25-dihydroxy derivative (Fig. 9.18), active in the regulation of Ca^{2+} and phosphate metabolism.

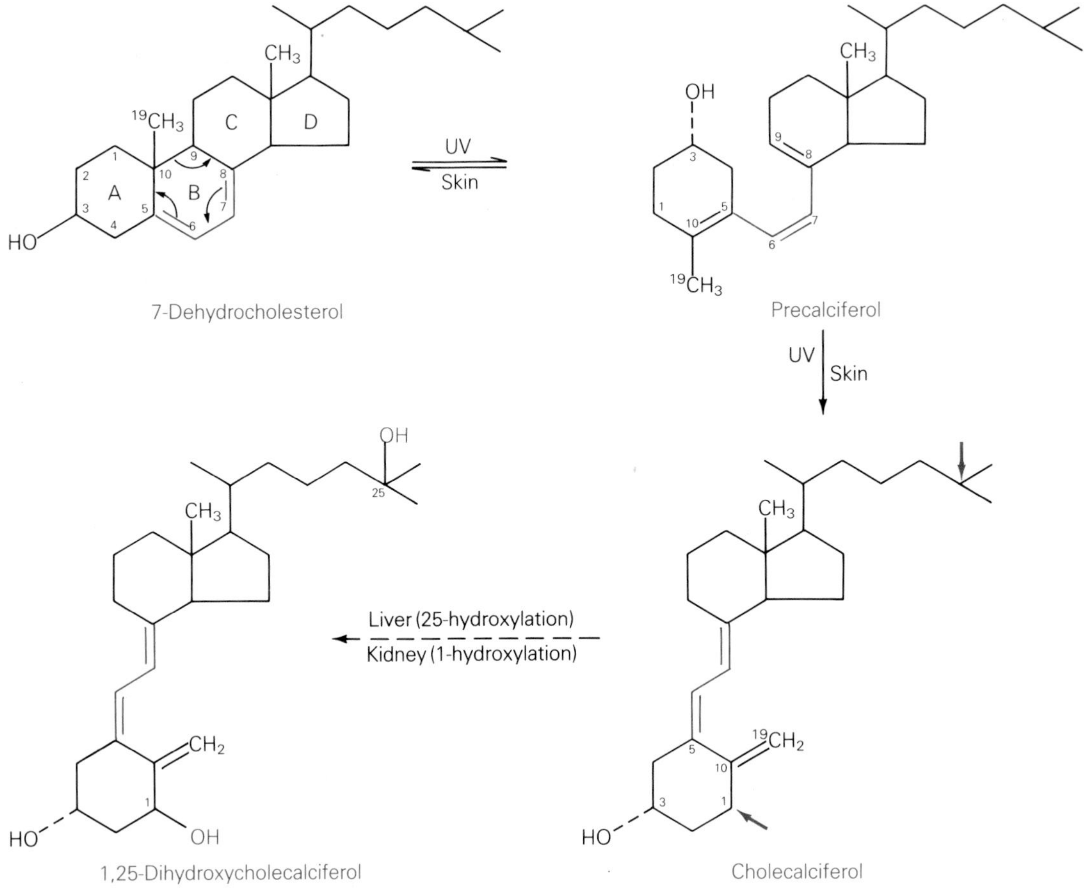

Fig. 9.18 Formation of cholecalciferol (vitamin D₃) and the active hormone. R, the side-chain of cholesterol.

Mono-, di- and tetraterpenes (C_{10}, C_{20} and C_{40} products)

Other isoprenoid derivatives are also potent in their biological activity but sometimes produced in relatively small amounts. Included in such a grouping are the C_{10} monoterpenes and the C_{20} gibberellins, diterpenes of complex cyclic structure which act as plant growth hormones. In addition, phytol is a diterpene derived from **geranylgeranyl-PP** (see Fig. 9.12) but this compound is widely distributed in nature. Three out of the four isoprene residues are reduced. A whole range of C_{40} carotenoid pigments also exists. These latter products include the hydrocarbon **carotenes** (see Box 9.3) plus the oxygenated or hydroxylated **xanthophylls**.

Carotenoids are formed initially by condensation of two geranylgeranyl-PP residues in a head-to-head manner to give the colourless **phytoene**. The linear chain of phytoene is susceptible to stepwise desaturation, generating substances with an extensive conjugated double bond system including β-carotene. These structures give rise to pronounced ultraviolet and visible light absorption spectra and hence carotenoids have intense colours. Carotenoids in general, therefore, act as photoreceptors and light-protecting agents in the grana of chloroplasts. These pigments are also responsible for some of the attractive colouring found in fruits and, to some extent, in seeds such as maize (corn) or flowers.

Tetraterpenes retain much of their linear structure and do not undergo extensive cyclization reactions as the di- or triterpene (sterol) groups. The

☐ **Monoterpenes** are derived directly from geranyl-PP without further condensation. They are readily isolated from plant extracts and frequently provide the basic fragrance underlying the attractive nature of perfumes and other essential oils. These derivatives include lavender, sandalwood, attar of roses, etc. In addition, many spices and flavour components found in fruit oils and leaves commonly used for culinary purposes are monoterpene compounds.

☐ Unlike squalene synthesis, the formation of **phytoene** does not utilize NADPH. During the head-to-head condensation reaction, two residues of pyrophosphate are released and a *cis* double bond is formed at the centre of the molecule:

Phytoene synthase

$$C_5 \rightarrow C_{10} \rightarrow C_{15} \rightarrow$$

$$R\cdots CH_2O\textcircled{P}\textcircled{P} + \textcircled{P}\textcircled{P}OCH_2\cdots R \longrightarrow$$
Geranylgeranyl-PP (C_{20})

$$R\cdots \overset{H}{\underset{|}{C}}\!\!=\!\!\overset{H}{\underset{|}{C}}\cdots R + 2\,PP_i$$
Phytoene (C_{40})

central double bond in phytoene prevents similar folding patterns.

Polyisoprene

Polyisoprenes include several materials of commercial importance, for example rubber and gutta percha.

RUBBER (POLY-CIS-ISOPRENE) is a linear chain polymer, containing some 1000–5000 isoprene residues. It is produced in the form of latex of the rubber tree, *Hevea brasiliensis*, by the action of an enzyme with very broad specificity, from isopentenyl-PP. Intermediates include *cis*-geranyl-PP, *cis,cis*-farnesyl-PP and many others. The prenyl transferase concerned possesses a different specificity at its active site from that involved in squalene and carotenoid biosynthesis. A different H atom is eliminated stereospecifically and all the double bonds formed are in the *cis* configuration. This structure permits free rotation around the —CH$_2$—CH$_2$— single bond to occur and gives rubber its characteristic properties (see Box 9.5).

GUTTA PERCHA (POLY-TRANS-ISOPRENE) is produced in the sap of the guayule plant (*Parthenium argentatum*) but in this case all the double bonds are in the *trans* form. The prenyl transferase system used removes the same H atom as that in squalene biosynthesis to give all-*trans* intermediates. Gutta percha is a hard, non-elastic substance which has been used as an insulating material in cables. Free rotation around the —CH$_2$—CH$_2$— regions is hindered by the presence of methyl groups on the same side of the *trans* double bond:

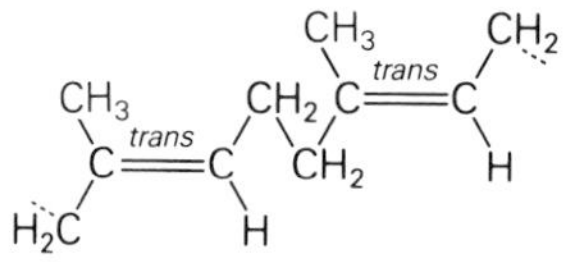

Box 9.5
Rubber

Rubber is produced in the latex of certain Angiosperms. Latex is simply a colloidal suspension of rubber particles in the watery sap. It is found in good yield in the bark of the tropical tree, *Hevea brasiliensis*, which is the commercial source cultivated in plantations. What a waste of energy for the plant (a great deal of ATP is used up in the synthesis of rubber) but how useful for us! An important feature of 'natural' rubber is its characteristic elastic properties. It is capable of returning to its original shape after extensive stretching. These properties are attributable to the structure of rubber and arise from the *cis* configuration. The hydrogen and methyl groups are found on the same side of the double bond and this positioning permits free rotation around the –CH$_2$–CH$_2$– single bond:

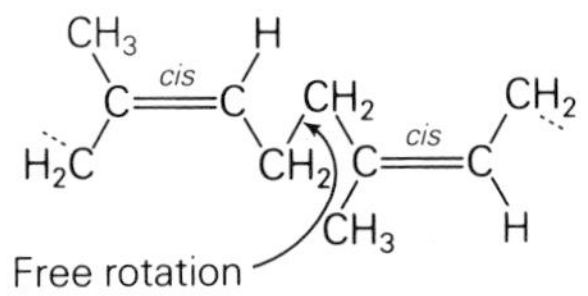

Simmonds, N.W. (1983) The rubber tree. *Biologist* **30**, 153–7. A synoptic history which describes the exploitation of natural rubber.

9.5 Biosynthesis of porphyrins

Porphyrin-containing compounds form a wide range of natural products of great biological importance. They include the oxygen-binding haem proteins, haemoglobin and myoglobin, the cytochromes, enzymes such as catalase and peroxidase, the photosynthetic pigment chlorophyll, and the corrin ring system of vitamin B_{12}. The haem proteins in general are essential components for cell respiration and energy generation in all organisms.

Porphyrins are all derived biosynthetically, at least in part and indirectly, from acetyl CoA but the more immediate precursors are succinyl CoA and glycine. Like cholesterol, therefore, these compounds are also assembled from key products of intermediary metabolism through a series of condensation steps. Moreover, metabolic blocks in ***porphyrin*** synthesis exist due to genetic lesions, and give rise to aberrant metabolism of intermediates in humans with consequent clinical symptoms (see Box 9.6).

Chemically, porphyrins are complex ring products, with structures comprising four substituted pyrrole units linked together by methene (—CH=) bridges. The basic planar skeleton is termed a porphin:

Porphin

Various side-chain substituents at the peripheral positions give rise to different classes of tetrapyrrole intermediates (see Fig. 9.19). Uroporphyrinogen, for example, is a precursor of haem and possesses four acetate and four propionate groups. In porphyrins themselves, a large number of conjugated double bonds are present which make these compounds absorb light strongly. They are therefore highly coloured. Porphyrins form ***chelates*** by co-ordination of suitable metal ions with the four nitrogen atoms.

Formation of the linear tetrapyrrole intermediate

The two major substrates for porphyrin biosynthesis are glycine and succinyl CoA. Succinyl CoA is an intermediate of the TCA cycle and its carbon atoms originate therefore from acetyl CoA. Like cholesterol, the initial breakthrough in determining the course of the pathway lay in the use of a 'heavy' isotope as tracer; this time ^{15}N was used. As this is not radioactive, incorporation into products was detected by mass spectrometry.

Shemin demonstrated in feeding studies that the nitrogen from [^{15}N]-glycine was incorporated into the nitrogen atoms of haem much more effectively than from other ^{15}N-labelled amino acids. Shortly afterwards, the involvement of the α-carbon of glycine, but not the carboxyl group, was confirmed using the ^{14}C-labelled substrates. Related degradative studies with methyl-labelled ([2-^{14}C]-) and carboxyl-labelled ([1-^{14}C])-acetate were performed. These studies showed that the former was incorporated into most of the remaining carbon atoms of the ^{14}C-labelled porphyrin from haem (24 out of 26), whereas [2-^{14}C]acetate was incorporated into the other two. These initial results suggested an involvement of succinyl CoA.

It was then established that the first (and committed) step in the biosynthesis of porphyrins is a condensation reaction, catalysed by a pyridoxal phosphate-dependent synthetase, between glycine and succinyl CoA. The

□ The porphyrin ring structure is somewhat complex to recall and even more formidable to draw correctly. A simple guide to assist the reader to draw the formula correctly is as follows:

1. first draw all the C–C and C–N bonds of the porphin ring system as single bonds;
2. next draw alternate (conjugated) double bonds starting anywhere but omitting two nitrogen atoms;
3. fill in the missing double bonds on appropriate rings;
4. finally, assign hydrogen atoms to the two nitrogen atoms with the 'missing' valency.

All double bond structures which can be drawn are equivalent because of resonance.

Exercise 6

Trace the pattern of labelling from [1-^{14}C]- and [2-^{14}C]acetyl CoA into the carbon atoms of succinyl CoA after one and two rounds of the TCA cycle. Hint: it may be helpful to consult *Energy in Biological Systems*, Chapter 4.

Porphyrin: from the Greek word meaning 'red' and relates to the port-wine colour, characteristic of urine from patients suffering from porphyrias.

Chelate: a term derived from the Greek word for 'crab's claw'. It generally refers to the co-ordinate bonds made between metal ions and two or more negatively-charged atoms or lone pairs of electrons on the same organic molecule.

product is δ-**aminolaevulinate** (ALA). This condensation takes place in two stages with the intermediate formation of an enzyme-bound (Schiff's base) derivative of α-amino-β-adipate which, on decarboxylation and release from the enzyme, gives ALA:

$$\text{Succinyl CoA} + \text{Glycine} \xrightarrow[\text{ALA synthetase}]{H^+ \quad CoASH \quad CO_2} \text{ALA}$$

ALA synthetase is located in mitochondria. The reaction is inhibited in a feedback manner by the end-product of this pathway, haem. Two molecules of ALA then condense, with loss of two molecules of water, to generate **porphobilinogen**. This product, with a single pyrrole ring, has one acetate (A) and one propionate (P) side-chain:

$$\text{ALA} \xrightarrow{\text{ALA dehydrase}} \text{Porphobilinogen} + 2\,H_2O + H^+$$

Further condensation of four porphobilinogen molecules then takes place in a head-to-tail manner to give an enzyme-bound tetrapyrrole. The linear product (aminomethyl-bilane) is released, after loss of ammonium ions, when each pyrrole ring is linked by a methylene ($-CH_2-$) group. Thus, three ammonium ions are lost at this stage, with a fourth eliminated later on ring closure to **uroporphyrinogen III**:

Uroporphyrinogen III

Haem formation

The subsequent stages of the pathway leading to **haem** formation are given in Fig. 9.19. The reaction sequence goes through the intermediate formation of uroporphyrinogen III (which is the first cyclic tetrapyrrole made), coproporphyrinogen III and protoporphyrin IX. These products are characterized by the nature of their side-chains. The reactions involve various side-chain modifications, coupled with oxidation of the methylene bridges. Finally, Fe^{2+} ions are inserted to give haem.

The first reaction shown in Figure 9.19 generates an asymmetric product in which an acetate (A) and propionate (P) residue from one end of the linear tetrapyrrole have become inverted. This requires the presence of two enzymes, porphobilinogen deaminase and a cosynthetase, since the deaminase alone gives the symmetrical I isomer. During the conversion of the biologically-active product, uroporphyrinogen III, to coproporphyrinogen III, all the acetate side-chains are decarboxylated to become methyl (M)

☐ Porphyrinogens are structures containing four substituted pyrrole rings connected by methylene (–CH₂–) bridges. In the intermediate, uroporphyrinogen III, the substituents are acetate (A) and propionate (P) residues present in the order A, P; A, P; A, P; P, A. Porphyrins are oxidized products whose structures are based on porphin, where each ring is linked by a methene (–CH═) bridge.

4 Phorphobilinogen

Aminomethyl-bilane

Reference Warren, M.J. and Scott, A.I. (1990) Tetrapyrrole assembly and modification into ligands of biologically functional cofactors. *Trends in Biochemical Sciences*, **15**, 486–91. Marvellous overview of the topic, all in less than six pages!

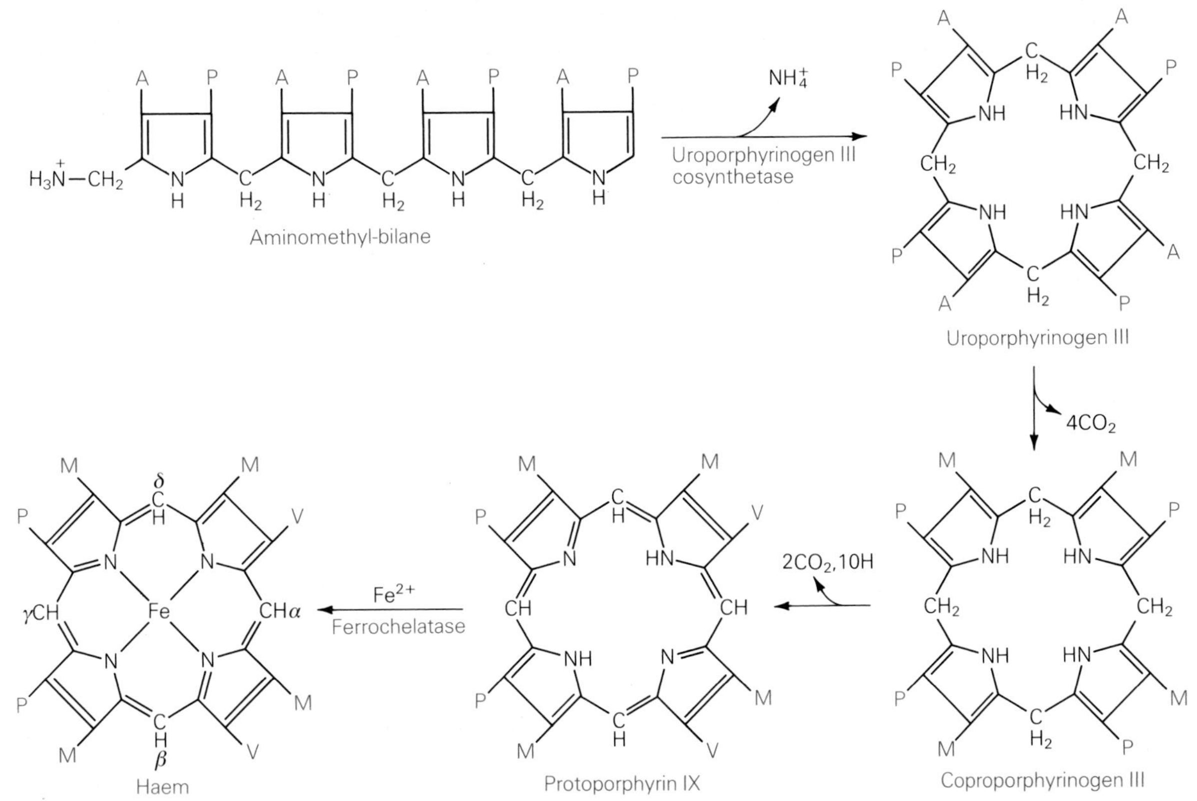

Fig. 9.19 Pathway for the cyclization of tetrapyrrole and formation of haem. A, $-CH_2CO_2^-$ (acetate); P, $-CH_2CH_2CO_2^-$ (propionate); M, $-CH_3$ (methyl); V, $-CH=CH_2$ (vinyl).

groups:

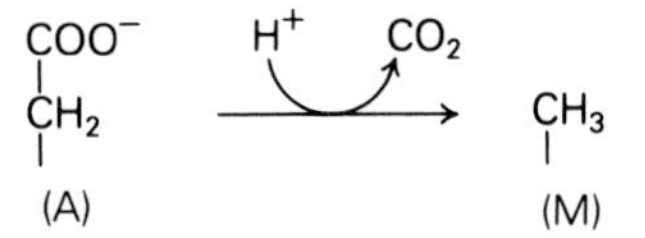

In the next step, all the methylene bridges are desaturated to methene groups, while two of the propionate side-chains are decarboxylated and desaturated to vinyl (V) groups:

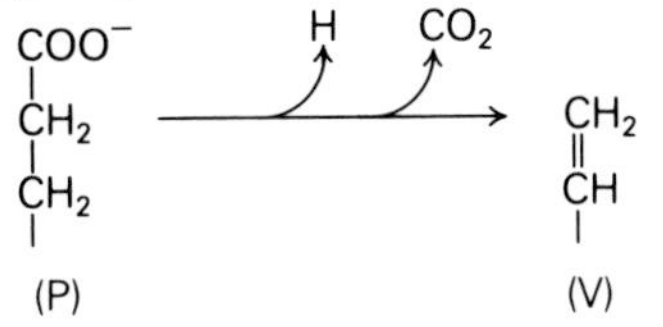

Thus, protoporphyrin IX, the only substituted porphin found in biological systems, is formed. Finally, insertion of an Fe^{2+} chelated to each of the four nitrogen atoms, is introduced by the enzyme ferrochelatase (protohaem ferrolyase). Haem is thus completed. The insertion of Fe^{2+} occurs by displacement of a proton from each of two opposing nitrogen atoms to generate a square-planar complex.

The fifth co-ordination position, available on the other side of the iron, is occupied in haemoglobin and myoglobin by a histidine residue within the globin polypeptides. In oxyhaemoglobin and oxymyoglobin, the remaining sixth position of the reduced Fe^{2+} is attached to a molecule of oxygen (Fig. 9.20). This position is free in the deoxy forms. In most cytochromes, both these co-ordination positions of the iron are bound covalently to amino acid residues.

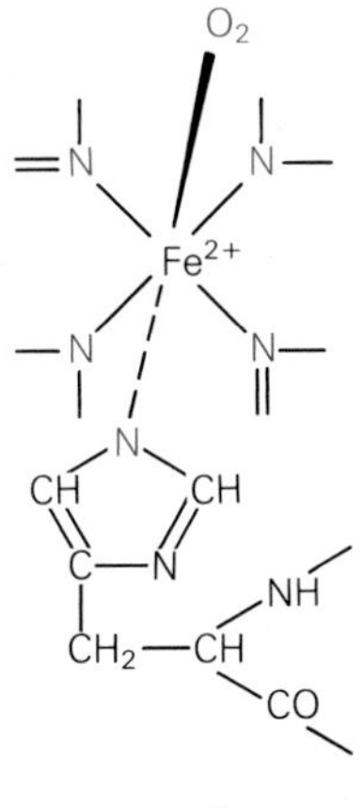

Fig. 9.20 Attachment of Fe^{2+} to the four N atoms of the haem pyrrole rings, the imidazole N of a histidine residue and O_2 in oxyhaemoglobin or oxymyoglobin.

Reference Granick, K.S. and Beale, S.I. (1978) Hemes, chlorophylls and related compounds: biosynthesis and metabolic regulation. *Advances in Enzymology* **46**, 33–203. Rather old, but still an excellent and comprehensive review of tetrapyrrole metabolism.

Reference Maines, M.D. and Kappas, A. (1977) Metals as regulators of heme metabolism. *Science* **198**, 1215–21. Includes a discussion of the role, biosynthesis and degradation of haem proteins.

Porphyria is the name given to a group of hereditary diseases arising from disorders of porphyrin biosynthesis. The absence of the cosynthetase is one such case. This deficiency causes the production of the tetrapyrrole, uroporphyrinogen I, on cyclization of the linear intermediate. This symmetrical product, with substituents positioned in the order A, P; A, P; A, P; A, P, and related derivatives are physiologically inactive. They accumulate in cells and may be excreted in the urine, giving it a port-wine colour.

This disorder is known as erythropoietic porphyria and results in erythrocytes being destroyed prematurely. The skin of these individuals retain porphyrins and is extremely sensitive to exposure to light. Serious lesions such as ulceration and even loss of extremities result, as does excessive hairiness. These symptoms may have been the basic cause of the problem ascribed to 'werewolves' and 'vampires' in the Middle Ages. Perhaps they learned to alleviate their symptoms by drinking blood?

Acute intermittent porphyria results from another defective gene causing a decrease in uroporphyrinogen III synthetase, with a corresponding compensatory increase in the ALA synthetase. This gives rise to an overproduction of porphobilinogen and ALA. These accumulate in the liver and enhanced excretion into the urine is observed; abdominal pain ensues associated with neurological symptoms. The apparent madness and symptoms periodically suffered by King George III may well have been attributed to such a condition.

The disease may be aggravated by certain drugs such as barbiturates and oestrogens, or other chemicals including dioxin. These exert their toxic effect by inducing synthesis of ALA synthetase and hence even more ALA is produced.

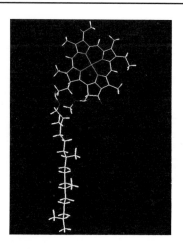

Fig. 9.21 Computer-drawn model of chlorophyll *a*. Courtesy of Dr C. Freeman, Polygen, University of York, UK.

9.6 Biosynthesis of chlorophyll

Chlorophylls are the green pigments found in all plant photosynthetic tissues (Fig. 9.21). They are located in thylakoid membranes stacked in the grana of chloroplasts. Chlorophyll *a* differs from chlorophyll *b* by a single substituent in one of the pyrrole rings. The closely-related bacteriochlorophyll is present in chromatophores within the cell membrane of true photosynthetic bacteria.

Chlorophyll functions in clusters and acts as a vital component in capturing light energy in order to excite and transfer electrons. The extensive conjugated ring system present renders it ideal for this purpose. The molecule is a tetrapyrrole derivative with an isoprenoid side-chain; in this case, a phytyl (C_{20}) derivative is attached (Fig. 9.22). The source of phytol is the C_{15} compound, farnesyl-PP, which is extended by further condensation with one molecule of isopentenyl-PP to give geranylgeranyl-PP (see Fig. 9.11). This C_{20} product then undergoes partial reduction.

The pathway leading to the synthesis of two metalloporphyrins, haem and chlorophyll, are common up to the intermediate, protoporphyrin. This was

Photochemotherapy is a newly developing treatment for certain cancers. This therapy uses the combination of a photosensitizing drug and laser light to destroy tumours. The drug in clinical use is Photofrin II, a mixture of porphyrins, which has the additional beneficial property of concentrating in tumour cells relative to the surrounding healthy tissue. In the absence of light it is harmless and will not damage tumour or healthy cells. However, the drug which strongly absorbs light may be activated by diverting laser light into tissue containing it. Toxic products are created which rapidly destroy the cells. The technique offers advantages over conventional chemotherapy in the treatment of solid tumours, since normal tissue damage (with its consequent undesirable side-effects) can be minimized.

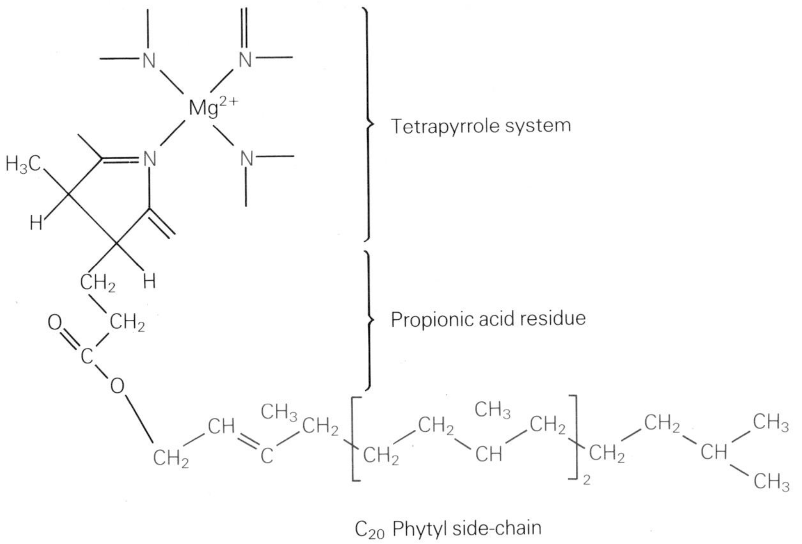

Fig. 9.22 Attachment of the phytyl side-chain through a propionate ester bond to one of the pyrrole rings in chlorophyll.

established in studies with mutants of the alga, *Chlorella* spp. Conversion of this compound into chlorophyll requires, among other steps, insertion of Mg^{2+}, in contrast to Fe^{2+}, for haem pigments and cytochromes. In addition, esterification of a propionate residue with phytol takes place. This confers lipid solubility and biological activity on the product. The presence of the isoprenoid side-chain permits chlorophyll to be anchored in hydrophobic regions within thylakoid membranes.

See *Cell Biology*, Chapter 5

9.7 Porphyrin degradation: bile pigment formation

As with cholesterol, haem needs to undergo degradation by means of attack by an oxygenase enzyme to give more active and soluble products. In the process, iron is released and may be reutilized by the body.

When red blood cells die after about 120 days in the circulation, the haem reacts with an endoplasmic reticulum monooxygenase in the liver or spleen. Oxygenation occurs initially at one of the methene carbon atoms of the protoporphyrin IX ring at the α-bridge position. Haem is cleaved to give **biliverdin**, together with carbon monoxide and iron as Fe^{3+}. NADPH is used as a co-reductant and the overall reaction requires two molecules of oxygen. Subsequently, biliverdin is converted to **bilirubin** by reduction with NADPH. Formation of these two **bile pigments** which are linear tetrapyrroles is shown in Fig. 9.23.

Bilirubin is transported to the liver after complexing with serum albumin, conjugated with glucuronate to give a more soluble form which is excreted into the bile. Bile pigments are ultimately eliminated from the body, after further modification, via the urine and faeces. Excess material may remain uncomplexed in the system if there is impaired liver function or blockage of the bile duct. The pigments then accumulate in the skin giving rise to the characteristic yellow colour of jaundice.

□ Various colorations are frequently observed after subcutaneous bruising. This situation results in the rupture of blood vessels and release of haemoglobin from damaged erythrocytes. This produces an initial purple colour at the site of the injury, followed by a blue-green appearance as **biliverdin** is formed and, eventually, the yellow colour of **bilirubin**. On recovery, normal skin colour returns as the bilirubin is transported to the liver and excreted as the glucuronide.

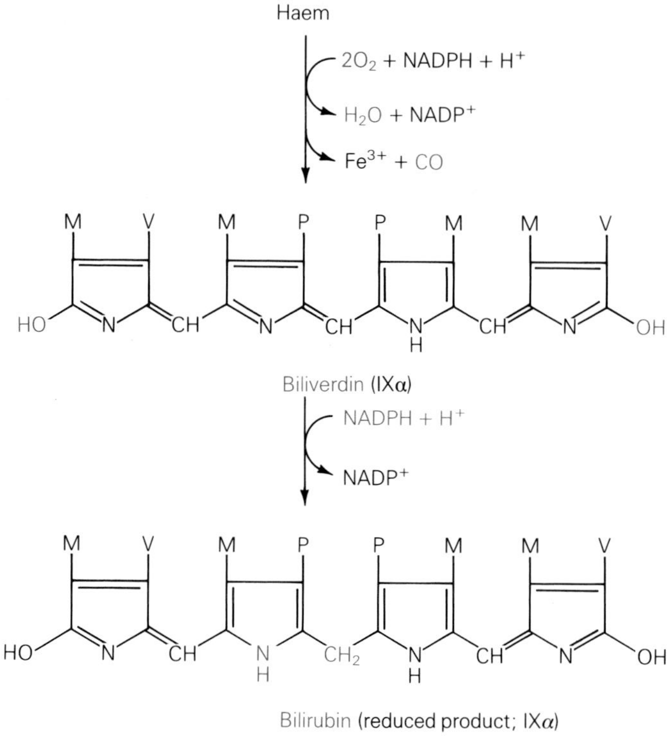

Haem
2O2 + NADPH + H+
H2O + NADP+
Fe3+ + CO
M V M P P M M V
HO N CH N CH N H CH N OH
Biliverdin (IXα)
NADPH + H+
NADP+
M V M P P M M V
HO N CH N H CH2 N H CH N OH
Bilirubin (reduced product; IXα)

Fig. 9.23 Formation of bile pigments from haem.

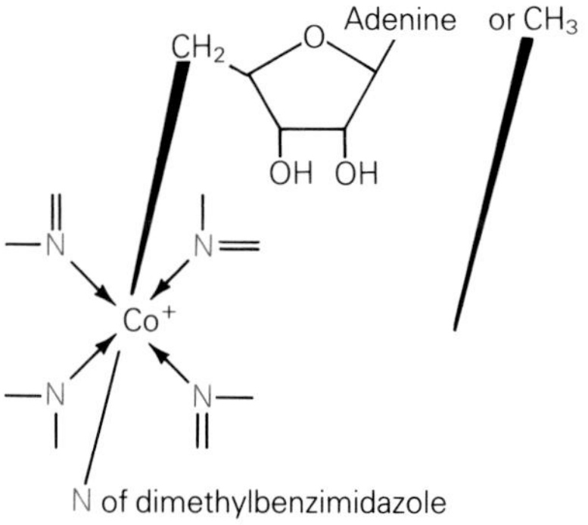

CH2
O
Adenine
or CH3
OH OH
N
N
Co+
N
N
N of dimethylbenzimidazole

Fig. 9.24 Attachment of the cobalt atom to the N of the corrin ring, the N of dimethylbenzimidazole and C.

9.8 Vitamin B_{12} and pernicious anaemia

Vitamin B_{12} (cobalamin) is a remarkably complex structure and has many novel features (see also Box 6.4). It too contains a tetrapyrrole ring system, which is not quite a porphyrin (or even a porphin) since one methene bridge is missing; rather it is termed a 'corrin'. In addition, a dimethylbenzimidazole ring is present which is bonded *N*-glycosidically to ribose 3-phosphate, that is, as a ribonucleotide.

The central cobalt atom is complexed to the five nitrogen atoms concerned, leaving a sixth co-ordination site available for a further ligand. This is occupied by CN in cyanocobalamin, the commercially available form of the vitamin. Biochemically, B_{12} may react as a 5′-deoxyadenosyl derivative (Fig. 9.24) as the coenzyme in rearrangement reactions, including the conversion of L-methylmalonyl CoA to succinyl CoA. The reactivity of coenzyme B_{12} relates essentially to the lability of the Co–C bond; this enables groups such as 'CH_3' to migrate between adjacent carbon atoms of the substrate.

Vitamin B_{12} is found mainly in liver and egg yolks. It is made by certain microorganisms including some which inhabit the intestines.

□ Absorption of **vitamin B_{12}** from the intestine depends upon the presence of a glycoprotein. This so-called '**intrinsic factor**' is normally present in gastric secretion and binds to B_{12}. The resultant complex is capable of binding to a receptor in the lining of the ileum. It may then be absorbed across the membrane and eventually released into the plasma. Pernicious anaemia generally results from a deficiency of this factor rather than a dietary deficiency of vitamin B_{12}.

Reference Stadtman, T.C. (1971) Vitamin B_{12}. *Science* **171**, 859–67. A fairly detailed account of the role of B_{12} in various metabolic processes for the student who would like more information.

9.9 Overview

Formation of cholesterol and other polyisoprenoid products is catalysed by a long, co-ordinated series of enzymes located in the cytosol and endoplasmic reticulum. Overall synthesis uses acetyl CoA as starting point in a reductive condensation sequence, which is subject to strict control exerted mainly at the level of HMG CoA reductase. NADPH is used as reductant throughout.

Cholesterol formation takes place primarily in the liver. It is transported to other tissues as LDL-cholesterol where recognition by an LDL-cholesterol receptor facilitates uptake. Cholesterol is a significant component in the lipid bilayer of plasma membranes of eukaryotic cells. It is also converted into bile acids and steroid hormones in appropriate tissues. Much effort has been expended in developing drugs which can regulate its formation in those situations where excess is made.

Porphyrins too are formed by condensation reactions from the key intermediates, succinyl CoA and glycine. Haemoglobin, cytochromes, catalase and vitamin B_{12} are vitally important biological molecules formed by this route.

The formation of both polyisoprenoids and porphyrins is important in ensuring the normal metabolism of cells. Absence of functional enzymes or errors in control mechanisms give rise to excess or aberrant products leading to serious clinical disorders.

1. Reaction pathway from [1-^{14}C]acetate ('c'-labelled):

$$CH_3^*CO_2Na \rightarrow CH_3^*COSCoA \rightarrow CH_3^*COCH_2^*COSCoA \rightarrow$$

$$\underset{\underset{CH_2^*CO_2H}{|}}{\overset{\overset{OH}{|}}{CH_3^*C}}CH_2^*COSCoA \longrightarrow \underset{\underset{CH_2^*CO_2H}{|}}{\overset{\overset{OH}{|}}{CH_3^*C}}CH_2^*CH_2OH \longrightarrow \underset{\underset{CH_2}{||}}{CH_3^*C}CH_2^*CH_2O\textcircled{P}\textcircled{P} \longrightarrow$$

$$\underset{CH_3}{\overset{CH_3}{C}}=CH\overset{*}{C}H_2O\textcircled{P}\textcircled{P} \cdots\cdots \rightarrow \underset{CH_3}{\overset{CH_3}{\overset{*}{C}}}=CH\overset{*}{C}H_2 \Big| CH_2\overset{CH_3}{\overset{|}{C}}=CH\overset{*}{C}H_2 \Big| CH_2\overset{CH_3}{\overset{|}{\overset{*}{C}}}=CH\overset{*}{C}H_2O\textcircled{P}\textcircled{P} \longrightarrow$$

Farnesyl-PP (C$_{15}$)

Squalene $\cdots\cdots\rightarrow$ Cholesterol

Similarly, for [2-^{14}C]acetate ('m'-labelled).

Homogenates of liver or cytosol plus microsomal preparations from liver (or yeast) were incubated in buffer with appropriately labelled acetate plus suitable cofactors. Lipids were extracted from these preparations with chloroform-methanol mixtures; squalene and cholesterol were then resolved from other lipids, purified and degraded chemically.

2. Use ^{14}C-labelled alanine which readily passes through the cell membrane. It is metabolised in part as follows:

^{14}C-alanine $\rightarrow$ ^{14}C-pyruvate $\rightarrow$ ^{14}C-acetyl CoA $\rightarrow$ ^{14}C-citrate (mitochondria) $\rightarrow$

$$^{14}\text{C-citrate (cytosol)} \xrightarrow[\text{[Hydroxycitrate} \ominus]}{\text{ATP citrate lyase}} {}^{14}\text{C-acetyl CoA} \begin{array}{l} \nearrow {}^{14}\text{C-labelled fatty acids} \\ \searrow {}^{14}\text{C-labelled cholesterol} \end{array}$$

Isolate the ^{14}C-labelled lipids, resolve fatty acids and cholesterol from each other and purify; both would be labelled in control homogenates. Addition of hydroxycitrate to the system inhibits ATP citrate lyase and prevents formation of cytosolic acetyl CoA, essential for both fatty acid and cholesterol biosynthesis. Thus neither product is formed under these conditions.

3. In each of the three pathways, enzymes catalysing physiologically unidirectional reactions are important.
(a) Ketone bodies: sequential use of HMG CoA synthase and ligase generate acetoacetate with loss of coenzyme A. High [NADH]/[NAD$^+$] in mitochondria in starved/diabetic state favours formation of β-hydroxybutyrate.
(b) Cholesterol: sequential use of HMG CoA synthase and reductase generate mevalonate with loss of coenzyme A. High [NADPH]/[NADP$^+$] in cytosol during fed state assists reduction. Reactions in direction of synthesis are also encouraged by loss of pyrophosphate (and its exothermic conversion to phosphate) in the condensation reactions which follow. (See also Exercise 4.)
(c) Palmitate: hydrolysis of ATP in formation of malonyl CoA; subsequent loss of CO$_2$ favours condensation step.

Again, high [NADPH]/[NADP$^+$] in cytosol during fed state assists reduction reactions.

4. These derivatives are effectively activated by this means since two molecules of ATP are required for their synthesis. Each condensation reaction results in loss of pyrophosphate, with immediate hydrolysis by pyrophosphatase into two residues of inorganic phosphate. The ΔG value for the overall process of condensation is, therefore, highly negative (exergonic) and drives the reaction sequence in the forward direction of synthesis.

5. See Fig. 9.14: lanosterol is converted into cholesterol by ER enzymes after (a) oxidation of three methyl groups to carboxyl groups and their subsequent loss as CO$_2$; (b) reduction of the side-chain double bond with NADPH; (c) migration of the double bond at C-8 to C-5.

Formation of vitamin D takes place through the intermediacy of 7-dehydro-cholesterol, which is attacked by ultra violet light (ring B). See Fig. 9.18 and related text for further details.

6. From [1-^{14}C]acetyl CoA:

$$CH_3^*COSCoA \longrightarrow \underset{\underset{CH_2CO_2H}{|}}{\overset{\overset{CH_2^*CO_2H}{|}}{HOCCO_2H}} \longrightarrow \underset{\underset{HOCCO_2H}{|}}{\overset{\overset{CH_2^*CO_2H}{|}}{CHCO_2H}} \longrightarrow \underset{\underset{COCO_2H}{|}}{\overset{\overset{CH_2CO_2H}{|}}{CH_2}} \longrightarrow$$

citrate　　　isocitrate　　[γ-^{14}-CO$_2$H]α-ketoglutarate

$$\underset{CH_2COSCoA}{\overset{CH_2^*CO_2H}{|}} \longrightarrow \underset{CH_2CO_2H}{\overset{CH_2^*CO_2H}{|}} \cdots\cdots\rightarrow \underset{CH_2^*CO_2H}{\overset{COCO_2H}{|}}$$

succinyl CoA　　　succinate　　　oxaloacetate
　　　　　　　　(label randomized)

Both labelled carboxyl groups are lost as ^{14}CO$_2$ in the next round of the cycle at the level of isocitrate dehydrogenase and α-ketoglutarate dehydrogenase.

From [2-^{14}C]acetyl CoA:

$$^{**}CH_3COSCoA \longrightarrow \underset{\underset{CH_2CO_2H}{|}}{\overset{\overset{^{**}CH_2CO_2H}{|}}{HOCCO_2H}} \xrightarrow{\text{as above}} \underset{^*CH_2CO_2H}{\overset{^*CH_2CO_2H}{|}} \cdots\cdots\rightarrow \underset{^*CH_2CO_2H}{\overset{^*COCO_2H}{|}}$$

　　　　　　　　　citrate　　　　succinate　　　oxaloacetate
　　　　　　　　　　　　　　(label randomized)

No ^{14}CO$_2$ is lost in round 1.

$$\xrightarrow{\text{round 2}} \underset{\underset{CH_2CO_2H}{|}}{\overset{\overset{^{**}CH_2CO_2H}{|}}{HOCCO_2H}} \cdots\cdots\rightarrow \underset{^*CH_2CO_2H}{\overset{^{**}CH_2CO_2H}{|}} \longrightarrow$$

further randomization and loss of half of 'original' label as ^{14}CO$_2$ in round 3.

Thus, the two CO$_2$ molecules lost at each round of the TCA cycle originate from the oxaloacetate formed in the 'previous' round and not from the C atoms of the 'incoming' acetyl CoA.

FILL IN THE BLANKS

1.   Polyisoprenoids are widely found in nature and are formed initially from _____ CoA and _____ CoA. _____ , a C_6 precursor, and _____ , the active biological isoprene unit, also act as intermediates in this biosynthetic process. Isoprenoid products include _____ , the major sterol in vertebrate tissues, and from this _____ acids and many _____ hormones are made.

The reactions leading to sterol synthesis take place in the _____ of the cell and many enzymes are located in the _____ _____ . The reducing equivalents required are supplied by _____ . Intermediates in the process are the hydrocarbon _____ , which has _____ carbon atoms, and _____ the first sterol formed. Formation of this sterol requires _____ oxygen.

Most of the cholesterol formed in the body is made in the _____. The hormones _____ and _____ play important roles in the reaction sequence and exert their control at the level of the enzyme _____ _____ _____ . This enzyme may also be inhibited by _____ inhibitors of its normal substrate.

Choose from: acetoacetyl, acetyl, bile, cholesterol, competitive, cytosol, endoplasmic reticulum, glucagon, HMG CoA reductase, insulin, isopentenyl-PP, lanosterol, liver, mevalonate, molecular, NADPH, squalene, steroid, 30.

MULTIPLE-CHOICE QUESTIONS

2.   Mevalonate is a key intermediate concerned with isoprenoid biosynthesis and is formed by reduction of HMG CoA. Which of the following statements is/are correct?

A. mevalonate is a branched-chain C_5 compound
B. the reducing equivalents required for mevalonate formation are supplied by NADH
C. HMG CoA undergoes two separate reductive steps to generate mevalonate
D. HMG CoA may also be utilized for ketone body formation in liver mitochondria
E. mevalonate reacts with ATP at the carboxyl group
F. synthesis of HMG CoA from acetyl CoA requires three condensation steps

3.   Biosynthesis of isoprenoid-derived products utilizes the reactive intermediate isopentenyl-PP. Which of the following statements is/are correct?

A. isopentenyl-PP is reactive because it contains a branched methyl group
B. isopentenyl-PP initially reacts with its isomeric dimethylallyl derivative
C. condensation reactions leading to longer-chain products always involve dimethylallyl-PP
D. pyrophosphate is released in the reaction described in B
E. the C_{15} product, farnesyl-PP, cannot undergo further condensation in liver cytosol
F. farnesyl-PP undergoes further condensation to longer-chain products in mitochondria and chloroplasts

4. Tetrapyrroles form the basic structure of many important biological compounds. Which of the following statements concerning their metabolism or structure is/are correct?

A. the major precursors of porphyrins are succinyl CoA and lysine
B. haem, cytochrome c, catalase and porphobilinogen are all tetrapyrrole compounds
C. the pyrrole rings in uroporphyrinogen III are linked by methene bridges
D. another name for haem is Fe-protoporphyrin IX
E. the Fe atom in all haem-containing compounds may be linked to O_2
F. conversion of haem to biliverdin requires an input of NADPH and O_2, and release of Fe^{3+}

SHORT-ANSWER QUESTIONS

5. Name three important anabolic enzymes, apart from HMG CoA reductase, whose amount would be reduced in liver cells during starvation or uncontrolled diabetes owing to lack of insulin.

6. Why is the structure of mevalonate particularly suitable for its function as precursor of isoprenoids?

7. Insertion of oxygen into hydrocarbon chains or ring systems is frequently performed throughout nature by monooxygenase or dioxygenase enzymes. This procedure renders the substrate more chemically active or water-soluble. Name some important examples of this phenomenon.

8. Why is the porphyrin ring structure suited to accept metal ions? How is the protein moiety in such compounds as (a) myoglobin or (b) cytochrome c associated with the Fe in the porphyrin prosthetic group? (c) How does the association in cytochrome oxidase differ from that found in other cytochromes?

ESSAY QUESTION

9. Discuss the importance of NADPH in the biosynthesis of cholesterol in mammalian liver.

Answers to Questions

Chapter 1

1. *Biosynthesis* is synonymous with anabolism which describes the metabolic processes whereby organisms build up complex molecules from simpler ones. Some organisms can build all their cellular material from CO_2, water and a source of nitrogen, but on the whole animal cells cannot do this. The macromolecules of the cell are built up from monomer units in an energy requiring process which usually involves ATP or another nucleoside triphosphate. This process is a series of condensation reactions to produce long chain molecules.

2. In general, the majority of the energy released by living cells comes from oxidative processes such as the catabolism of carbohydrates and fats obtained from or stored food materials. Plant cells differ in that they obtain energy from sunlight, but they still store carbohydrates and fats for use in periods of darkness. The formation of the monomer building blocks of cellular macromolecules takes place either *de novo*, and requires reducing power: alternatively these units are obtained from food materials. The coupling together of monomers to form macromolecules is neither oxidative nor reductive but is described as a condensation reaction requiring the removal of water.

3. C, D, E.

4. E.

5. C, E.

6. C, D.

7. When two monosaccharides link to form a disaccharide (glycosidic link formation) and when two amino acids link to form a disaccharide (peptide bond formation). Reverse of condensation is hydrolysis.

8. Highly specific, controllable, highly efficient, work at relatively low temperatures.

9. For making small molecules from simple or inorganic precursors a whole molecule such as a sugar or an amino acid has to be built up. The pathways for doing this are typically complex, and ATP and often reducing power (NADPH) are required. Formation of macromolecules from monomers involves condensation and ATP (or some other nucleoside triphosphate) which typically form a compound with the monomewr to be added. 'Information' is also usually required to ensure the correct order of monomer units.

10. Amino acids for proteins, and for information specifying the order is in the sequences of bases in nucleic acids. Ribonucleoside triphosphates for RNA. The information specifying the order of the bases is in DNA: for DNA, usually another DNA molecule specifies the order and the monomer units are in the form of deoxynucleoside triphosphates. For polysaccharides the monomers are monosaccharides but the order is specified by the enzymes catalysing the addition of these units.

11. It is not feasible to mix two amino acids and ATP is aqueous solution and expect to get a dipeptide. Even if it happened (which it does not) it would be unsatisfactory. Monomer units (here amino acids) are added one at a time to a growing polymer (polypeptide) in a specified order. The monomer is 'activated' in that it carries its 'energy' with it so that the reaction proceeds, and an enzyme directs and catalyses the process. In the case of amino acids ATP is originally used in reactions that produce amino acyl-tRNA complexes which are the ultimate amino acid residue donors and they are activated in

the sense that they are fairly unstable and readily donate their amino acids to the growing polypeptide. Similar, but not identical, reactions are used for the biosynthesis of polysaccharides and nucleic acids.

12. Pure water is $1000/18 = 55.5 \, \mathrm{mol \, dm^{-3}}$, therefore cells have a water concentration of $0.7 \times 55.5 = 38.9 \, \mathrm{mol \, dm^{-3}}$.

13. Identify monomer units, how order is specified, how energy is used to drive the condensation reactions.

14. Cells store information as a sequence of base residues in a DNA or RNA molecule. A certain code is used with specific rules and there are elaborate mechanisms for decoding and using the information as well as for ensuring fidelity when the information is copied (e.g. prior to cell division). Human ingenuity has devised a multitude of ways of storing information from the printed word and the gramophone record (which stores an analogue of the sound information) to braille. Modern information storage is almost entirely digital, e.g. pits on a CD or binary information in magnetic media such as tapes or floppy disks. Digital information is better in many respects. Apart from being in the right form for modern electronics to handle (at great speed), it is easier to copy and error-checking is easier (why is this?). Is biological information storage analogue or digital.

15. The building-up of anabolism on the whole require energy because more complicated structures are being built from simple ones. In detail, the reaction:

$$\text{monosaccharide} + \text{monosaccharide} \rightarrow \text{disaccharide} + H_2O$$

for example, is endergonic or requires energy (hydrolysis releases the same amount of energy). Therefore ATP is required: the hydrolysis of ATP is an exergonic process and therefore coupling this to the synthesis of a disaccharide produces an overall reaction that *will* proceed:

monosaccharide + monosaccharide →

disaccharide + H_2O $\Delta G^{0'} \sim$

$+ 23.1$ kal mol^{-1}

$ATP + H_2O \rightarrow$

$ADP + P_i$ $\Delta G^{0'} \sim - 30$ kcal mol^{-1}.

The net reaction, i.e. the sum of these two equations has $\Delta G^{0'} - 7$ kcal mol^{-1}.

16. The two major problems are that the macromolecule may be insoluble (e.g. collagen, cellulose) and therefore difficult to transport in aqueous media or through membranes, and that all the enzymes and other necessary ingredients and participants, as well as the energy supply or the activated monomers must be outside the cell membrane. This presents a number of problems compared with the tightly controlled environment inside the cell. The process is not impossible. Cellulose of plant cell walls is probably built in this way but collagen is exported from animal cells as a soluble precursor, procollagen, to be made insoluble in the extracellular matrix.

17. The structures must be attractive to one another and must have appropriate complementary shapes so that numerous weak, non-covalent bonds may form and hold the structure together. Mutually attractive structures are the phospholipids of the cell membrane (hydrophobic) and membrane proteins. The amphoteric nature of the lipids keeps them in the correct orientation in the lipid bilayer. Ionic and other bonds such as hydrogen bonds are hydrophilic and are responsible for holding other structures together. (Research: look up why sickle cell haemoglobin in such a disaster.) The point of having many weak, non-covalent forces is that these require little or no energy to make or break and thus supra-molecular structrures from easily (also breaks down easily) when required.

Chapter 2

1. In C_3 plants the first stable product of CO_2 fixation is 3-phosphoglycerate which is formed by carboxylation of the C_5 acceptor molecule ribulose 1,5-bis-phosphate by the enzyme ribulose-1,5-bisphosphate carboxylase and subsequent hydrolysis of the transient C_6 intermediate. The 3-phosphoglycerate produced is phosphorylated to produce 1,3-bisphosphoglycerate and reduced by glyceraldehyde 3-phosphate dehydrogenase using NADPH as electron donor to produce glyceraldehyde 3-phosphate. Some of the glyceraldehyde 3-phosphate is converted to the hexose fructose 6-phosphate, but most of it is recycled in a complex series of reactions to produce the C_5 sugar phosphate ribulose 5-phosphate, which is then phosphorylated to regenerate the CO_2 acceptor ribulose 1,5-bisphosphate. These reactions are collectively called the Calvin cycle.

2. The enzyme ribulose-1.5-bis-phosphate carboxylase, which catalyses the first step in CO_2 fixation, comprises small regulatory and large catalytic subunits. The small subunits are synthesized in the cytosol whilst the large subunits are synthesized in the chloroplast stroma. Assembly of the small and large subunits into an active oligomer requires the presence of a binding protein. The active complex is subject to control by a number of factors. For example NADPH binds and allo-sterically activates. Also the pH of the stroma rises when the thylakoid membranes take up H$^+$ ions to a pH that is optimum for ribulose-5-bisphosphate carboxylase activity. Other Calvin cycle enzymes are activated by reduction of disulphide bonds by the small redox protein thioredoxin. This protein in turn receives its electrons from ferredoxin. Since the reduced form of the latter protein accumulates in the light most of the Calvin cycle enzymes are strongly stimulated by light.

3. C_4 and CAM plants differ from C_3 species in that oxaloacetate and not the C_3 molecule 3-phosphoglycerate is the first stable CO_2 fixation product. This C_4 compound results from the carboxylation of phosphoenolpyruvate by the enzyme phosphenolpyruvate carboxylase. In C_4 plants this reaction takes place in the mesophyll cells, which surround the bundle sheath cells. The oxaloacetate is then reduced to malate which is transported to the bundle sheath cells and decarboxylated. Fixation of the released CO_2 is then by the enzyme ribulose-1,5-bisphosphate carboxylase using ribulose 1,5-bisphosphate as CO_2 acceptor. This scheme of CO_2 fixation means that CO_2 becomes concentrated in the bundle sheath cells and the energetically wasteful process of photorespiration is suppressed. The special organization of mesophyll and bundle sheath cells is called Krantz anatomy.

4. A, D
5. B, D
6. A, C
7. Similarities: CO_2 acceptor is phosphoenolpyruvate, C_4 acid is transported and then decarboxylated. Ultimate CO_2 acceptor is ribulose 1,5-bisphosphate. Differences: C_4 plant leaves have specialized anatomy, separation of C_4 enzymes into specialized mesophyll and bundle sheath cells.

CAM plants separate enzyme activites in time, not spatially.

8.

Enzyme	Regulatory factor
Rubisco	NADPH (activates)
	Mg^{2+} (activates)
	pH (increased pH activates)
	CA1P (inhibitor)
Glyceraldehyde-3-phosphate dehydro-genase	Thioredoxin

(other examples also possible)

9. Distinguish between C_3 and C_4 by comparing compensation point, $^{13}C/^{12}C$ discrimination or first product of CO_2 fixation. In C_3 plants supplied with $^{14}CO_2$ the first product would be [^{14}C]-3-phosphoglycerate; in C_4 plants [^{14}C]-oxaloacetate would be the first compound found.

10. (a) 2;
(b) 1;
(c) 2;
(d) 6×3 ATP + 6×2 NADPH.
11. (a) 18 ATP + 12 NADPH;
(b) 2 ATP/CO_2 transported;
(c) 18 ATP + 6×2 ATP (C_4 pathway) + 12 NADPH.

Chapter 3

1. The pathway of gluconeogenesis from lactate involves <u>four</u> reactions not encountered in glycolysis. These are catalysed by the enzymes <u>pyruvate carboxylase</u>, <u>phophoenolpyruvate carboxykinase</u>, <u>fructose bisphosphatase</u>, and <u>glucose-6-phophatase</u>, respectively.

The conversion of two molecules of lactate to one molecule of glucose requires the hydrolysis of <u>four</u> molecules of ATP and <u>two</u> of GTP. The ATP is required in the reactions catalysed by <u>pyruvate carboxylase</u> and <u>phospho-glycerate kinase</u>, while the GTP is required in the <u>phosphoenolpyruvate carboxykinase</u> -catalysed reaction.

The three major non-carbohydrate sources of glucose production in liver are <u>lactate</u>, <u>amino acids</u> and <u>glycerol</u>.

2. A, B, D.

3. B, D.

4. B, D, E.

5. (a) False. (b) False. (c) False. (d) False. (e) True.

6. In liver and kidney, during starvation or after muscular exercise.

7. The metabolism of glucose to lactate proceeds with a high negative standard free energy change and is irreversible. In gluconeogenesis, the three irreversible enzyme reactions of glycolysis are bypassed. The overall synthesis of glucose from lactate is also irreversible, but requires an input of metabolic energy from ATP and GTP.

8. The pyruvate carboxylase-catalysed reaction is mitochondrial. The mito-chondrial membrane is impermeable to oxaloacetate. Oxaloacetate can leave the mitochondria via one of two shuttle systems:

(i) by intramitochondrial reduction to malate, efflux of malate and reconversion to oxaloacetate in the cytosol. This pathway exports reducing power from the mitochondria and is involved in gluconeogenesis from malate and from other substrates whose metabolism does not generate cytosolic NADH directly.

(ii) via intramitochondrial transamination to aspartate, efflux of aspartate and cytosolic transamination of the aspartate back to oxaloacetate. This does not results in the transfer of reducing equivalents across the mitochondrial membrane and is involved in the metabolism of lactate.

9.

$$\text{Asparagine} \longrightarrow \text{Aspartate} \longrightarrow \text{Fumarate}$$
$$\text{NH}_4^+ \qquad\qquad \text{Urea}$$
$$\text{NAD}^+ \quad \text{NADH} + \text{H}^+$$
$$\longrightarrow \text{Malate} \longrightarrow \text{Oxaloacetate}$$
$$\downarrow$$
$$\downarrow$$
$$\downarrow$$
$$\text{Glucose}$$

Mitochondrial reactions are not involved. The oxaloacetate is generated in the cytosol. The required NADH comes from the cytosolic malate dehydrogenase reaction.

10. In the conversion of lactate to glucose, oxaloacetate is transferred out of the mitochondria as aspartate (see answer to Question 8). Aminooxyacetate inhibits the transamination of oxalo-acetate to aspartate and hence inhibits gluconeogenesis. With pyruvate as the only substrate, NADH must be generated in the cytosol. Oxaloacetate leaves the mitochondria as malate, transamination is not involved and aminooxyacetate does not inhibit glucose production.

11. If lactate is the only energy source, then lactate must be converted to pyruvate and the pyruvate must be oxidized via the TCA cycle. Oxidation of pyruvate to carbon dioxide gives 15 ATP molecules/pyruvate via oxidative phosphorylation. In addition the NADH generated by the lactate dehydrogenase reaction will be reoxidized to give three ATP molecules. Therefore the oxidation of one lactate gives 18 ATP. Effectively six ATP molecules are required for the conversion of two molecules of lactate to one of glucose. Thus to generate this ATP, 1/3 molecules of lactate must be oxidized. The ratio of glucose formed to lactate consumed is therefore 1 : 2.33.

Chapter 4

1. The control of glycogen biosynthesis is <u>regulated</u> by the concentrations of certain effector molecules including <u>Glc 6-P</u> and <u>AMP</u>, as well as by hormones such as <u>glucagon</u>, <u>insulin</u> and <u>adrenalin</u>. The enzymes involved, <u>glycogen synthetases</u>, can exist in <u>two</u> inter-convertible forms only one of which is active. The change from active to inactive state or <u>vice versa</u> involves a <u>covalent</u> modification to the protein and this control is co-ordinated with the process of glycogen <u>breakdown</u>.

2. Sucrose is a <u>disaccharide</u> composed of <u>Glucose</u> and <u>Fructose</u> residues, and is produced by photosynthetic <u>cells</u> of green plants and is subsequently <u>transported</u> round the plant. Two enzymes are known to catalyse sucrose synthesis: <u>sucrose</u> synthetase and <u>sucrose 6-phosphate</u> synthetase. One of these produces sucrose 6-phosphate in photosynthetic tissues and sucrose formation from this compound is promoted by the action of <u>sucrose-6-phosphate</u> phosphatase. In <u>non-photosynthetic</u> tissues, in contrast, the enzyme <u>sucrose synthetase</u> tends to catalyse the formation of UDP-Glc and <u>fructose</u> rather than sucrose <u>synthesis</u>.

3. When the periodic extracellular polysaccharides are being synthesized, a typical process involves transfer of <u>single</u> sugars to form <u>oligosaccharide</u> units which are subsequently transferred themselves. The process occurs <u>within</u> the cell, and the oligosaccharides are then <u>transported</u> to the <u>external</u> face of the cytoplasmic membrane to be added to the growing <u>polysaccharide</u>. However, other pathways exist and some poly-saccharides are produced by the action of <u>enzymes</u> in the periplasmic space. <u>Lipid</u> phosphoryl sugars play a major role in the biosynthesis of many bacterial poly-saccharides although the assembly of bacterial <u>cellulose</u> does not require a lipid intermediate.

4. B, C, D.

5. B, D, E.

6. D.

7. B, E.

8. The cytosol is an aqueous solution, at about pH 7, of many our M_r compounds comprising intermediates of metabolic pathways. The extracellular space varies in pH as it is in contact with the environment. Only extremely alkaline conditions, lethal to the cell, solubilize otherwise insoluble polysaccharides.

There are two reasons why a poly-saccharide is insoluble. One is that it is strongly bound, probably through covalent bonds, to an insoluble polymer. For example, some β-glucans in yeast cell walls which, from their structure, might be though to be soluble, are insoluble because they are probably bound to insoluble chitin.

Examples of the second reason are chitin and cellulose, insoluble because the –OH and >N–H groups are H-bonded to similar groups in other polymers and not to solubilizing water molecules. They form large aggregates leaving few residues for solvation.

Some polysaccharides, like the agars and carrageenans in the cell walls of algae, form intermolecular complexes that, far from resisting association with water, trap it to form gels. The cell would be incapacitated if this occurred in the cytosol.

9. (b) The C-1 of the transferred glucosyl residue is activated through carbonium ion formation and is received by the oxygen of the C-4 hydroxyl group.

10. The pathway is a combination of Figs 4.2 and 4.8. Synthesis is driven by the hydrolysis of pyrophosphate.

The rate of glycogen synthesis is controlled by glycogen synthetase in mammals and UDP-Glc pyrophosphorylase in bacteria. The reason for this is the difference in the use of UDP-Glc in the difference cells.

11. Fructose-6-phosphate, may exist both in the ring structure and the open-chain form. It is the open form that is aminated by glutamine to glucosamine 6-phosphate or isomerized to mannose 6-phosphate and glucose 6-phosphate:

Fructose 6-phosphate can thus give rise to nucleotide diphosphoglucose, mannose and N-acetylglucosamine which in turn give rise to other sugar diphosphonucleotides (Figs 4.5–7).

12. See Fig. 4.6. There are two differences between bacterial bactoprenols and eukaryotic dolichols. The structures of lipids in the yeast intermediates have more, and a greater range of, isoprenyl units. In addition, the isoprenyl unit attached to the diphospho group is unsaturated.

13. Two types of sugar residues, xylose and galactose, require three transferases. The first transferase attaches a xylosyl residue to the protein. In the first of two galactosyl transfers the acceptor is a xylosyl residue and in the second it is a galactosyl residue. Thus two transferases are required despite the common donor of UDP-Gal.

14. Transfer made on the cytoplasmic side of the membrane. Sugar nucleotides, in common with other charged molecules, cannot cross the cytoplasmic membrane except on specific transport proteins. In this case the charged diphospho group is shepherded across in the enveloping lipophilic environment of a polyisoprenyl moiety.

See Fig. 4.19 for the assembly of the trisaccharide pyrophosphate.

15. (a) The reducing end is not free, but bound to a pyrophosphate group. Both

added unit and partially assembled polysaccharide are anchored to the cytoplasmic membrane by the lipid carrier. The incoming unit cannot break free and seek the non-reducing end which might be at a considerable distance into the periplasm.

16. On the outer surface of the cytoplasmic membrane (Fig. 4.21). The strategy for assembly is the same as for the biosynthesis of lipopolysaccharide.

17. One such change is the internal bridging of galactose residues between C-3 and C-6 which alters the conformational structure of the ring and promotes association with similar polymers to form a double helix. The consequence of this is that polysaccharides like algal iota carrageenan and agarose form molecular networks that trap water to form protective gels.

Another example, on the surface of mammalian cells, is the inversion of configuration at C-5 of a –CH$_2$OH, changing an incorporated residue of D-glucuronate to L-iduronate thus changing the viscosity of the extra-cellular proteoglycan complex.

Both bridging and inversion are catalysed by enzymes with the consequence that the cell, in controlling the activity of these enzymes, can rapidly and easily alter the physical properties of the extracellular matrix without removing existing polysaccharides and replacing them with new ones; an energetically expensive process.

Fructose 6-P

Glucose 6-P

Mannose 6-P

Glucosamine 6-P

Chapter 5

1. Glutamate synthase catalyses the reaction:

L-glutamine + α-oxoglutarate + NADH + H$^+$ = α-L-glutamate + NAD$^+$

Nitrogenase is generally detected using the reaction:

C_2H_2 = + H_2 = C_2H_4

Pyridoxal phosphate is a cofactor concerned with the transfer of amino groups in transamination reactions. A root nodule bacteria such as *Rhizobium* requires a carbon source in the form of sucrose or its hydrolysis products from its leguminous host. A heterocyst supplies nitrogen in the form of glutamine to the adjacent vegetative cells.

2. D
3. C
4. A
5. B
6. (i) anaerobes, compartmentation as in heterocysts, (ii) oxygen-binding proteins, leghaemoglobin, (iii) high rate of oxidative metabolism as in *Azotobacter*.
7. Glutamine is the corepressor for nitrogen repression. It cannot be synthesized at a sufficient rate in the mutant.
8. O_2/H_2O, $FAD/FADH_2$, $NAD^+/NADH$, Ferredoxin Fe^{3+}/Fe^{2+}.
9. Fe^{2+}, Mo^{2+}, Mg^{2+}.
10. Glutamate dehydrogenase 36% V_{max}, glutamine sythetase 95% V_{max}.

Chapter 6

1. One-carbon units are used in various biosynthetic processes. Purine synthesis involves incorporation of two formyl groups from tetrahydrofolate derivatives, while thymidine and serine arise from hydroxymethyl groups. 5-methyl tetrahydrofolate donates the methyl group of methionine, which is important as it is the origin (via S-adenosyl-methionine) of methyl groups in a wide range of cell constituents. Most of the one-carbon units in the cell arise from the hydroxymethyl group of serine, which is formed by transfer from serine to tetrahydrofolate with the formation of glycine. Serine, in turn, is formed from 3-phosphoglycerate, an intermediate of the glycolytic pathway, by successive oxidation, transamination and dephosphorylation reactions.
2. Glycine is a precursor of a large number of non-protein cell constituents including purines, essential in coenzymes and nucleic acids, and of various excretion products, such as hippurate, formed from benzoate. Any glycine ingested in the diet that is not required for such purposes is broken down via an enzyme system that involves four different proteins, one of which, the H-protein is absent in the congenital condition called non-ketotic hyperglycinaemia. This condition is characterized by high concentrations of glycine in the blood. Apart from CO_2, the

other product of this system is methylenetetrahydrofolate, which can be used for biosynthetic purposes, such as the synthesis of purines, or can be oxidized to CO_2 via the enzymes methylenetetrahydrofolate dehydrogenase, cyclohydrolase and formyltetrahydrofolate dehydrogenase. The electron acceptor for the last-mentioned enzyme is $NADP^+$.
3. Glutamate is the parent compound from which the following amino acids are derived: glutamine, γ-aminobutyrate, arginine and proline. Ornithine not only gives rise to arginine, but also to putrescine and the C_4 portions of spermidine and spermine. Acetylation of the amino group of glutamate is an essential preliminary to ornithine formation to prevent the spontaneous cyclization of glutamate semialdehyde to give Δ'-pyrroline carboxyate, the precursor of proline.
4. Different polyamines have two, three or four positive charges. They consist of C_3 and C_4 portions joined by secondary amino groups. The C_3 portion arises from decarboxylated S-adenosyl-methionine, while the C_4 portion arises by a similar process from ornithine. Inhibitors of polyamine synthesis have clinical application.
5. B, D, E
6. A, E
7. A, B, E
8. O-methyl compounds: lignin, pectin, some methylated proteins, rRNA, tRNA, chlorophyll.
N-methyl compounds: adrenalin, betaine, creatine, phosphatidylcholine, sarcosine, some methylated proteins, DNA (e.g. N^6-methyladenine), rRNA, tRNA.
C-methyl compounds: cyclopropane fatty acids, ergosterol (and other methylated steroids), DNA (e.g. 5-methylcytosine).
S-methylmethionine is an S-methyl compound!
9. Serine + Tetrahydrofolate → Glycine + NAD^+ + 5,10-Methylene-tetrahydrofolate
Glycine + NAD^+ + Tetrahydrofolate → CO_2 + 5,10-Methylenetetra-hydrofolate + NADH + NH_4^+
5,10-Methylenetetrahydrofolate + $NAD(P)^+$ → 5,10-Methenyltetra-hydrofolate + NAD(P)H + H^+ (twice)
5,10-Methenyltetrahydrofolate + H_2O →

10-Formyltetrahydrofolate + H^+ (twice)
10-Formyltetrahydrofolate + $NADP^+$ + H_2O → CO_2 + Tetrahydrofolate + NADPH + H^+ (twice)
Overall: Serine + $5NAD(P)^+$ + $3H_2O$ → $3CO_2$ + $5NAD(P)H$ + $4H^+$ + NH_4^+

Chapter 7

1. AMP or adenosine 5′-mono-phosphate, contains the base adenine or aspartate, the sugar ribose and a phosphate. The 6-amino group is derived from 6-aminopurine, as is one of the ring nitrogens. Two of the purine ring carbon atoms originate from N^{10}-formyl tetra-hydrofolate. Amino acids contributing to the ring skeleton include glycine and glutamine.
2. The *de novo* metabolic route for producing purine nucleotides involves assembly of the purine ring on the ribose sugar. The necessary enzymes are located in the cytosol of the cell. Several amino acids supply atoms for the purine ring, and these include aspartate, glutamine and glycine. An initial product is IMP from which AMP and GMP are derived subsequently. The amino acid serine supplies one-carbon units via formyl tetrahydrofolate.
3. A, D.
4. C, D, E.
5.
Aspartate + Glutamine + Ribose 5-phosphate + HCO_3^- + 2ATP

$$UMP \xrightarrow[ATP]{} UDP \xrightarrow[ATP]{} UTP \xrightarrow[Glutamine + ATP]{} CTP$$

Thus five ATP molecules are required for each molecule of CTP synthesised.
6.

$$Ribose\ 5\text{-phosphate} \xrightarrow[]{ATP} PRPP \longrightarrow$$
$$AMP = 2ATP$$

Glycine + ATP, Glutamine + ATP, ATP, Aspartate + ATP

⟶ IMP

Glutamine + ATP
IMP ⟶ XMP ⟶ GMP
$AMP + PP_i = 2ATP$

Thus six ATP molecules are required for IMP synthesis and a further two for GMP formation, an overall value of 8ATP.

7. Radioactive labelling is a key experimental technique in metabolic studies. The constancy of the specific activity ratio when comparing ^{14}C-labelled cytosine with labelled ribose and with deoxyribose strongly suggests that the ribonucleotide is not cleaved prior to reduction but stays intact. This finding led to the search for a ribonucleotide reductase enzyme system. Subsequently ribonucleoside diphosphate reductase was identified.

8. Hydroxyurea inhibits ribonucleotide reductase preventing the formation of DNA precursors. Cells accumulate at the G_1/S border. After a suitable time interval washing the cells and resuspending them in fresh medium (diluting out the hydroxyurea) allows the cell population to enter the S phase and to replicate DNA in synchrony.

High concentrations of deoxy-thymidine can also be used to achieve synchrony. Uptake by cells and phosphorylation (dT → dTMP) raises the concentration of dTTP regulating (allosterically inhibiting) ribonucleotide reductase. This leads to a marked diminution of dCDP formation from CDP, and the consequent lack of dCTP diminishes DNA synthesis with a build-up of cells at the G_1/S border. DNA synthesis is slowed rather than stopped so to achieve a high degree of synchrony there is the need for a double deoxythymidine block.

9. 8-Azaguanine is converted to the 'fraudulent' purine ribonucleotide by the HGPRTase salvage pathway. Thus this analogue can be incorporated into mRNA and this leads to marked impairment of polysomal structure and function in translation (R.S. Rivest, D. Irwin and H.G. Mandel (1982) *Advances in Enzyme Regulation* **20**, 351–373). Thus cells containing hypoxanthine-guanine phosphoribosyltransferase (enzyme E) lethally impair their protein synthetic machinery. Only mutants lacking this salvage pathway (c.f. Lesch–Nyham syndrome) survive.

Fused mutant myeloma-spleen cells survive and flourish in the HAT medium because the hybridoma possesses HGPRTase (E) from the spleen partner. This means that hypoxanthine (H) in the medium can be salvaged to meet the purine nucleotide needs of cell growth and division. The presence of

Aminopterin (A), a highly effective inhibitor of folate reduction, blocks the *de novo* purine biosynthetic pathway and thus the salvage pathway is vital (and is missing in the selected myeloma cells). The presence of Aminopterin also blocks *de novo* thymidine nucleotide formation. Cells take up deoxythymidine (T) from the medium and thymidine kinase enables them to use this as a source of dTMP leading to DNA synthesis. HAT is an excellent example of a selective medium, it has proved of great importance in monoclonal biotechnology.

Chapter 8

1. Fatty acids are produced by the <u>fatty acid synthase</u> complex in the cytosol. The major product of this enzyme is <u>palmitate</u>. The production of <u>malonyl CoA</u> is stimulated by <u>carbon dioxide</u> and requires <u>biotin</u>. The function of the latter is to activate <u>acetyl CoA</u>. Long-chain, C_{20-24}, fatty acids are produced by <u>elongation</u> reactions in <u>mitochondria</u> or <u>endoplasmic reticulum</u>. These two elongation systems differ in their cofactor requirements, which are <u>NADH</u> and <u>NADPH</u>, and sources of carbon, <u>acetyl CoA</u> and <u>malonyl CoA</u>, respectively. A further series of reactions produces <u>unsaturated</u> fatty acids. There are <u>four</u> families of these fatty acids. Each contain one or more <u>double</u> bonds. The <u>four</u> families of acids are called <u>palmitoleate</u>, <u>oleate</u>, <u>linoleate</u> and <u>linolenate</u>.

2. B, E, G.
3. C, D.
4. A, B, C.
5. A.
6. A, B.
7. $^{14}CO_2$ would be used as the source of the carboxy group of malonate but would not be incorporated into the fatty acid. During the formation of malonyl CoA one of the ^{2}H atoms would be lost from the methyl group of acetyl CoA. A further ^{2}H atom would be lost during the dehydration step. Therefore each malonyl CoA molecule would incorporate one ^{2}H atom; two atoms of ^{2}H would be incorporated from the primer acetyl CoA. A ratio of ^{2}H ^{14}C of 9 : 0 would be found.

In the second experiment, again the carboxyl group of malonyl CoA would be lost and there would be no incorporation

of $^{14}CO_2$. Only one acetyl CoA would be incorporated into the fatty acid giving a ratio of ^{2}H : ^{14}C of 2 : 0.

In the third experiment a ratio of ^{2}H : ^{14}C of 2 : 7 would be found. The ^{14}C could be measured by scintillation spectrometry, but the non-radioactive ^{2}H would have to be determined by a weighing method or by mass spectrometry.

8.

	(a) palmitate	(b) palmitoleate
ATP	1	1
NADPH	14	14†
acetyl CoA*	8	8

* Assuming malonyl CoA is synthesized from acetyl CoA.
† 1 mole of NADH or NADPH generated during the unsaturation step.

9. First, elongation to the C_{20} unsaturated arachidonate occurs. Arachidonate reacts with oxygen catalysed by a cyclooxygenase introducing a five-membered peroxide containing ring. This is reduced to the prostaglandin.

10. (a) albumin, (b) VLDL, (c) LDL.

Chapter 9

1. Polyisoprenoids are widely found in nature and are formed initially from <u>acetyl</u> CoA and <u>acetoacetyl</u> CoA. <u>Mevalonate</u>, a C_6 precursor, and <u>isopentenyl-PP</u>, the active biological isoprene unit, also act as intermediates in this biosynthetic process. Isoprenoid products include <u>cholesterol</u>, the major sterol in vertebrate tissues, and from this <u>bile</u> acids and many <u>steroid</u> hormones are made.

The reactions leading to sterol synthesis takes place in the <u>cytosol</u> of the cell and many enzymes are located in the <u>endoplasmic reticulum</u>. The reducing equivalents required are supplied by <u>NADPH</u>. Intermediates in the process are the hydrocarbon <u>squalene</u>, which has <u>30</u> carbon atoms, and <u>lanosterol</u>, the first sterol formed. Formation of this sterol requires <u>molecular</u> oxygen.

Most of the cholesterol formed in the body is made in the <u>liver</u>. The hormones <u>insulin</u> and <u>glucagon</u> play important roles in the reaction sequence and exert their control at the level of the enzyme <u>HMG CoA reductase</u>. This enzyme may also be inhibited by <u>competitive</u> inhibitors of its normal substrate.

2. C, D.

3. B, D, E, F.

4. D, F.

5. ATP citrate lyase (provision of acetyl CoA); glucose-6-phosphate dehydrogenase, 6-phosphogluconate dehydrogenase, malic enzyme (provision of NADPH); acetyl CoA carboxylase; fatty acid synthase. Also, kinases from glycolysis.

6. Primary alcohol group → Ⓟ → Ⓟ Ⓟ (potential release of PP_i); carboxyl group/tertiary alcohol group → concerted decarboxylation → C_5 product containing nucleophilic methylene group → isomeric derivative capable of electrophilic reaction with first product.

7. Phenylalanine → tyrosine; catabolism of tyrosine; hydroxylation of ring systems or hydrocarbon chains in detoxification of, or tolerance towards, drugs; hydroxylation of ring systems or hydrocarbon chains in bacterial degradation.

8. Four co-ordination positions are provided in the planar porphyrin molecule by a lone pair of electrons on each nitrogen atom. The fifth and sixth co-ordination positions on the Fe are available for further bonding to a polypeptide and, in some cases, oxygen.

The remaining co-ordination positions in the following examples are filled by amino acid residues and/or O_2:

(a) myoglobin: imidazole N of histidine plus O_2 (only Fe^{2+});

(b) cytochrome *c*: imidazole N of histidine and S of methionine;

(c) cytochrome oxidase (aa_3): 'amino acid' plus O_2 ($\rightarrow Cu^{2+/+}$).

Glossary

A

Aldolase: *an enzyme that catalyses the transfer of a C_3 unit from a ketose sugar to an aldose sugar.*

Allyl: *the grouping $CH_2{=}CH{-}CH_2{-}$ as in allyl chloride: $CH_2{=}CH{-}CH_2Cl$.*

B

Bacteroid: *the name used for the irregularly-shaped or branched form of bacteria such as Rhizobium as it occurs in the root cells during nodule formation.*

Betaine: *beet leaves may contain up to 3% of this compound. Betaine is $(CH_3)_3N^+CH_2COO^-$, but in fact there is a family of such 'biogenic amines' which can serve as methyl donors in methylation reactions.*

Bundle sheath cells: *large parenchyma cells, surrounding a vascular bundle in photosynthetic tissue, characteristic of C_4 plants.*

C

Carboxylation: *the addition of a CO_2 unit to an organic compound.*

Carnitine: *previously called 'vitamin B_T', identified as an essential component of the meal worm, Tenebrio molitor. Vertebrates can synthesize carnitine and thus it is not a vitamin for them.*

Cellulose: *a long, unbranched polymer consisting only of D-glucose units linked $\beta1{-}4$.*

Chelate: *a term derived from the Greek word for 'crab's claw'. It generally refers to the coordinate bonds made between metal ions and two or more negatively-charged atoms or lone pairs of electrons on the same organic molecule.*

Cholesterol: *an isoprenoid. The name is derived from the Greek words chole, meaning bile (appreciable amounts of cholesterol are found in bile), and stereos, solid.*

Chylomicrons: *Chyle is the milky fluid appearing in the lacteals of the lymphatic vessels of the small intestine when fat is being absorbed from the gut. From the Greek, chylos, juice. The milky appearance results from the presence of microscopic globules of lipid, micro small.*

D

De novo: *means synthesized 'from new' rather than being formed from other organic compounds. However, its meaning is often broader. One could speak of de novo synthesis of cholesterol from acetate units, for example.*

Diabetes: *A groups of diseases characterized by a lack of insulin from the pancreas or an insensitivity to insulin (deficiency of receptors?). This results in a high and potentially fatal blood glucose level. The word diabetes comes from the Greek word for 'siphon' because untreated diabetic individuals excrete large volumes of urine.*

Diazotroph: *The suffix -troph, as in autotroph, auxotroph, diazotroph, heterotroph, prototroph and phototroph, all relate to the mode of nutrition, and is derived from the Greek word meaning nourish. A diazotroph is an organism which is able to use N_2 as its sole source of nitrogen. The N_2 is reduced to ammonia and then incorporated into organic nitrogen.*

Dinitrogen: *used to distinguish N_2 from nitrogen in its more general sense in nitrogen metabolism, nitrogen repression, etc. In the same way, the terms dihydrogen and dioxygen should also be used, but this would be cumbersome and is usually unnecessary since there is unlikely to be any ambiguity.*

1,3-Diol: *A structure with two alcohol groups separated by a methylene group.*

$$\underset{\underset{|}{}}{-}\overset{\overset{OH}{|}}{C}-CH_2-\overset{\overset{OH}{|}}{C}-$$

F

Feedback inhibition: *the inhibition of the first enzyme of a metabolic pathway by the end-products of the pathway.*

Fixation: *we talk about 'fixing' CO_2 or N_2. The idea is that these free gaseous substances become 'fixed' or incorporated into more solid organic compounds.*

Folic acid: *part of the B-vitamin complex. It was first isolated from 4 tons of spinach leaves in 1941, as an acid that stimulated the growth of Streptococcus faecalis. From the Latin, folium, leaf.*

G

Glucocorticoids: *a category of naturally-occurring as well as synthetic steroids and steroid-like compounds that have effects on glucose metabolism. Example: hydrocortisone.*

Gluconeogenesis: *literally, the generation of new glucose. The term refers to the set of enzyme-catalysed reactions by which glucose is generated from non-carbohydrate precursors. The term is derived from several Greek words which literally mean 'building new glucose'.*

Glycogen: *a branched polymer consisting only of D-glucose molecules.*

Glycolysis: *the process by which sugars, usually glucose, are catabolized to provide energy. The process may take place either aerobically or anaerobically.*

Glycosyltransferases: *enzymes involved in transferring a sugar molecule from a donor (normally a nucleoside diphosphate sugar) to an acceptor molecule:*

NDP-sugar $+$ acceptor $\rightarrow$

NDP $+$ acceptor-sugar

H

Hepatic: *pertaining to the liver (Greek hepar, the liver).*

Homocysteine: *'homo-' means that this is the higher homologue of cysteine, that is, the compound of similar structure but having an additional methylene group.*

β-hydroxyketone: *a structure containing ketone and alcohol functional groups separated by a methylene group.*

$$-\overset{\overset{OH}{|}}{C}-CH_2-\underset{\underset{\beta}{}}{C}{=}O$$

I

Isozyme: *a contraction of **isoenzyme**. Two enzymes may be regarded as isoenzymes if they catalyse the same reaction but differ in their amino acid sequence.*

K

Ketone bodies: *these are acetoacetate, β-hydroxybutyrate and acetone. They are chemical compounds rather than bodies and unfortunately β-hydroxybutyrate is not even a ketone. However, the name has stuck and they form a closely related metabolic group.*

Kranz anatomy *refers to the arrangement of the bundle sheath cells in C_4 plants. Kranz is German for 'wreath', and refers to the ring-like arrangement of the photosynthetic cells around the vein.*

L

Lecithin: *the old name for the phospholipid now called phosphatidylcholine. This lipid is an important component of most biological membranes.*

Light reaction of photosynthesis: *that part of the photosynthetic process whereby light is trapped. It certainly consists of more than a single reaction. The light reaction may be distinguished experimentally from all the subsequent reactions, called the dark reaction.*

M

Macromolecule: *means large molecule, but what does that mean? The division is hard to define. Most people would call insulin, a small protein of M, about 6000, a macromolecule. Anything smaller they would not.*

Mesophyll cells: *the cells of the internal parenchyma of a leaf: they are the major photosynthetic cells.*

Molar growth yield: *an index of the efficiency of a substrate to support growth. It is defined as the increase in dry weight of biomass per mole of substrate consumed.*

mRNA: *the common abbreviation for messenger RNA, which forms the intermediary between DNA and protein synthesis. It carries the information specifying how to join up amino acids in a specific sequence to form a particular polypeptide.*

Muramic acids (muramates): *and teichoic acids (teichoates) are compounds found in bacterial cell walls, from the Latin* murus, *a wall, Greek* teichos, *a wall.*

N

NTP: *the abbreviation for 'nucleoside triphosphate'; similarly, dNTP means deoxynucleotide triphosphate. These abbreviations are used when the exact nucleoside triphosphate participating in a reaction (e.g. ATP, UTP) is either unknown or may be any of four.*

P

Peroxisomes: *small organelles bounded by a single membrane and containing catalase and peroxidases.*

Photorespiration: *generally regarded as 'wasteful'. It occurs when Rubisco picks up O_2 instead of CO_2 and produces glycolate.*

Polyamines: *a group of small basic compounds, so named because they carry several amino groups. In fact, putrescine and cadaverine are diamines. Spermidine and spermine were given their names because they where first 'discovered' in semen.*

Porphyrin: *a group of polyisoprenoid compounds. From the Greek word meaning 'red' and relates to the port wine colour, characteristic of urine from patients suffering with porphyrias.*

Prostaglandins: *complex derivatives of arachidonate. Originally classified as ether-soluble (the E series), or phosphate-soluble (the F series, from the German fosphat). This is a peculiar nomenclature because it is not based on the structure of the compound.*

Prosthetic group: *used to describe a group attached to a protein which is not part of the polypeptide backbone, and is thus not an amino acid residue. A prosthetic group is important in the functioning of the protein, and it is not easily dissociated from the protein.*

R

Residue: *a convenient way of describing a monomer incorporated into a polymer.*

S

Self-assembly: *furniture that you have to assemble from the parts yourself, is known as self-assembly furniture. Self-assembling macromolecular complexes come together themselves (without any human assistance!) because their charges, shapes and hydrophobicities are complementary.*

Semiconservative: *in making a new DNA double helix it would be possible to envisage a replica of the molecule being formed. One helix would be two 'old' strands and the other, two 'new' strands. In fact, the new molecules each consist of one new and one old strand, so that half of each molecule is* conserved *in the replication process.*

Serine: *an amino acid, first isolated from sericin, the protein associated with silk. It is usually removed from the silk fibres (fibroin) by boiling with alkali.*

Sterol: *compounds composed of four fused rings with an hydroxyl at C-3, e.g. cholesterol.*

Sulphate: *a term widely used in describing plant and mammalian polysaccharides such as algal fucan sulphate, human chondroitin 4- and 6-sulphates, although not strictly correct. Such groups are better thought of as sulphuryl radicals attached to the oxygen atom of the hydroxyl group.*

Sulphuryl   Sulphate

T

Transketolase: *an enzyme that catalyses the transfer of a C_2 unit from a ketose sugar to an aldose sugar.*

Triacylglycerols: *when glycerol is esterified with three molecules of long-chain fatty acid, the product is called a triacylglycerol. The former term was 'triglyceride'.*

tRNA: *the abbreviation for transfer RNA, a family of small RNA molecules that participate in protein synthesis, bringing amino acids to the ribosome–mRNA complex for addition to the growing polypeptide.*

Turnover number: *a measure of the catalytic activity of an enzyme. It is defined as the number of moles substrate transformed to product per mole of enzyme per second.*

U

α-β Unsaturated ketone: *structure containing a ketone functional group immediately adjacent to a carbon–carbon double bond.*

Index

Page references to Figures and Tables are in **bold**. References to Boxes and Side-notes are indicated by (B) and (S) respectively, after the page numbers.